Blackwell Maths 2

Mark Riddington B.Sc.

Martin Grier B.Sc.

Woodroffe School, Lyme Regis

Basil Blackwell

Basil Blackwell Publisher Ltd.
108 Cowley Road
Oxford OX4 1JF

First published 1981 by Sorrett Publishing Pty Ltd.
© 1981 B.J. Lynch, L.P. Picking, J.L. Anders and M.J. Coffey

This edition published 1984 by Basil Blackwell Publisher Ltd.
Adaptations © 1984 M. Riddington and M. Grier

ISBN 0 631 90620 7

Printed in Great Britain
at Blantyre Printing Company, Glasgow

Contents

Symbols

$=$	is equal to
$\neq$	is not equal to
$\approx$	is approximately equal to
$<$	is less than
$>$	is greater than
$\leq$	is less than or equal to
$\geq$	is greater than or equal to
$\therefore$	therefore
$\in$	is an element of
$\notin$	is not an element of
$\mathscr{E}$	the universal set
ϕ	the empty or null set
$\{\ \}$	notation for a set
$\leftrightarrow$	equivalence of two sets
$\subset$	is a subset of
$\not\subset$	is not a subset of
$\cap$	intersection
$\cup$	union
$n(A)$	cardinal number of set A
A'	complement of set A
$\llcorner$	right angle
$\overline{AB}$	line segment or interval AB
$\mathrm{d}\overline{AB}$	length measure of segment AB
$\triangle AOB$	triangle AOB
$\angle AOB$	angle AOB or magnitude of angle AOB
$\perp$	is perpendicular to
$\parallel$	is parallel to
(x, y)	ordered pair, Cartesian coordinates of a point
$a : b$	ratio of a to b
$\%$	percent
$\overrightarrow{AB}$	vector AB

Preface

Blackwell Maths 2 is the second in a series of five texts, the first three of which are designed to cover the full ability range in the first three years of the secondary school. The last two volumes will be aimed specifically at the middle of the ability range; that is those pupils who at present follow a course of study with CSE as the final examination.

Emphasis is placed on the mastering of basic skills and their application in order to provide a sound basis for subsequent studies. New topics are introduced assuming little or no prior knowledge of that topic, other topics are extended from the first book after revision. The graded exercises contain questions suitable for a wide range of ability. Thus it is not expected that a pupil will attempt every question. Chapter 14 could serve to provide further work for pupils who finish exercises quickly but all pupils should be encouraged to attempt the exercises in this chapter. We have assumed that calculators will be available at times for some of the more difficult computations.

The authors wish to acknowledge the help and encouragement of their colleagues at Woodroffe School.

Arithmetic 1

1.1 Whole Numbers

Exercise 1 A

1. What is the value of the 3 in each of the following:
   - (a) 32
   - (b) 43
   - (c) 309
   - (d) 137
   - (e) 2304
   - (f) 9403
   - (g) 31 782
   - (h) 83 019

2. Write the following groups of numbers in ascending order:
   - (a) 294, 942, 429, 924, 249
   - (b) 362, 236, 326, 623, 263
   - (c) 1056, 5106, 1506, 5016, 1065
   - (d) 2608, 6802, 2806, 6280, 2860
   - (e) 8780, 7880, 8087, 7088, 8807
   - (f) 9413, 4931, 3914, 4391, 3419

3. Write i) the highest possible number
   ii) the lowest possible number
   using the following digits:
   - (a) 2, 6, 3, 0, 1
   - (b) 4, 2, 7, 0, 9
   - (c) 0, 5, 7, 2, 8
   - (d) 6, 4, 0, 2, 3
   - (e) 7, 0, 9
   - (f) 4, 2, 6
   - (g) 3, 5, 8, 0
   - (h) 2, 7, 5, 0
   - (i) 2, 6, 3, 5, 0
   - (j) 8, 3, 9, 5, 2

The result of adding numbers is called the 'sum'.
The result of subtracting two numbers is called the 'difference'.

Exercise 1 B

Calculate the following:

1. 235 + 416
2. 723 + 132
3. 521 + 762 + 1042
4. 816 + 2403 + 705
5. 4560 + 815 + 9206 + 41
6. 86 + 5201 + 7678 + 143
7. 16 402 + 2 345 + 920 + 1006
8. 24 014 + 30 230 + 4210 + 11 017
9. 52 + 18 605 + 976 + 7412
10. 370 + 4082 + 73 + 26 809
11. 426 − 135
12. 572 − 369
13. 1047 − 462
14. 3420 − 386
15. 9408 − 7625
16. 1495 − 1287
17. 15 460 − 10 825
18. 23 762 − 19 853
19. 37 005 − 29 518
20. 40 602 − 28 739

The result of multiplying two numbers together is called the *product*.
The result of dividing two numbers is called the *quotient*.

Exercise 1C

1. Calculate the following:
   (a) 416×3
   (b) 226×8
   (c) 204×6
   (d) 810×7
   (e) $18\,046 \times 5$
   (f) $12\,520 \times 9$
   (g) 623×10
   (h) 834×10
   (i) 1457×12
   (j) 413×11
   (k) $725 \div 5$
   (l) $942 \div 3$
   (m) $12\,768 \div 6$
   (n) $20\,466 \div 9$
   (o) $648 \div 9$
   (p) $512 \div 8$
   (q) $5376 \div 3$
   (r) $2982 \div 3$
   (s) $2484 \div 8$
   (t) $7503 \div 6$

2. Find the product of the following:
   (a) 56×20
   (b) 87×40
   (c) 492×30
   (d) 675×60
   (e) 36×59
   (f) 42×27
   (g) 159×23
   (h) 216×45
   (i) 301×36
   (j) 402×23
   (k) 560×78
   (l) 830×35
   (m) 2708×52
   (n) 3017×43

3. Calculate the following:
   (a) 18×10
   (b) 29×100
   (c) 30×100
   (d) 10×731
   (e) $620 \div 10$
   (f) $456 \div 10$
   (g) $5220 \div 100$
   (h) $9015 \div 100$
   (i) 1000×34
   (j) 100×100
   (k) $9200 \div 1000$
   (l) $870 \div 100$

In calculating a problem which has more than one operation (i.e. $+$, $-$, $\times$, $\div$ or $(\ )$), it is very important that an agreed order of performing the operations is used.
Remembering that 'of' means '$\times$':

Order of Operations:

1. Brackets $(\ \ )$
2. Of, Division, Multiplication $\div$, $\times$
3. Addition, Subtraction $+$, $-$
Work the problem from left to right.

This is easily remembered using the word BODMAS.

Example 1.1

Evaluate:
(a) $95 + 64 \div 8 \times 3$
(b) $18 \times 7 - (31 - 9) \div 11$

Solution

(a) $\qquad 95 + 64 \div 8 \times 3$ Working from left to right we calculate the division first,

$= 95 + \quad 8 \quad \times 3$ then the multiplication,

$= 95 + \qquad 24$ finally the addition.

$= \qquad 119$

(b) $18 \times 7 - (31 - 9) \div 11$ Work the brackets first.

$= 18 \times 7 - \quad 22 \quad \div 11$ From left to right, the multiplication is next,

$= \quad 126 \quad - \quad 22 \quad \div 11$ then the division,

$= \quad 126 \quad - \quad\quad 2$ finally the subtraction.

$= \quad\quad\quad 124$

Exercise 1D

Evaluate the following:

1. $6 \times 8 + (4 - 3) \times 7$
2. $9 \div 3 - (11 - 9) \times 1$
3. $94 + 16 \div 4 - 100 \div 4$
4. $24 - 42 \div 7 + 5 \times 6$
5. $32 \div (7 - 3) \times 12$
6. $56 \times (3 + 5) \div 7$
7. $(2 \times 24) - 81 \div 9 + 3$
8. $72 \div 9 \times (22 - 14)$
9. $3 + 14 \times 7 \div (13 - 11)$
10. $(44 \div 11) \times 16 - 23$
11. $(7 + 23) - 3 \times (20 - 15)$
12. $(19 + 46) \times 3 + (39 - 24)$
13. $(14 \div 7) \times (6 + 15) - 3$
14. $(35 \div 5) \times (42 - 16) + 15$
15. $6 \times 5 \div 3 + 15 \div 5$
16. $5 \times 8 \div 20 - 16 \div 8$
17. $3 + 9 \div 3 \times 6 - 4$
18. $5 + 36 \div 2 - 55 \div 11$
19. $8 - 7 \times 6 \div 21 - 6$
20. $12 - 56 \div 8 + 13 \times 5$

1.2   Factors and Primes

> Whole numbers which multiply together to give a certain number are said to be *factors* of that number.

Thus, for example,

$\quad\quad 4 \times \quad 3 = 12 \quad \therefore$ 4 and 3 are factors of 12

and $2 \times \quad 6 = 12 \quad \therefore$ 2 and 6 are factors of 12

and $1 \times 12 = 12 \quad \therefore$ 1 and 12 are factors of 12.

Thus the factors of 12 are 4, 3, 2, 6, 1 and 12.

Similarly,

$\quad\quad 2 \times 8 = 16 \quad \therefore$ 2 and 8 are factors of 16

and $\quad 4 \times 4 - 16 \quad \therefore$ 4 is a factor of 16

and $16 \times 1 = 16 \quad \therefore$ 16 and 1 are factors of 16.

Thus the factors of 16 are 2, 8, 4, 16 and 1.

Consider the number 5.

$\quad\quad\quad 5 \times 1 = 5 \quad$ 5 and 1 are factors of 5 and, in fact, they are the *only* factors of 5.

$\quad\quad\quad$ 5 is a special type of number known as a *prime number*.

> A number which has only itself and 1 as factors is known as a *prime number*. E.g., 2, 3, 5, 7, 11, 13, 17 are prime numbers.
>
> *N.B.* Remember 1 is not a prime number.

Consider the factors of 12 : 4, 3, 2, 6, 1 and 12.

Now consider the factors of 12 which are also prime numbers, that is 3 and 2. Thus we call 3 and 2 *prime factors* of 12.

> The *prime factors* of a number are the factors which are prime numbers.

Consider the factors of 12 and 42

$$12 = 3 \times 4$$
$$= 2 \times 6$$
$$= 1 \times 12 \qquad \text{the factors of 12 are } 4, ③, ②, ⑥, ①, 12$$
$$42 = 6 \times 7$$
$$= 3 \times 14$$
$$= 2 \times 21$$
$$= 1 \times 42 \qquad \text{the factors of 42 are } ⑥, 7, 14, ③, ②, 21, ①, 42$$

Comparing both sets of factors we can see that 1, 2, 3 and 6 are common to both 12 and 42. Thus we call them *common factors*.

The common factor, 6, is called the *highest common factor* (*H.C.F.*) of these two numbers.

> *Common factors* are factors which are common to two or more numbers. The *highest common factor* (*H.C.F.*) is the largest factor common to the numbers.

Example 1.2

(a) Find the factors of 60.
(b) Find the common factors to 18 and 45, and state the *H.C.F.*
(c) Find the prime factors of 28.

Solution

(a) $60 = 2 \times 30$
$$= 3 \times 20$$
$$= 4 \times 15$$
$$= 5 \times 12$$
$$= 6 \times 10$$
$$= 1 \times 60$$

Thus the factors of 60 are 2, 30, 3, 20, 4, 15, 5 12, 6, 10, 1, 60.

(b) $18 = 2 \times 9 \qquad\qquad\qquad 45 = 3 \times 15$
$$= 3 \times 6 \qquad\qquad\qquad\quad = 5 \times 9$$
$$= 1 \times 18 \qquad\qquad\qquad\quad = 45 \times 1$$

$(2, ⑨, ③, 6, ①, 18) \qquad\qquad (③, 15, 5, ⑨, 45, ①)$

The common factors of 18 and 45 are 3, 9 and 1.

Thus the *H.C.F.* of 18 and 45 is 9.

(c) $28 = 2 \times 14$
$$= 4 \times 7$$
$$= 1 \times 28$$

Therefore the factors are 2, 14, 4, 7, 1, 28; the prime numbers are 2, 7. So the prime factors are 2 and 7.

Exercise 1E

1. Find the factors of each of the following numbers:

(a)   10 (b)   8
(c)   25 (d)   65
(e)   48 (f)   32
(g) 150 (h) 144
(i) 360 (j) 220

2. Find the prime factors for each of the numbers in question 1.

3. Find the common factors for the following pairs of numbers:

(a) 30, 45 (b) 12, 48
(c) 16, 36 (d) 18, 42
(e) 100, 60 (f) 75, 175
(g) 8, 50 (h) 14, 26
(i) 240, 600 (j) 320, 140

4. State the highest common factor for each of the pairs of numbers in question 3.
5. Find the prime factors of the following:

   (a) 26

   (b) **44**

   (c) 56

   (d) 72

   (e) 81

   (f) 125

   (g) 108

   (h) 120

   (i) 67

   (j) 43

1.3 Fractions

When you first learnt about fractions, they were considered to be *part of one whole*, but we have since learnt that there are many types of fractions, some of which are greater than 1.

Thus, we needed a more general statement to cover all fractions. For example, $\frac{2}{3}$ is written in fraction form, with one number (*numerator*) over another number (*denominator*). This can also be written as $2 \div 3$. In general

> A fraction is one number divided by another number expressed in the form $\frac{a}{b}$ where
>
> a is the *numerator* and b is the *denominator*.

Exercise 1F

1. Express the following as mixed numbers:

   (a) $\frac{4}{3}$

   (b) $\frac{5}{4}$

   (c) $\frac{12}{7}$

   (d) $\frac{9}{8}$

   (e) $\frac{16}{5}$

   (f) $\frac{24}{9}$

   (g) $\frac{18}{10}$

   (h) $\frac{26}{18}$

   (i) $\frac{34}{6}$

   (j) $\frac{42}{5}$

   (k) $\frac{56}{8}$

   (l) $\frac{48}{4}$

2. Express the following as vulgar fractions:

   (a) $1\frac{1}{2}$

   (b) $1\frac{1}{3}$

   (c) $1\frac{5}{7}$

   (d) $1\frac{2}{9}$

   (e) $2\frac{1}{12}$

   (f) $2\frac{3}{8}$

   (g) $4\frac{6}{11}$

   (h) $5\frac{12}{15}$

   (i) $3\frac{11}{16}$

   (j) $4\frac{7}{12}$

   (k) 5

   (l) 3

3. Complete the following to make true statements:

   (a) $\dfrac{1}{2} = \dfrac{\square}{4} = \dfrac{7}{\square}$

   (b) $\dfrac{2}{3} = \dfrac{\square}{15} = \dfrac{18}{\square}$

   (c) $\dfrac{5}{12} = \dfrac{\square}{60} = \dfrac{60}{\square}$

   (d) $\dfrac{6}{14} = \dfrac{18}{\square} = \dfrac{\square}{112}$

   (e) $\dfrac{7}{\square} = \dfrac{42}{60} = \dfrac{\square}{110}$

   (f) $\dfrac{11}{\square} = \dfrac{55}{60} = \dfrac{\square}{144}$

4. Using the signs $<$, $>$ or $=$, make the following statements true:

   (a) $\frac{3}{8}$ $\frac{1}{4}$

   (b) $\frac{5}{6}$ $\frac{7}{12}$

   (c) $\frac{2}{3}$ $\frac{11}{15}$

   (d) $\frac{1}{2}$ $\frac{13}{16}$

   (e) $\frac{9}{24}$ $\frac{3}{8}$

   (f) $\frac{15}{50}$ $\frac{3}{10}$

5. Express in simplest form:

   (a) $\frac{5}{15}$

   (b) $\frac{4}{20}$

   (c) $\frac{8}{14}$

   (d) $\frac{6}{9}$

   (e) $\frac{36}{90}$

   (f) $\frac{50}{75}$

   (g) $\frac{18}{45}$

   (h) $\frac{20}{55}$

   (i) $\frac{52}{117}$

   (j) $\frac{48}{144}$

   (k) $2\frac{20}{48}$

   (l) $1\frac{6}{21}$

> When adding or subtracting fractions, the denominators must be the same, that is, they must have *common denominators*.

Exercise 1G

1. Find the lowest common multiple for the following groups of numbers:
   (a) 6 and 7
   (b) 5 and 8
   (c) 4 and 6
   (d) 8 and 10
   (e) 12 and 10
   (f) 6 and 8
   (g) 15 and 9
   (h) 14 and 12
   (i) 18 and 10
   (j) 12 and 15
   (k) 20, 12 and 5
   (l) 10, 8 and 6
   (m) 24, 8 and 18
   (n) 30, 6 and 12
   (o) 27, 18 and 24
   (p) 15, 9 and 40

2. Simplify the following:
   (a) $\frac{1}{3} + \frac{4}{9}$
   (b) $\frac{1}{4} + \frac{3}{8}$
   (c) $\frac{7}{8} - \frac{1}{2}$
   (d) $\frac{3}{5} - \frac{8}{15}$
   (e) $1\frac{2}{7} + \frac{8}{21}$
   (f) $\frac{6}{20} + 2\frac{1}{5}$
   (g) $3\frac{7}{8} - 2\frac{9}{24}$
   (h) $2\frac{4}{5} - 1\frac{6}{15}$
   (i) $\frac{3}{5} + \frac{4}{9}$
   (j) $\frac{2}{3} + \frac{7}{10}$
   (k) $\frac{5}{8} - \frac{1}{3}$
   (l) $\frac{8}{15} - \frac{1}{4}$
   (m) $\frac{11}{12} + \frac{5}{8}$
   (n) $\frac{7}{16} + \frac{3}{10}$
   (o) $\frac{4}{9} - \frac{2}{21}$
   (p) $\frac{5}{6} - \frac{4}{27}$
   (q) $1\frac{7}{18} + 2\frac{3}{10}$
   (r) $3\frac{2}{6} + 1\frac{9}{20}$
   (s) $1\frac{3}{24} - \frac{5}{9}$
   (t) $1\frac{3}{16} - \frac{7}{12}$
   (u) $\frac{5}{6} + \frac{7}{16} + \frac{15}{24}$
   (v) $\frac{7}{10} + \frac{3}{8} + \frac{5}{6}$
   (w) $1\frac{1}{3} + \frac{7}{10} - \frac{4}{15}$
   (x) $\frac{5}{14} + 1\frac{3}{7} - \frac{5}{12}$
   (y) $3\frac{11}{18} + 1\frac{5}{12} - 2\frac{3}{4}$
   (z) $2\frac{7}{10} + 1\frac{4}{30} - 1\frac{3}{4}$

Exercise 1H

1. Find the product of the following:
   (a) $\frac{2}{3} \times \frac{5}{7}$
   (b) $\frac{1}{4} \times \frac{3}{5}$
   (c) $\frac{3}{5} \times \frac{7}{8}$
   (d) $\frac{2}{3} \times \frac{8}{11}$
   (e) $\frac{5}{8} \times \frac{4}{9}$
   (f) $\frac{2}{5} \times \frac{7}{8}$
   (g) $\frac{3}{4} \times \frac{8}{14}$
   (h) $\frac{5}{6} \times \frac{7}{10}$
   (i) $\frac{14}{15} \times \frac{3}{8}$
   (j) $\frac{8}{9} \times \frac{6}{15}$
   (k) $\frac{6}{7} \times \frac{21}{30}$
   (l) $\frac{9}{14} \times \frac{35}{36}$
   (m) $\frac{12}{21} \times \frac{15}{18}$
   (n) $\frac{7}{20} \times \frac{15}{20}$
   (o) $1\frac{1}{2} \times \frac{5}{6}$
   (p) $1\frac{4}{5} \times \frac{11}{12}$
   (q) $2\frac{3}{5} \times 1\frac{1}{4}$
   (r) $1\frac{1}{5} \times 2\frac{7}{8}$
   (s) $3\frac{2}{3} \times 2$
   (t) $4\frac{6}{8} \times 3$
   (u) $2\frac{11}{12} \times 3\frac{3}{5}$
   (v) $3\frac{3}{10} \times 1\frac{7}{15}$

2. Find the quotients of the following:
   (a) $\frac{1}{2} \div \frac{1}{3}$
   (b) $\frac{1}{4} \div \frac{1}{5}$
   (c) $\frac{2}{5} \div \frac{4}{9}$
   (d) $\frac{3}{8} \div \frac{1}{4}$
   (e) $\frac{7}{10} \div \frac{3}{5}$
   (f) $\frac{9}{14} \div \frac{2}{7}$
   (g) $\frac{6}{15} \div \frac{15}{21}$
   (h) $\frac{8}{9} \div \frac{12}{15}$
   (i) $1\frac{3}{4} \div \frac{7}{8}$
   (j) $1\frac{5}{6} \div \frac{7}{9}$
   (k) $2\frac{2}{5} \div 1\frac{1}{9}$
   (l) $3\frac{8}{9} \div 1\frac{1}{4}$
   (m) $3\frac{12}{20} \div 2$
   (n) $2\frac{6}{15} \div 6$
   (o) $2\frac{5}{16} \div 1\frac{11}{20}$
   (p) $1\frac{5}{12} \div 2\frac{2}{18}$

3. Simplify the following:
   (a) $\frac{3}{4} \times \frac{1}{2} \div \frac{1}{3}$
   (b) $\frac{4}{5} \times \frac{1}{3} \div \frac{2}{5}$
   (c) $\frac{3}{8} \times \frac{4}{3} \div \frac{2}{3}$
   (d) $\frac{6}{10} \times \frac{7}{8} \div \frac{3}{5}$
   (e) $\frac{5}{12} \div \frac{2}{6} \times \frac{1}{2}$
   (f) $\frac{4}{5} \div \frac{2}{3} \times \frac{1}{8}$
   (g) $\frac{7}{8} \div \frac{5}{7} \times \frac{3}{4}$
   (h) $\frac{11}{12} \div \frac{1}{4} \times \frac{5}{6}$
   (i) $1\frac{4}{5} \times 1\frac{1}{2} \div \frac{3}{4}$
   (j) $1\frac{2}{3} \times 1\frac{7}{8} \div \frac{5}{9}$

1.4 Decimals

Decimals give another way of writing fractions and mixed numbers.

In the decimal number system the *place value* system of units, tens, hundreds, etc. is maintained, but a decimal point is placed after the unit column to separate the whole numbers from the numbers less than one. Then the place value columns become tenths, hundredths, thousandths, etc.

E.g.   the number

Tens	Units	.	$\frac{1}{10}$	$\frac{1}{100}$	$\frac{1}{1000}$
1	2	.	5	2	6

reads as twelve point five two six

1.4.1 Addition and Subtraction

When adding and subtracting decimals it is simplest to work vertically, with all decimal points directly below each other forming a separate column. This helps to keep the place value distinct.

Exercise 1I

1. Using the signs $<$, $>$ or $=$, complete the following statements:

   (a) 0.5   5
   (b) 6   0.6
   (c) 2.01   2.1
   (d) 3.2   3.02
   (e) 6.2   6.05
   (f) 8.09   8.1
   (g) 3.010   3.001
   (h) 5.003   5.030
   (i) 0.062   0.0062
   (j) 0.0051   0.051
   (k) 0.0038   0.003
   (l) 0.007   0.0073
   (m) 0.52   0.526
   (n) 0.978   0.98
   (o) 0.405   0.045
   (p) 0.207   0.027

2. Write the following in ascending order:

   (a) 3.62, 36.2, 0.362
   (b) 4.06, 40.6, 400.6
   (c) 23.04, 23.4, 2.34, 2.034
   (d) 17.2, 1.702, 17.02, 1.72
   (e) 54.03, 5.43, 50.403, 5.403
   (f) 6.027, 60.27, 60.027, 6.27
   (g) 1.493, 1.943, 1.394, 1.439
   (h) 1.276, 1.762, 1.672, 1.267
   (i) 0.503, 0.526, 0.507, 0.500
   (j) 0.72, 0.725, 0.702, 0.7005
   (k) 0.010, 0.011, 0.110, 0.111
   (l) 0.2, 0.202, 0.022, 0.220

3. Evaluate the following:

   (a) $4.26 + 5.73$
   (b) $8.41 + 2.95$
   (c) $3.76 - 2.19$
   (d) $5.03 - 2.93$
   (e) $0.42 + 1.065$
   (f) $3.02 + 0.504$
   (g) $10.4 - 3.63$
   (h) $12.06 - 4.397$
   (i) $100.5 + 0.736 + 21.82$
   (j) $86.59 + 0.032 + 18.3$
   (k) $27.8 - 2.69$
   (l) $13.5 - 4.821$
   (m) $0.02 + 5.7 + 208.062$
   (n) $5.86 + 39.247 + 0.3$
   (o) $0.76 - 0.086$
   (p) $0.32 - 0.068$
   (q) $10.1 + 1.01 + 101 + 0.101$
   (r) $0.605 + 60.5 + 6.05 + 605$
   (s) $8.45 + 43.8 - 12.5$
   (t) $83.7 + 9.28 - 13.2$
   (u) $0.54 + 3.6 - 0.207$
   (v) $4.2 + 0.17 - 1.416$
   (w) $30.05 - 14.732 + 22.39$
   (x) $6.4 - 3.592 + 7.82$
   (y) $302.5 - 30.25 + 3.025$
   (z) $46.07 - 40.67 + 4.607$

1.4.2   Multiplication and Division

Example 1.3
Calculate:
(a) 6.2 × 10   (b) 0.05 × 1000
(c) 10.326 × 100

Solution
(a) 6.2 × 10 ———1 zero   Move decimal point
      = 62.   1 place to the *right*.

      = 62
(b) 0.05 × 1000 ——3 zeros   Move decimal point
      = 0050.   3 places to the right.

      = 50.0
      = 50

(c) 10.326 × 100 ——2 zeros   Move decimal point
      = 1032.6   2 places to the right.

      = 1032.6

Example 1.4
Calculate:
(a) 5.6 ÷ 100   (b) 3600 ÷ 1000
(c) 0.25 ÷ 10

Solution
(a) 5.6 ÷ 100   so move decimal point 2 places to the left.
      = .056
      = 0.056

(b) 3600 ÷ 1000   so move decimal point 3 places to the left.
      = 3.600
      = 3.6

(c) 0.25 ÷ 10   so move decimal point 1 place to the left.
      = .025
      = 0.025

Exercise 1J
Evaluate:

1. 2.6 × 10   2. 3.1 × 10
3. 42.6 × 100   4. 53.8 × 100
5. 13.4 ÷ 10   6. 16.8 ÷ 10
7. $\dfrac{106.7}{100}$ 8. $\dfrac{210.3}{100}$
9. 1.76 × 1000   10. 2.04 × 1000
11. $\dfrac{4.32}{100}$ 12. $\dfrac{8.61}{100}$
13. 0.065 × 10   14. 0.103 × 10
15. 0.34 ÷ 10   16. 0.49 ÷ 10
17. 407.1 × 100   18. 820.3 × 100
19. $\dfrac{281.6}{1000}$ 20. $\dfrac{347.2}{1000}$
21. 0.03 × 1000   22. 0.07 × 1000
23. $\dfrac{0.6}{100}$ 24. $\dfrac{0.8}{100}$
25. 42.8 × 100 ÷ 1000   26. 17.9 × 1000 ÷ 10

Example 1.5

Calculate:

(a) 2.65×1.4

(b) $4.68 \div 0.6$

Solution

(a)
$$\begin{array}{r} 265 \\ \times\ 14 \\ \hline 1060 \\ 2650 \\ \hline 3710 \end{array}$$

Multiply, temporarily ignoring the decimal points.

There is a total of 3 decimal places in the question so there will be 3 decimal places in the answer. The answer is not complete until the correct number of decimal places has been shown.

Thus, $2.65 \times 1.4 = 3.710$ (3 decimal places in the answer).

(b) In $4.68 \div 0.6$ the divisor (0.6) is a decimal which makes the division awkward; it is asking 'How many times is 0.6 contained in 4.68?'

Thus we make the divisor a whole number by multiplying by 10, (i.e. $0.6 \times 10 = 6$).
We have multiplied 0.6 by 10 so we must also multiply 4.68 by 10, (i.e. $4.68 \times 10 = 46.8$).
Now the question reads $46.8 \div 6$.

$$\begin{array}{r} 7.8 \\ \hline 6\overline{)46.8} \end{array}$$
$$\therefore\ 4.68 \div 0.6 = 7.8$$

Exercise 1 K

Calculate the following:

1. 3.6×9
2. 4.7×8
3. 3.54×4
4. 83.2×6
5. 42×0.3
6. 72×0.2
7. 0.5×23
8. 0.5×2.3
9. 0.5×0.23
10. 5×0.023
11. 72.6×18
12. 39.5×13
13. 16.5×3.4
14. 23.9×6.8
15. 3.62×8.5
16. 4.49×7.3
17. 26.5×0.46
18. 43.2×0.71
19. 2.1×4.083
20. 6.7×1.709
21. 0.871×0.34
22. 0.927×0.18
23. 0.02×0.008
24. 0.05×0.0015
25. $5.6 \div 8$
26. $8.4 \div 7$
27. $0.15 \div 5$
28. $0.54 \div 6$
29. $1.6 \div 0.4$
30. $4.5 \div 0.9$
31. $28 \div 0.7$
32. $96 \div 0.4$
33. $10 \div 0.8$
34. $21 \div 0.8$
35. $21.7 \div 0.4$
36. $3 \div 0.08$

1.4.3   Decimals to Fractions

Most decimals can be expressed in fraction form. Because of the place value system of tenths, hundredths, thousandths, etc., found in decimals, the conversion can be made easily into the appropriate denominator and then simplified.

For example

	$\frac{1}{10}$	$\frac{1}{100}$
$0\cdot$	4	2

Thus there are $\frac{4}{10} + \frac{2}{100}$

$$= \frac{40}{100} + \frac{2}{100} \qquad L.C.D. = 100$$

$$= \frac{42\,^{21}}{100\,_{50}}$$

$$= \frac{21}{50}$$

Note that $0.42 = \frac{42}{100}$ since there are digits in the decimal places up to the hundredths column.

Exercise 1L

Express the following as fractions in simplest form:

1. 0.2
2. 0.4
3. 0.08
4. 0.05
5. 0.12
6. 0.26
7. 0.005
8. 0.003
9. 0.072
10. 0.053
11. 0.325
12. 0.625
13. 1.7
14. 2.3
15. 2.65
16. 3.15
17. 1.048
18. 1.062
19. 3.808
20. 2.302

1.4.4   Fractions to Decimals

Any fraction can be expressed in decimal form by dividing the denominator into the numerator. We have already defined a fraction as one number, the numerator, divided by another number, the denominator; thus, to express in decimal form we should carry out the division. Thus, for example,

$$\frac{3}{4} = 3 \div 4$$

$$\begin{array}{r} 0.75 \\ 4\overline{)3.000} \end{array}$$

So $\frac{3}{4} = 0.75$

If we write $\frac{2}{9}$ as a decimal

$$\begin{array}{r} 0.22222\ldots \\ 9\overline{)2.00000\ldots} \end{array}$$

we find $\frac{2}{9} = 0.2222\ldots$
Similarly, $\frac{1}{3}$ as a decimal gives

$$\begin{array}{r} 0.33333\ldots \\ 3\overline{)1.00000\ldots} \end{array}$$

so $\frac{1}{3} = 0.3333\ldots$
Both these decimals continue without end and so are called *non-terminating*.

> A decimal which continues without an end is called a non-terminating decimal.

Also, in each of the above, the digit repeats indefinitely. This is called a *recurring* decimal and in such cases we place a dot over the repeating digit instead of writing them out more fully.
Thus $\frac{2}{9} = 0.2222\ldots = 0.\dot{2}$
and $\frac{1}{3} = 0.3333\ldots = 0.\dot{3}$
In fact, all fractions which become non-terminating decimals will repeat. Sometimes, however, we have to take the division for a large number of decimal places before we can see the pattern:

For $\frac{4}{7}$,
$$\frac{0.571428571\ldots}{7)4.000000000\ldots}$$
$\frac{4}{7} = 0.571428\,571428\ldots$
$\qquad\qquad$ repeating pattern

To indicate the recurring digits in this case we usually place a dot above the first and last digit of the recurring pattern:
$\frac{4}{7} = 0.\dot{5}7142\dot{8}$

If one or more digits are repeated continuously the decimal is called a recurring decimal.
If one or two digits recur a dot is placed over these digits.
If three or more digits recur a dot is placed on the first and last recurring digit.

Example 1.6
Write the following decimals in recurring decimal form:
(a) 0.444... (b) 0.6222...
(c) 0.313131... (d) 0.8423423...

Solution
(a) $0.444\ldots = 0.\dot{4}$ $\qquad$ since the 4 is repeating
(b) $0.6222\ldots = 0.6\dot{2}$ $\qquad$ since the 2 is repeating
(c) $0.313131\ldots = 0.\dot{3}\dot{1}$ $\qquad$ since the pattern 31 is repeating
(d) $0.8423423\ldots = 0.8\dot{4}2\dot{3}$ $\qquad$ since the pattern 423 is repeating
The dot shows that the decimal goes on continuously in a pattern.

Exercise 1 M
1. Write the following decimals in recurring decimal form:
   (a) 0.777... (b) 0.333...
   (c) 0.818181... (d) 0.040404...
   (e) 0.6020202... (f) 0.7363636...
   (g) 0.418418... (h) 0.915915...
   (i) 0.04321321... (j) 0.23627627...

2. Write the following as recurring decimals. Use a calculator where necessary:
   (a) $\frac{1}{3}$ (b) $\frac{2}{3}$
   (c) $\frac{4}{9}$ (d) $\frac{2}{9}$
   (e) $\frac{5}{6}$ (f) $\frac{1}{6}$
   (g) $\frac{5}{11}$ (h) $\frac{7}{11}$
   (i) $\frac{7}{15}$ (j) $\frac{8}{15}$
   (k) $\frac{7}{18}$ (l) $\frac{4}{33}$
   (m) $\frac{5}{22}$ (n) $\frac{3}{22}$
   (o) $\frac{21}{37}$ (p) $\frac{5}{27}$

3. Which of the following are recurring decimals:
   (a) $\frac{3}{8}$ (b) $\frac{7}{100}$
   (c) $\frac{3}{5}$ (d) $\frac{7}{8}$
   (e) $\frac{3}{11}$ (f) $\frac{7}{9}$
   (g) $\frac{3}{10}$ (h) $\frac{7}{11}$
   (i) $\frac{3}{7}$ (j) $\frac{7}{12}$

1.5   Rounding off

Decimals are used extensively in everyday life. Probably the major uses arise through money and measurement in the metric system.

When performing any of the operations with decimals more decimal places than required may result. For example £6.758 has no meaning. Therefore it is necessary to introduce a rule to terminate decimals after a specified number of decimal places.

Example 1.7

(a) Round-off the following decimals to one decimal place

  (i) 6.73   (ii) 12.48   (iii) 0.55

(b) Round-off the following decimals to two decimal places

  (i) 6.486   (ii) 3.7816   (iii) 3.78123

Solution

(a) (i) 6.73 lies between 6.70 and 6.80

6.73 is nearer to 6.70, therefore 6.73 ≈ 6.7 to one decimal place.

(ii) 12.48 lies between 12.40 and 12.50

12.48 is nearer to 12.50 therefore 12.48 ≈ 12.5 to one decimal place.

(iii) 0.55 lies between 0.50 and 0.60

0.55 is in the middle of 0.50 and 0.60. In this case the larger decimal is taken, so 0.55 ≈ 0.6 to one decimal place.

(b) (i) 6.486 lies between 6.480 and 6.490

6.486 is nearer to 6.490 therefore 6.486 ≈ 6.49 to two decimal places.

(ii) 3.7816 lies between 3.78 and 3.79

3.7816 is nearer to 3.78 therefore 3.7816 ≈ 3.78 to two decimal places.

(iii) 3.78123 lies between 3.78 and 3.79

3.78123 is nearer 3.78 therefore 3.78123 ≈ 3.78 to two decimal places.

The last two examples show that it does not matter how many decimal places are given, it is only the specified decimal place and the next place to the right that are important.

Example 1.8

Express 14.8385 to

(a) one decimal place

(b) the nearest hundredth

(c) three decimal places

Solution
(a) 14.8385   83 is nearer 80

one decimal place.   To one decimal place 14.83 ≈ 14.8
(b) 14.8385   38 is nearer 40

hundredth column.   To the nearest hundredth 14.835 ≈ 14.84
(c) 14.8385   85 is midway, so use 90

three decimal places   To three decimal places 14.8385 ≈ 14.839

Exercise 1 N

1. Express the following decimals correct to two decimal places:
   (a) 4.834   (b) 1.641
   (c) 6.978   (d) 2.887
   (e) 14.055   (f) 28.065
   (g) 4.8319   (h) 2.8427
   (i) 7.0842   (j) 17.0413
   (k) 0.98932   (l) 0.27642

2. Round off the following decimals to the number of decimal places indicated in the bracket:
   (a) 4.87 (1)   (b) 12.843 (2)
   (c) 0.0475 (3)   (d) 6.79 (1)
   (e) 0.9408 (3)   (f) 0.8736 (3)
   (g) 18.687 (2)   (h) 4.0649 (3)
   (i) 2.94 (1)   (j) 17.63 (1)
   (k) 3.863 (1)   (l) 9.877 (1)
   (m) 24.938 (1)   (n) 0.145 (1)
   (o) 7.9327 (2)   (p) 73.0645 (3)
   (q) 0.00762 (3)   (r) 2.0419 (2)
   (s) 14.83333 (2)   (t) 4.9128 (2)

3. Write the following decimals to the nearest tenth:
   (a) 12.63   (b) 4.71
   (c) 4.07   (d) 16.08
   (e) 0.928   (f) 0.739
   (g) 36.4732   (h) 2.592
   (i) 0.8563   (j) 104.8617
   (k) 1.97   (l) 9.96

1.6 Averages
To find the average of two numbers, we add them together and divide by two.

Example 1.9
Last week the Smith family used 12 pints of milk. This week they used 18 pints. Find the average number of pints used per week during the fortnight.

Solution
Total number of pints $= 12 + 18$
$$= 30$$
Average $= 30 \div 2$
$$= 15 \text{ pints per week.}$$

To find the average of three numbers we again add them up but this time we divide by three.

Example 1.10

Form 2A has 28 pupils, 2B and 2C have 31 each. What is the average form size?

Solution

$$\text{Total number of pupils} = 28 + 31 + 31$$
$$= 90$$
$$\text{Average} = 90 \div 3$$
$$= 30 \text{ pupils per class.}$$

> The average of a set of numbers is found by adding the numbers together and dividing by the total number of elements in the set.

Example 1.11

(a) Betty Bright found that her examination results were as follows:

English	98%
History	94%
Geography	82%
Mathematics	99%
Science	87%

What is Betty's average mark in the examinations?

(b) If Betty's Art mark was 80%, how would this affect her average mark?

Solution

(a) $\text{Average mark} = \dfrac{\text{Sum of all the marks}}{\text{Total number of subjects}}$

$$= \frac{98 + 94 + 82 + 99 + 87}{5}$$

$$= \frac{460}{5}$$

$$= 92$$

$\therefore$ Betty's average mark is 92%.

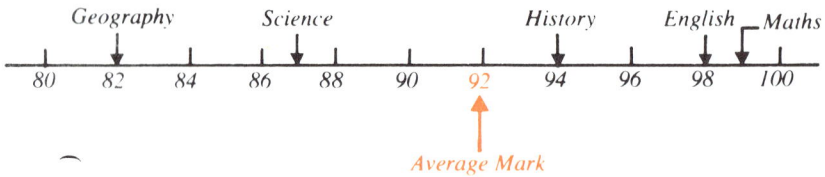

Fig 1.1

N.B.   The average of a set of numbers always lies between the largest and the smallest numbers.

(b) Average mark $= \dfrac{\text{Sum of all the marks}}{\text{Total number of subjects}}$

$= \dfrac{98 + 94 + 82 + 99 + 87 + 80}{6}$

$= \dfrac{540}{6}$

$= 90$

∴ Betty's average mark is now 90%.

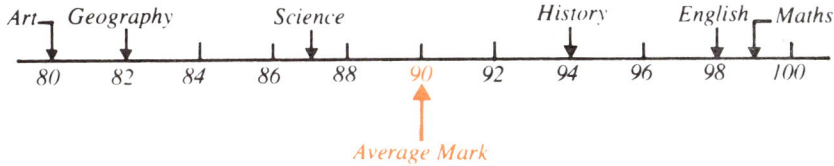

Average Mark

Fig 1.2

Betty's average mark has been lowered by her mark in Art.

Exercise 1O

1. Find the average of each of the following sets of numbers:
   (a) 6, 9, 8, 7   (b) 12, 15, 8, 3
   (c) 17, 14, 19, 6, 17   (d) 3, 15, 18, 27, 16
   (e) 9, 53, 84, 13, 63   (f) 4, 25, 16, 81, 13, 19
   (g) 7.2, 3.6, 4.7, 9.6, 8.5   (h) 3.2, 7.5, 6.9, 11.3, 8.05
   (i) 7.32, 5.1, 16.04, 24.1, 33.64   (j) 6.97, 9.86, 18.3, 5.74, 27.3

2. In the examinations Anne Smart obtained the following results:

English	83
French	69
History	72
Geography	67
Maths	95
Science	81
Art	75
Music	87

   What was Anne's average mark?

3. Jenny and Jill were comparing their examination results and found that their marks were as follows:

	Jenny	*Jill*
English	95	96
French	90	94
General Studies	92	94
Maths	99	99
Science	96	89
Craft	97	50

   (a) Calculate the average of each student.
   (b) Why are the averages so different?

4. The takings for a school tuck shop for one week are as follows:

Monday	£5.20
Tuesday	£6.33
Wednesday	£7.92
Thursday	£4.18
Friday	£10.72

Find the average takings per day.

5. In the last seven matches the school hockey team scored 5, 2, 7, 9, 0, 8 and 4 goals. What is the average score for the seven matches?

6. On a get-fit campaign I ran an average of 2 km each day. How far did I run in a week?

7. The pay of a part-time worker averages £60 per week. How much did he receive for six weeks' work?

2.1 Definitions

> A set is a collection of things of a particular kind and is presented by a pair of braces, { }, which means 'the set of'.

A set can be named by a capital letter.

E.g. W = 'the set of' days in the week.

  or W = {days in the week}.

The members of a set are called the *elements* of that set.

Thus, the elements of W are Monday, Tuesday, Wednesday, Thursday, Friday, Saturday and Sunday.

Similarly, if S = {1, 3, 5, 7, 9} then we could say, 3 *is an element of* {1, 3, 5, 7, 9} or 3 is an element of S.

Using set notation for 'is an element of' $(\in)$, we could write

  $3 \in S$

Also, we know that 8 is not an element of S.

Thus we would write $8 \notin S$.

The elements of a set can be either *described* or *listed*.

For example, A = {prime numbers between 1 and 23}

The elements of A are *described*.

    or A = {2, 3, 5, 7, 11, 13, 17, 19}

The elements of A are *listed*.

Note: the word '*between*' means that the numbers stated are *not* included in the set.

The number of elements in a set is called the *cardinal number* denoted by a small n in front of brackets containing the name of the set: $n(\ )$.

For example, B = {2, 9, 15, 24, 31, 48}

There are 6 elements in set B, thus we say

'The cardinal number of set B' = 6

In set notation, $n(B)$ = 6

The *universal set*, denoted by $\mathscr{E}$, is the set which contains all the elements that may be considered.

A *Venn Diagram* is a pictorial representation of all the sets being considered and is usually bound by the universal set.

E.g. $\mathscr{E}$ = {3, 6, 9, 12, 15, 18, 21, 24, 27, 30}

  and A = {multiples of 9}

∴ the multiples of 9 within the universal set are 9, 18 and 27.

∴ The Venn diagram would show:

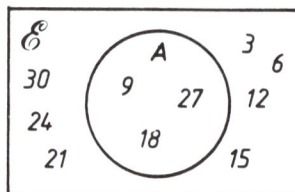

Fig 2.1

The Venn diagram shows that A is contained in $\mathscr{E}$; that is, all the elements of A are in $\mathscr{E}$.
Thus we say A 'is a subset of' $\mathscr{E}$ which can be denoted by $A \subset \mathscr{E}$ where $\subset$ is the symbol for 'is a subset of'.
Thus, a *subset* is a set which has all its elements contained within another set.
Consider the following Venn diagram:

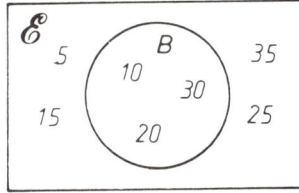

Fig 2.2

Fig 2.2 shows $\mathscr{E} = \{5, 10, 15, 20, 25, 30, 35\}$, that is, everything within $\mathscr{E}$

and $B = \{10, 20, 30\}$, everything within B.

The only region not considered separately is that outside B and inside $\mathscr{E}$.

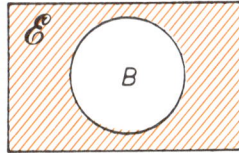

Fig 2.3

This region is called the *complement of B*, denoted by B', as it contains all elements in the universal set except those in B.
Thus: $B' = \{5, 15, 25, 35\}$

Example 2.1

$\mathscr{E} = \{1, 2, 3, 4, 5, 6, 7, 8, 9, 10, 11, 12, 13, 14, 15\}$
$A = \{\text{prime numbers}\}$
$B = \{3, 6, 9, 12\}$

(a) List the elements of A.
(b) State whether the following are, True or False?
   $9 \in A, 9 \in B, 9 \notin \mathscr{E}$.
(c) Find $n(A)$ and $n(B)$.
(d) State the subsets of B.
(e) Draw a Venn diagram of $\mathscr{E}$ and A and hence find A'.

Solution

(a) $A = \{\text{prime numbers}\}$
   $A = \{2, 3, 5, 7, 11, 13\}$ within the bounds of $\mathscr{E}$.
(b) 9 is not a prime number $\therefore$ $9 \in A$ is false.
   9 is contained in B $\therefore$ $9 \in B$ is true.
   9 is contained in $\mathscr{E}$ $\therefore$ $9 \notin \mathscr{E}$ is false.
(c) $n(A) = 6$ as there are 6 elements.
   $n(B) = 4$ as there are 4 elements.
(d) $B = \{3, 6, 9, 12\}$
The subsets of B are $\{3\}, \{6\}, \{9\}, \{12\}$
                $\{3, 6\}, \{3, 9\}, \{3, 12\}, \{6, 9\}, \{6, 12\}, \{9, 12\}$
                $\{3, 6, 9\}, \{3, 6, 12\}, \{3, 9, 12\}, \{6, 9, 12\}$
                $\{3, 6, 9, 12\}$ and $\{\ \ \}$
Note: the subsets of any set include the set itself and the null set.

(e)

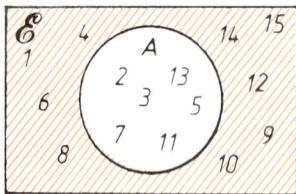

$A' = $ ▨

$A' = \{1, 4, 6, 8, 9, 10, 12, 14, 15\}$

Fig 2.4

2.1.1   Types of Sets

Consider $M = \{$cars displayed at the zoo$\}$.

As we do not see cars displayed at the zoo this set has no elements. Therefore this is called an *empty* set or a *null* set and is denoted by either

$M = \{\ \ \}$

or $M = \phi$

Sometimes it is not possible to *list* all the elements of a set as it is too large, for example, $\{$cities in the world$\}$.

But it can be represented by listing a part of the set only, as follows:

$\{$ Melbourne, Paris, London ... Munich$\}$

Note: The dots show there are more elements in the set than those already listed, also the element listed before the set is closed is to show that the set does not go on indefinitely. Thus it is called a *finite set*.

Similarly, it is not possible to list all the elements of a set such as $\{$whole numbers$\}$.

But this can be shown also by listing part of the set as follows:

$\{1, 2, 3, 4, 5 \ ... \ \}$.

Note: In this case the dots indicate that the set continues indefinitely as there are an infinite number of elements. This is called an *infinite set*.

Two sets which contain *exactly the same elements* are called *equal* sets.

E.g. $C = \{$the set of factors of 12$\}$

and $D = \{1, 2, 3, 4, 6, 12\}$

$\therefore C = D$

Two sets which have the *same cardinal number* (the same number of elements) are called *equivalent* sets.

E.g. $X = \{4, 8, 12, 16\}$

$Y = \{$oranges, apples, grapes, pears$\}$

$n(X) = 4$

$n(Y) = 4$

$\therefore X$ is equivalent to Y

$X \leftrightarrow Y$

Example 2.2

(a) State whether finite or infinite: $\{$people over 3 feet in height$\}$.

(b) State whether the following is true or false: $\{$even numbers between 8 and 10$\} = \phi$.

(c) Is $T = \{2, 3, 7, 11, 5, 9\}$ equal or equivalent to $A = \{2, 3, 5, 7, 11, 13\}$?

Solution

(a) Consider $\{$people over 3 feet in height$\}$. Although this is a very large set it cannot continue indefinitely so it is a finite set.

(b) As there are no elements in $\{$even numbers between 8 and 10$\}$ it is true to say $\{$even numbers between 8 and 10$\} = \phi$.

(c) For $T = \{2, 3, 7, 11, 5, 9\}$ and $A = \{2, 3, 5, 7, 11, 13\}$,

A contains different elements from T so they are not equal. However, $n(T) = 6$ and $n(A) = 6$ so they are equivalent, $A \leftrightarrow T$.

Summary of Set Notation

Symbol		Meaning
{ }	. . .	'the set of'
A, B, etc.	. . .	capital letters name a set
$\in$	. . .	'is an element of'
$\notin$	. . .	is not an element of
$n(A)$	. . .	the cardinal number of set A
{1, 2, 3 . . . 10}	. . .	a finite set
{1, 2, 3 . . .}	. . .	an infinite set
{ }, ϕ	. . .	empty set or null set
$A = B$	. . .	equal sets have the same elements
$A \leftrightarrow B$	. . .	equivalent sets have the same cardinal numbers
$\mathscr{E}$	. . .	the universal set
$\subset$	. . .	is a subset of
A'	. . .	the complement of set A; all elements in $\mathscr{E}$ except those in A.

Exercise 2A

1. List the following sets:
   (a) {digits}
   (b) {letters of the alphabet}
   (c) {odd numbers between 6 and 26}
   (d) {even numbers between 8 and 30}
   (e) {factors of 18}
   (f) {factors of 24}
   (g) {whole numbers between 7 and 7000}
   (h) {whole numbers from 101 to 202}
   (i) {multiples of 4}
   (j) {multiples of 6}
   (k) {fractions equivalent to $\frac{1}{2}$}
   (l) {fractions equivalent to $\frac{2}{3}$}

2. Describe the following sets:
   (a) {a, e, i, o, u}
   (b) {$+, -, \times, \div$}
   (c) {12, 14, 16, 18, 20, 22, 24}
   (d) {23, 25, 27, 29, 31, 33, 35, 37}
   (e) {1, 15, 3, 5}
   (f) {1, 12, 2, 6, 3, 4}
   (g) {81, 82, 83, 84 . . . 160}
   (h) {206, 207, 208, 209 . . . 1006}
   (i) {25, 30, 35, 40, 45 . . .}
   (j) {24, 32, 40, 48 . . .}
   (k) {0.1, 0.2, 0.3 . . . 1}
   (l) {2.4, 2.6, 2.8, 3.0 . . . 4.8}

3. Write the following in set notation, listing the elements when possible:
   (a) the set of odd numbers between 6 and 20;
   (b) the set of factors of 24;
   (c) P is equal to the set of even numbers;
   (d) M is equal to the set of months in the year;
   (e) June is an element of M;
   (f) Saturday is not an element of M;
   (g) the cardinal number of set M equals twelve;
   (h) set T is equal to the null set;
   (i) set Q is equivalent to set B;
   (j) the universal set is equal to the set of multiples of 3 between 2 and 40;
   (k) T is equal to the set of whole numbers 6, 9 and 12; T is a subset of the universal set;
   (l) the complement of set T is equal to the set of 3, 15, 18, 21, 24, 27, 30, 33, 36, and 39.

4. $A = $ {whole numbers between 4 and 16}
   $B = $ {the last 8 letters in the alphabet}
   $C = $ {multiples of 6 between 0 and 40}
   (a) Using the above information, state whether the following are true or false:
   (i) $7 \in A$
   (ii) $24 \in C$
   (iii) $t \in B$
   (iv) $16 \in A$
   (v) $r \in B$
   (vi) $18 \in A$
   (vii) $24 \in A$ and C
   (viii) $12 \in A$ and C
   (ix) $z \notin B$
   (x) $4 \notin A$

(xi)   12 ∉ A   (xii)   15 ∉ C
(xiii) 6 ∉ B   (xiv) x ∉ C
(b) Complete the following statements to make them true.
   (i)   7 ∈ ____   (ii)   30 ∈ ____
   (iii)   y ∈ ____   (iv)   14 ∈ ____
   (v)   12 ∉ ____   (vi)   6 ∉ ____
   (vii)   x ∉ ____   (viii) p ∉ ____
   (ix)   9 ∈ ____   (x)   24 ∈ ____
   (xi)   6 ∈ ____   (xii)   12 ∈ ____

5. Find the cardinal numbers of the following sets:
   (a) $A = \{2, 4, 6, 8, 10, 12, 14\}$
   (b) $C = \{a, b, c, d, e, f, g, h, i, j\}$
   (c) $M = \{\text{whole numbers from 1 to 100}\}$
   (d) $Q = \{\text{parts in question 5}\}$
   (e) $S = \{\text{different letters in 'intelligent'}\}$
   (f) $P = \{\text{whole numbers between 5 and 50}\}$
   (g) $T = \{\text{hours in a week}\}$
   (h) $B = \{\text{minutes in a day}\}$
   (i) $H = \{\text{toes on 16 people}\}$
   (j) $D = \{\text{legs on 20 spiders}\}$
   (k) $Z = \{\text{vowels in the word 'sky'}\}$
   (l) $J = \{\text{whales living in the jungle}\}$
   (m) $F = \{3, 6, 9 \ldots 42\}$
   (n) $W = \{2, 4, 6 \ldots 32\}$
   (o) $K = \{105, 110, 115 \ldots 455\}$
   (p) $E = \{640, 660, 680 \ldots 1420\}$

6. $P = \{\text{odd numbers less than 30}\}$
   $R = \{\text{factors of 15}\}$
   $T = \{1, 3, 5\}$
   (a) Using the above information, complete the following statements:
      (i) $\{25\} \subset$ ____   (ii) $\{17\} \subset$ ____
      (iii) $\{5, 15\} \subset$ ____   (iv) $\{15, 25\} \subset$ ____
      (v) $\{3\} \subset$ ____   (vi) $\{5\} \subset$ ____
      (vii) $\{13, 15, 17\} \subset$ ____ (viii) $\{3, 5, 7, 9\} \subset$ ____
      (ix) $\{1, 3, 5\} \subset$ ____   (x) $\{3, 5, 15\} \subset$ ____
      (xi) $R \subset$ ____   (xii) $T \subset$ ____
      (xiii) $\{15, 5, 1, 3\} \subset$ ____ (xiv) $\{3, 15, 1\} \subset$ ____
      (xv) $\phi \subset$ ____   (xvi) $\{\ \} \subset$ ____
   (b) List all the subsets of set T.
   (c) Draw a Venn diagram to show the relationship between P, R and T.

7. If $\mathscr{E} = \{2, 4, 6, 8, 10, 12, 14, 16, 18, 20, 22, 24\}$
   $A = \{4, 8, 12, 16, 20, 24\}$
   $B = \{6, 2, 18, 22\}$
   $C = \{10, 14\}$
   (a) draw a Venn diagram of
      (i) $\mathscr{E}$ and A (ii) $\mathscr{E}$ and B
      (iii) $\mathscr{E}$ and C
   (b) shade and list
      (i) A' (ii) B'
      (iii) C'

8. State whether the following sets are finite (F), infinite (I) or empty $\{\ \}$:
   (a) The set of multiples of 10 between 50 and 1000;
   (b) $\{5, 10, 15, 20 \ldots\}$;
   (c) $\{\text{all makes of cars in the world}\}$;
   (d) The set of prime numbers between 19 and 23;
   (e) $\{4, 8, 12 \ldots 404\}$;
   (f) $\{\text{6-wheeled bicycles}\}$;
   (g) The set of numbers greater than 10 020;
   (h) The set of words in the English language;
   (i) The set of whiskers on a snake;
   (j) $\{\text{fractions equivalent to } \frac{1}{2}\}$;
   (k) $\{\text{multiples of 7 between 22 and 27}\}$;
   (l) The set of factors of 1200.

9. State which of the following sets are equal and which are equivalent, using the symbols = or ↔ as applicable:
$A = \{1, 2, 3, 4 \ldots 21\}$
$B = \{$letters in the word 'God'$\}$
$C = \{$months in the year$\}$
$D = \{$numbers less than 12$\}$
$E = \{3, 9, 15, 21 \ldots 57\}$
$F = \{$consonants in the alphabet$\}$
$G = \{15, 30, 45, 60 \ldots 165\}$
$H = \{$letters in the word 'dog'$\}$
$I = \{7, 14, 21, 28 \ldots 84\}$
$J = \{$odd multiples of 3 less than 60$\}$

2.2   Use of Venn Diagrams

A Venn diagram, as stated before, is a pictorial representation of sets and the relationship between sets.

It is particularly useful when dealing with two or more sets, as the distribution of elements can be easily seen and thus interpreted.

For example, consider $S = \{1, 2, 3, 4, 5, 6, 7, 8\}$
and $T = \{2, 5, 7, 8\}$

To draw a Venn diagram we must first decide upon the relationship between S and T by comparing their elements.

All the elements of T are contained in S, thus the diagram should show T contained in S:

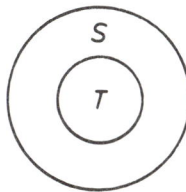

Fig 2.5

To complete the diagram the elements need to be placed in the appropriate sets, remembering that as T is contained in S those elements are already listed for S. Thus, the Venn diagram for S and T is

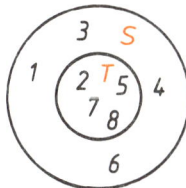

Fig 2.6

The diagram now shows us at a glance that
$T \subset S$
$T = \{2, 5, 7, 8\}$
$S = \{1, 3, 4, 6, 2, 5, 7, 8\}$
and $T' = \{1, 3, 4, 6\}$

2.2.1   Intersection of Sets

Consider the following sets:
$A = \{3, 6, 9, 12, 15, 18, 21, 24\}$
$B = \{4, 8, 12, 16, 20, 24\}$
Now, compare their elements, noting those which are common to both.
$A = \{3, 6, 9, 12, 15, 18, 21, 24\}$
$B = \{4, 8, 12, 16, 20, 24\}$
To represent these sets in a Venn diagram we would need to show an *overlap* of the sets containing the elements 12 and 24:

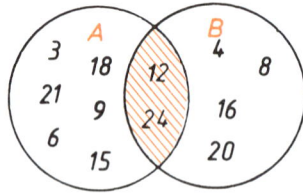

Fig 2.7

This *overlap* containing the common elements of both sets is called the *intersection* of A and B.
This is denoted by $A \cap B$
which says 'A *intersect* B' and therefore equals $\{12, 24\}$.
I.e. $A \cap B = \{12, 24\}$
Thus,

The set of elements which is contained in two or more sets, is called the intersection
of those sets. It is shown by an overlap of sets in a Venn diagram and denoted by the
symbol $\cap$.

E.g. $A \cap B$

Example 2.3
Given $P = \{2, 3, 5, 8, 9, 12, 15\}$
 $Q = \{4, 6, 7, 9, 14, 15\}$
 $R = \{5, 6, 9, 10, 12, 14, 16\}$
(a) Draw a Venn diagram to show the relationship between P, Q and R.
(b) Find (i) $P \cap Q$ (ii) $Q \cap R$ (iii) $R \cap P$ (iv) $P \cap Q \cap R$
Answer by listing the set and shading the appropriate region on a Venn diagram.

Solution
(a) To draw a Venn diagram we must first find any common elements to P, Q and R.
 $P = \{2, 3, 5, 8, 9, 12, 15\}$
 $Q = \{4, 6, 7, 9, 14, 15\}$
 $R = \{5, 6, 9, 10, 12, 14, 16\}$
9 is common to the three sets so we must show 3 sets overlapping.

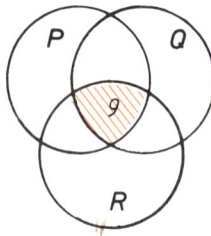

9 is contained in each of P, Q and R.

Fig 2.8

Note there are separate regions for the intersections of P and Q, Q and R, and R and P.
Considering each separately:
 $P = \{2, 3, 5, 8, 9, 12, 15\}$
and $Q = \{4, 6, 7, 9, 14, 15\}$

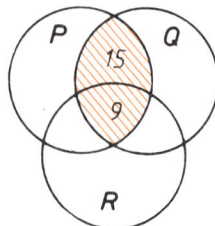

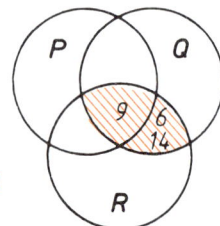

Fig 2.9

 $Q = \{4, 6, 7, 9, 14, 15\}$
and $R = \{5, 6, 9, 10, 12, 14, 16\}$

$R = \{5, 6, 9, 10, 12, 14, 16\}$
and $P = \{2, 3, 5, 8, 9, 12, 15\}$

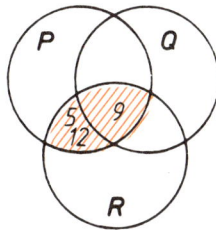

Fig 2.9 (cont)

Now putting this all together in one diagram we have:

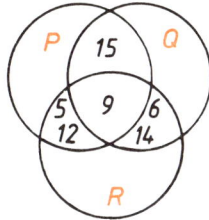

Fig 2.10

Complete the Venn diagram by filling in the sections of the sets which do not overlap, that is, the elements which are exclusive to each set.

$P = \{2, 3, 5, 8, 9, 12, 15\}$
$Q = \{4, 6, 7, 9, 14, 15\}$
$R = \{5, 6, 9, 10, 12, 14, 16\}$

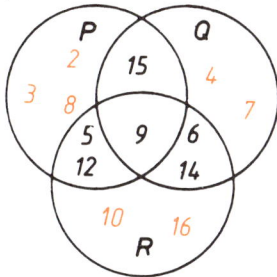

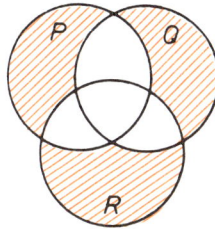

Fig 2.11

(b) (i) $P \cap Q$

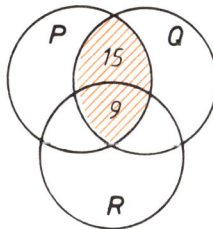

$\therefore P \cap Q = \{15, 9\}$

(ii) $Q \cap R$

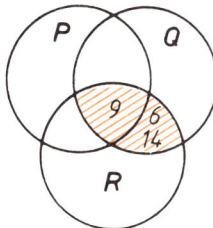

$\therefore Q \cap R = \{9, 6, 14\}$

(iii) $R \cap P$

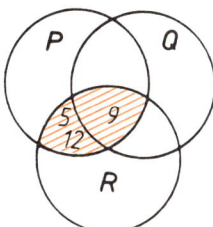

$\therefore R \cap P = \{9, 5, 12\}$

Fig 2.12

(iv) $P \cap Q \cap R$

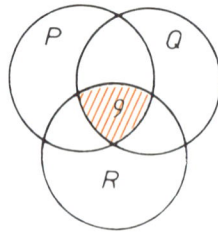

Fig 2.12 (cont)

$\therefore P \cap Q \cap R = \{9\}$

2.2.2   Union of Sets

The *union* of two or more sets, as the name suggests, is the set which contains all the elements of those sets.

Consider the following sets:

$$M = \{3, 4, 8, 10, 11, 16, 17\}$$
$$N = \{4, 6, 9, 11, 14, 15\}$$

The *union* of sets M and N is a representative of each different element of M and N put in one united set.

The union of M and N = $\{3, 4, 8, 10, 11, 16, 17, 6, 9, 14, 15\}$.

The symbol for the union of sets is $\cup$.

Thus $M \cup N = \{3, 4, 8, 10, 11, 16, 17, 6, 9, 14, 15\}$

Note: the elements 4 and 11, which are common to both M and N (i.e. $M \cap N = \{4, 11\}$) are only listed once.

A Venn diagram would show M intersecting N with 4 and 11 in the intersection, and the remaining elements in their respective sets.

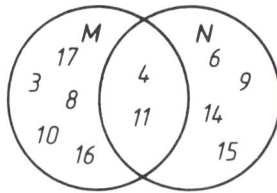

Fig 2.13

Thus the union of M and N is the entire shaded region.

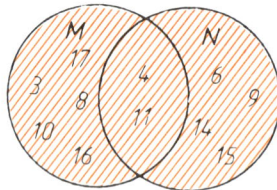

Fig 2.14

$M \cup N = \{3, 4, 6, 8, 9, 10, 11, 14, 15, 16, 17\}$.

> The union of two or more sets is the set which contains a representative of each of the elements in those sets. It is shown in a Venn diagram by shading the entire region of those sets and is denoted by the symbol $\cup$.
>
> E.g. $A \cup B$

Example 2.4

Draw a Venn diagram for the following sets and shade the region of the specified union.

(a) $A = \{4, 6, 8, 10, 12\}$
    $B = \{a, i, e, g\}$
    Find $A \cup B$

(b) $F = \{$positive whole numbers $< 6\}$
    $G = \{1, 2, 3, 4, 5, 6, 7, 8, 9, 10\}$
    Find $F \cup G$

(c) $\mathscr{E} = \{\text{odd numbers} < 30\}$
$X = \{3, 7, 15, 21, 25, 29\}$
$Y = \{1, 5, 7, 11, 15, 19, 23\}$
Find (i) $X \cup Y$
(ii) X'

Solution
(a) $A = \{4, 6, 8, 10, 12\}$
$B = \{a, i, e, g\}$
There is no intersection between A and B.
Thus the Venn diagram shows:

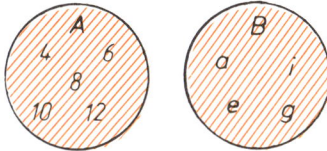

Fig 2.15

$\therefore A \cup B$ is the entire region of A and B (shaded red).
$A \cup B = \{4, 6, 8, 10, 12, a, i, e, g\}$.
(b) $F = \{\text{whole numbers} < 6\}$
$= \{1, 2, 3, 4, 5\}$
$G = \{1, 2, 3, 4, 5, 6, 7, 8, 9, 10\}$
All the elements of F are contained in G so $F \subset G$.
Thus the Venn diagram shows:

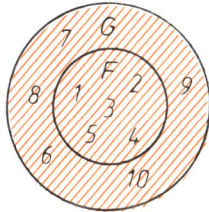

Fig 2.16

Thus the union, $F \cup G$, is the entire region of F and G (shaded red).
$\therefore F \cup G = \{1, 2, 3, 4, 5, 6, 7, 8, 9, 10\}$.
(c) $\mathscr{E} = \{\text{positive odd numbers} < 30\}$
$- \{1, 3, 5, 7, 9, 11, 13, 15, 17, 19, 21, 23, 25, 27, 29\}$
$X = \{3, 7, 15, 21, 25, 29\}$
$Y = \{1, 5, 7, 11, 15, 19, 23\}$
X and Y intersect as they have common elements 7 and 15.
Thus in the Venn diagram they overlap.

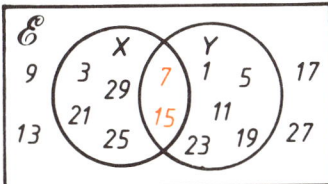

Fig 2.17

The remainder of elements in the universal set not already listed in X or Y are listed within the rectangle.

(i) In the Venn diagram, $X \cup Y$ is all of X and all of Y as shaded.

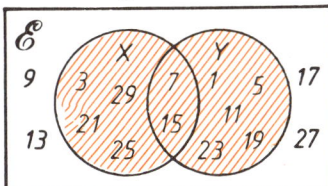

Fig 2.18

$X \cup Y = \{1, 3, 5, 7, 11, 15, 19, 21, 23, 25, 29\}$.

(ii) X' (the complement of X) is everything *except* the elements of X. Thus by shading the Venn diagram to show everything except X,

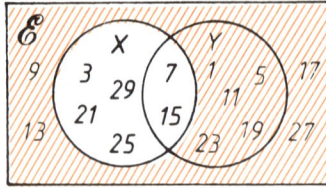

Fig 2.19

we find $X' = \{1, 5, 11, 19, 23, 9, 13, 17, 27\}$.

Note: the universal set ($\mathscr{E}$) contains all the elements that *may* be considered. As all the elements of $\mathscr{E}$ are not necessarily considered, the *union* of two sets is not necessarily the same as the universal set.

Exercise 2B

1. Using set notation, give the meaning of the following Venn diagrams. (If shaded, give the meaning of the shaded region.)

(a)

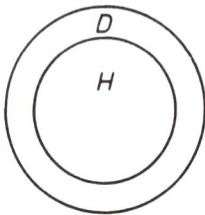

(b)

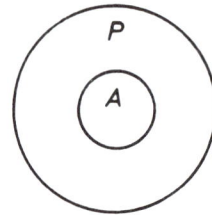

(c)

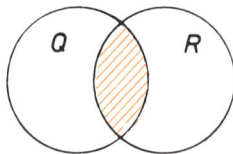

(d)

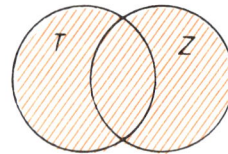

(e)

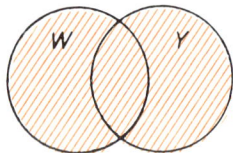

(f)

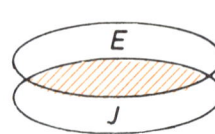

(g)

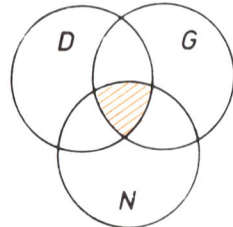

(h)

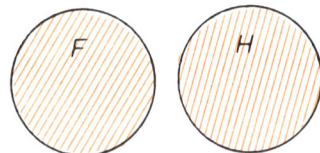

(i)

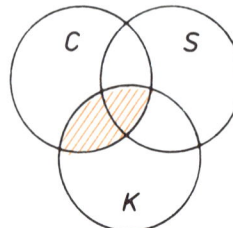

(j)

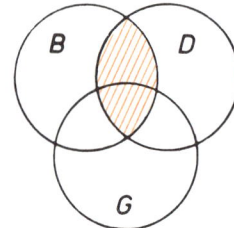

(k)

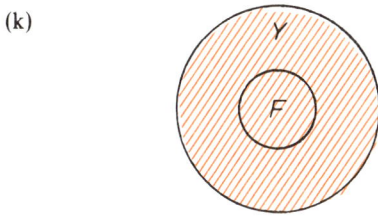

(l)

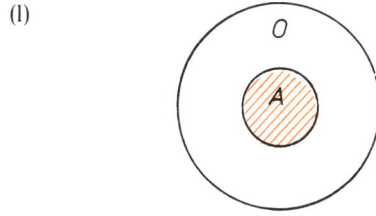

(m)

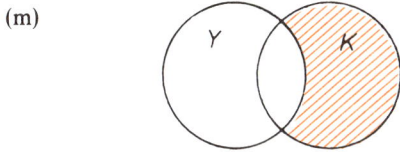

(n)

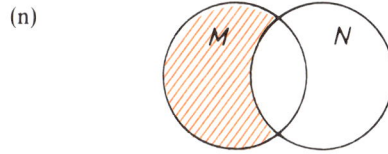

(o)

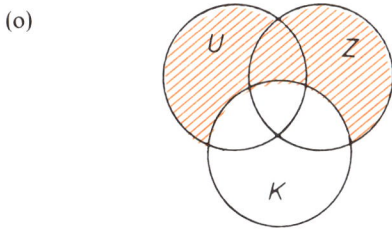

(p)

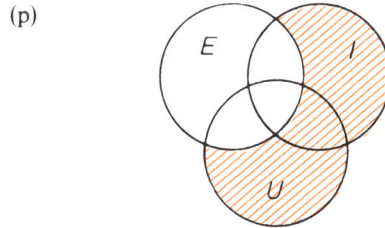

(q)

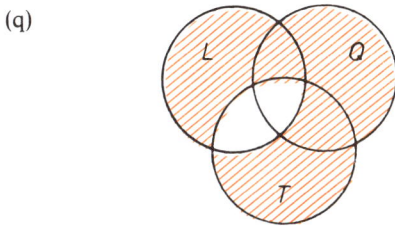

(r)

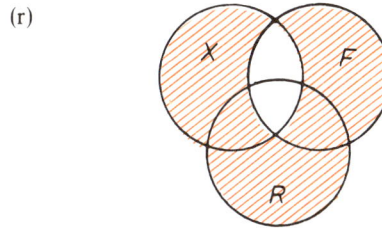

2. Draw Venn diagrams to show the relationship between the following sets:
   (a) Z = {animals at the zoo}
        Y = {Australian animals}
   (b) D = {dairy products}
        B = {eggs, bacon, toast, butter, coffee}
   (c) F = {Labradors, German Shepherds, Rottweilers}
        G = {dogs in the world}
   (d) T = {gum trees, oak trees, maple trees, silver birch trees}
        G = {trees}
   (e) A = {children under 10 years}
        S = {students in your form}
   (f) M = {land animals}
        Q = {whales}
   (g) C = {dogs}
        F = {four legged animals}
        R = {Rex, the guard dog}
   (h) R = {boys with red hair in Form 2}
        C = {children at school}
        F = {children in Form 2}
   (i) M = {students studying Maths}
        S = {students in your school}
        T = {students who play tennis}
   (j) C = {4 cylinder cars}
        R = {red cars}
        D = {cars with 2 doors}

3. Use the following sets to find the stated intersections and draw and shade a Venn diagram to show these intersections

$A = \{$natural numbers between 7 and 21$\}$

$B = \{2, 4, 6, 8, 10, 12\}$

$C = \{20, 24, 28, 32, 36, 40\}$

$D = \{$multiples of 10 less than 60$\}$

$E = \{$odd numbers between 8 and 27$\}$

$F = \{5, 10, 15, 20, 25, 30\}$

(a) $A \cap B$ (b) $A \cap C$

(c) $A \cap E$ (d) $A \cap D$

(e) $A \cap F$ (f) $C \cap D$

(g) $B \cap D$ (h) $D \cap F$

(i) $E \cap F$ (j) $C \cap F$

(k) $A \cap D \cap B$ (l) $A \cap E \cap F$

(m) $C \cap D \cap F$ (n) $B \cap D \cap F$

(o) $B \cap E$ (p) $C \cap E$

4. Using the sets A, C and F from question 2, draw a Venn diagram and hence list the following:

(a) $A \cap C \cap F$ (b) $A \cap F$

(c) $C \cap F$ (d) $A \cap C$

(e) A' (f) C'

(g) $(A \cap C)'$ (h) $(A \cap F)'$

(i) $(C \cap F)'$ (j) F'

(k) $(A \cap F \cap C)'$ (l) $A \cap (C \cap F)'$

(m) $C \cap (A \cap F)'$ (n) $F \cap (A \cap C)'$

5. Draw a Venn diagram showing the relationship of the following sets and use it to find the following unions. Show each union by shading the Venn diagram.

$P = \{$factors of 48$\}$

$T = \{$multiples of 6 less than 50$\}$

$F = \{4, 9, 12, 15, 16, 18, 20\}$

(a) $P \cup T$ (b) $T \cup F$

(c) $P \cup F \cup T$ (d) $F \cup P$

(e) P' (f) T'

(g) $(T \cup P)'$ (h) $(F \cup T)'$

(i) $(F \cup P)'$ (j) F'

(k) $P' \cup T$ (l) $T' \cup F$

(m) $F' \cup P'$ (n) $T' \cup P'$

2.3   Problem Solving Using Sets

Sets and Venn diagrams can be useful in solving certain problems.

Example 2.5

Draw a Venn diagram of the following situation and answer the questions:

(a) In a class of 30 boys, some play football (F), some play tennis (T), and some do swimming (S) or a combination of these. 3 boys do all three sports, 5 play both football and tennis only, 6 do swimming and football only; a total of 15 do swimming and 2 of the 14 boys who play tennis do swimming as well but not football. If all boys must play at least one sport, find:

(i)   how many play tennis only;

(ii) how many play football only;

(iii) how many do swimming only;

(iv) the total number of boys who play football;

(v)   how many do not play tennis;

(vi) how many do not swim.

Solution

The Venn diagram will show 3 intersecting sets F, T and S. The numbers placed in each region represent the number of elements in that region.

That is;
1. 3 do all sports $\therefore n(F \cap T \cap S) = 3$
2. 5 do F and T only $\therefore n(F \cap T) = 8$
3. 6 do S and F only $\therefore n(S \cap F) = 9$
4. 2 do S and T only $\therefore n(S \cap T) = 5$
5. Total of 15 in S $\therefore$ 4 in S only
6. Total of 14 in T $\therefore$ 4 in T only

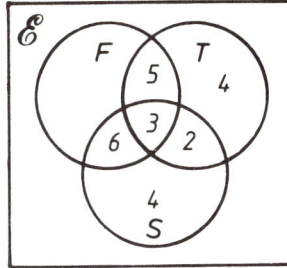

Fig 2.20

As there are 30 boys in the class and they must take at least one sport, those not already represented must be in F only. (i.e. play football only)

There is a total of 24 represented $(5 + 4 + 3 + 2 + 6 + 4)$.

Thus, the remaining 6 boys must be those who play football only.

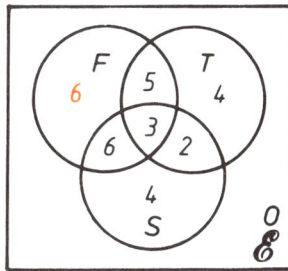

Fig 2.21

(i)   Those who play tennis but not football or swimming:

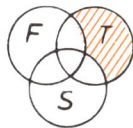

$\therefore$ tennis only $= 4$

(ii) Those who play football but not tennis or swimming:

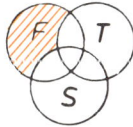

$\therefore$ football only $= 6$

(iii) Those who do swimming but not tennis or football:

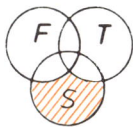

$\therefore$ swimming only $= 4$

(iv) Total number who play football:

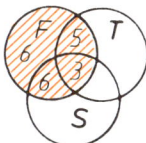

$\therefore$ football $= 6 + 5 + 3 + 6$
$\qquad\qquad = 20$

(v)   Those who do not play tennis, i.e. T':

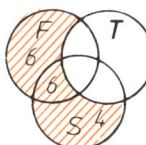

$\therefore T' = 6 + 6 + 4$
$\qquad = 16$

(vi) Those who do not swim, i.e. S':

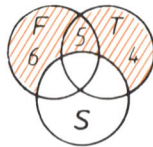

$$\therefore S' = 6 + 5 + 4$$
$$= 15$$

Exercise 2C

1. Draw a Venn diagram to represent the information given and solve the problem.

(a) In a class of 32 students, 18 do Drama and 22 are in the Choir. If 16 students do both Drama and Singing, how many do neither?

(b) In a factory which employs 85 workers, 52 catch the bus to work and 47 catch the train to work. If all the workers travel to work on either bus or train, how many travel to work on both the train and the bus?

(c) In a cricket club there are 18 people who can bowl and 30 who are batsmen; 7 of these members both bat and bowl. How many members does the club have?

(d) In a Form 2 class all the students learn at least one language. If 21 students do French, 14 do German and 6 do both, how many students are in the class?

(e) In a nursery they have 36 varieties of trees which are evergreens, and 55 varieties of trees which flower. If there is a total of 72 different varieties of trees, how many are evergreens which also flower?

(f) There are 56 members in a rugby club, all of whom participate in at least one other sport. 16 do swimming only, 12 play golf only, 11 play squash only, and 4 do both swimming and squash as well. If no member does all three sports as well as the rugby, and 7 play both squash and golf, how many play golf and swim?

2. (a)

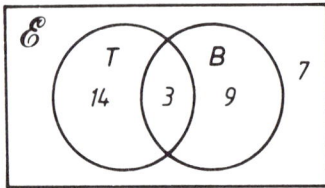

The Venn diagram represents a class of Form 2 girls, ($\mathscr{E}$), some of whom do ballet (B) and some tennis (T). The numbers represent the number of girls in each activity.

  (i)   How many girls in the class?
  (ii) How many girls do both ballet and tennis?
  (iii) How many girls learn ballet only?
  (iv) How many girls do neither tennis nor ballet?
  (v)   How many girls learn tennis?
  (vi) How many girls learn ballet?

(b)

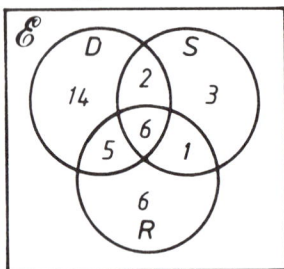

The Venn diagram represents a car park in which there are 42 cars, ($\mathscr{E}$). Some have four doors (D), some have seat covers (S), and some are red (R), or a combination of these. The numbers represent the number of cars in each category.

  (i)   How many cars are neither 4-door, red, nor have seat covers?
  (ii)   How many cars are 4-door, red, and have seat covers?
  (iii) How many cars are red and have seat covers, but not 4 doors?
  (iv)   How many cars are red and have seat covers?
  (v)   How many cars have seat covers and are 4-door?
  (vi)   How many cars are red?
  (vii) How many cars have seat covers but are neither red nor 4-door?
  (viii) How many cars are either 4-door, red or have seat covers?

Directed Numbers

3

3.1 Magnitude and Direction

On the Celsius temperature scale, water boils at 100°C and freezes at 0°C. However it can be even colder than 0°C. Temperatures below freezing point are called minus temperatures.
Thus 1°C below zero is called ‾1°C (minus 1°C)
and 10°C below zero is called ‾10°C (minus 10°C).

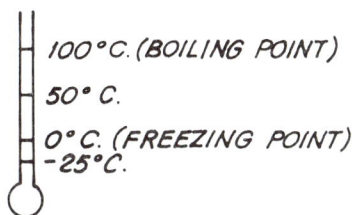

100°C. (BOILING POINT)

50° C.

0°C. (FREEZING POINT)
‑25°C.

Fig 3.1

The minus sign indicates that the temperature is below 0°C. In fact, many measurements require us to specify the direction as well as the amount or size so as to be more exact.
E.g.   9 km away, needs a direction of *N*, *S*, *E* or *W* etc. to indicate its exact position.

2 hours from noon, needs the direction of either before noon or after noon or the use of *A.M.* or *P.M.*

300 m from sea level, needs an indication as to whether it is above or below sea level.
Thus, if direction is not specified different meanings could be taken from that which was intended.
E.g.   The third street on the *right* is different from the third street on the *left*.

A *gain* of 12 kg is different from a *loss* of 12 kg.

419 km to the *west* is different from 419 km to the *east*.

An *increase* in temperature of 2°C is different from a *decrease* in temperature of 2°C.

Exercise 3A

1. Give the word which is the opposite to
   (a) down
   (b) left
   (c) increase
   (d) early
   (e) positive
   (f) negative

2. State the opposite of the following:
   (a) 5 steps up
   (b) 18 km east
   (c) 15 minutes late
   (d) win by 5 goals
   (e) a gain of 3 kg in weight
   (f) a profit of $6000
   (g) a fall in temperature of 6°C
   (h) the second door on the left
   (i) 3 hours before
   (j) 4 years in the past
   (k) 5°C below zero
   (l) 16°C above zero
   (m) positive
   (n) minus
   (o) +5°C
   (p) −16°C
   (q) −78°C
   (r) +98.6°C

We use + and − signs (positive and negative) to indicate either above zero or below zero respectively when talking about temperature. Likewise we can use + and − signs to indicate opposite directions with other quantities. If left is considered negative then right would be positive; if east is considered positive then west would be negative; if up is considered positive then down would be negative and so on.

Example 3.1

Rewrite the following using + or − signs given that 'right', 'east', 'up' and 'win' are considered positive (+).

(a) 3 steps to the left
(b) 6 km east
(c) 12 floors down
(d) Win by 15 points

Solution

(a) 3 steps to the left is negative 3 steps (− 3 steps).
(b) 6 km east is positive 6 km (+ 6 km).
(c) 12 floors down is negative 12 floors (− 12 floors).
(d) Win by 15 points is positive 15 points (+ 15 points).

Exercise 3B

1. Complete the following table. The first one is done for you.

NEGATIVE	POSITIVE
LOSS 2 kg	GAIN 2 kg
DOWN 9 steps	
	ABOVE S/L 2 km
BACKWARDS 6 steps	
	RISE 8°C
LOSS of £1400	
	EAST 6 km
LOSE by 6 points	
	INCREASE of £10
	NORTH 700 km
LEFT 9 paces	
	→
↓	

2. Using the table in question 1, rewrite each of the following using + or − to indicate direction:

(a) North 40 km
(b) Rise 14°C
(c) Down 4 floors
(d) Backwards 8 paces
(e) Increase of 10%
(f) Left 6 paces
(g) Lose 9 kg
(h) Profit of £2000
(i) Above sea level 14 m
(j) Up 7 steps
(k) Decrease of 14 g
(l) West 17 km
(m) Right 12 steps
(n) Positive 19

Example 3.2

On the east/west line below, follow the directions given and state the finishing point.

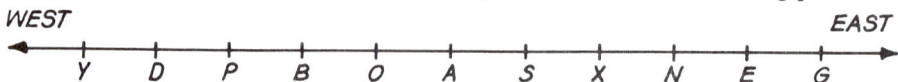

WEST   EAST

◄———+———+———+———+———+———+———+———+———+———+———
    Y   D   P   B   O   A   S   X   N   E   G

(a) Start at *O*, face east and move 5 units.
(b) Start at *O* and move + 5.

Solution

(a)

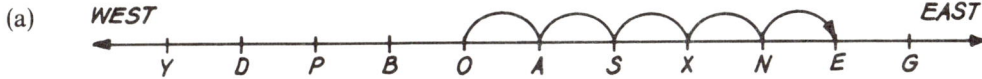

Finishing point *E*.

(b)

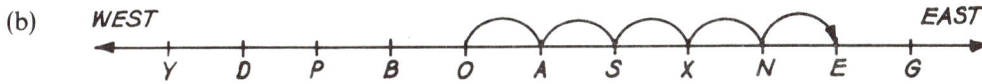

Finishing point *E*.

Exercise 3C

Using the east/west line in example 3.2, follow the directions given below and state the finishing point.

1. Start at *B*, face west and move 2 units.
2. Start at *A*, face east and move 4 units.
3. Start at *S*, face west and move 6 units.
4. Start at *P*, face east and move 3 units.
5. Start at *X*, face east and move 1 unit.
6. Start at *N*, face west and move 7 units.
7. Start at *O*, move $+4$ units.
8. Start at *Y*, move $+8$ units.
9. Start at *S*, move -5 units.
10. Start at *X*, move -1 unit.
11. Start at *B*, move $+3$ units.
12. Start at *A*, move -2 units.
13. Start at *P*, move $+6$ then -2.
14. Start at *S*, move -4 then $+5$.
15. Start at *O*, move $+5$ then -7.
16. Start at *X*, move -2 then -4.

> A *directed number* is a number for which direction $(+, -)$ as well as size is specified.

3.2   The Number Line

The number line, like the thermometer, can be extended to include numbers less than zero as well as greater than zero.

The numbers to the right of zero, as always, are positive and indicated by a $+$ director sign, while the numbers to the left of zero will therefore be negative, indicated by a $-$ director sign.

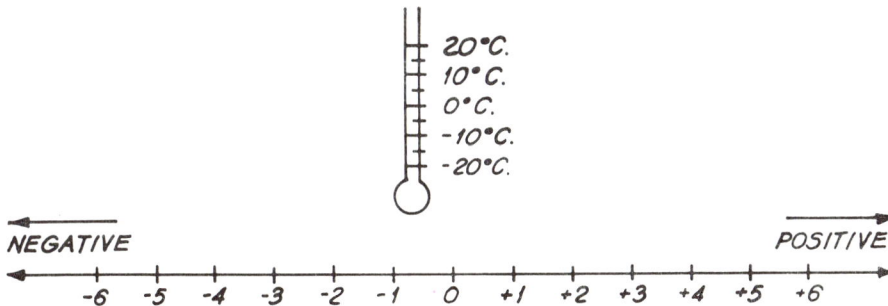

Fig 3.2

> A *number line* is used to show the position of a number as both distance and direction from zero and extends indefinitely in both positive and negative directions.

The number line in fig 3.2 can be used to compare the sizes of numbers. As you move in an easterly/positive direction along the number line, it can be seen that the numbers become larger. Similarly, as you move in a westerly/negative direction along the number line, it can be seen that the numbers decrease in size. Think of this in relation to the thermometer. As the temperatures rise and move in a positive direction it gets hotter (larger temperatures) and as they move down in a negative direction it gets colder (smaller temperatures).

Example 3.3

(i) Show the position of the following pairs of numbers on a number line.
(ii) State which is larger.

    (a) $^+6, ^+2$ (b) $^+5, ^-3$
    (c) $^+2, ^-5$ (d) $^-2, ^+1$
    (e) $^-4, 0$ (f) $^-1, ^-6$

Solution

(a) (i)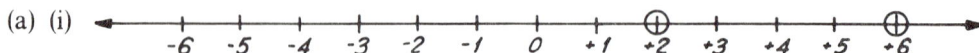

    (ii) $^+6$ is larger than $^+2$ because it lies to the right of $^+2$

(b) (i)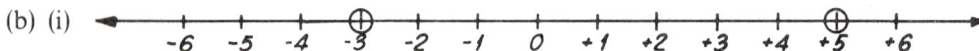

    (ii) $^+5$ is larger than $^-3$ because it lies to the right of $^-3$

(c) (i)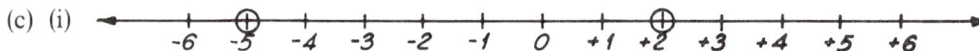

    (ii) $^+2$ is larger than $^-5$ because it lies to the right of $^-5$

(d) (i)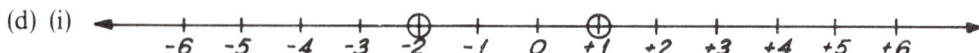

    (ii) $^+1$ is larger than $^-2$ because it lies to the right of $^-2$

(e) (i)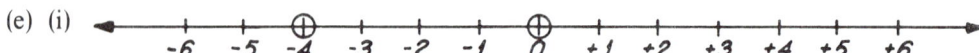

    (ii) 0 is larger than $^-4$ because it lies to the right of $^-4$

(f) (i)

    (ii) $^-1$ is larger than $^-6$ because it lies to the right of $^-6$

Exercise 3D

1. Mark the following groups of numbers on separate number lines, and state in which direction you are moving along the number line (easterly or westerly). Hence state whether the numbers are increasing or decreasing.

    (a) $^+6, ^+4, ^+1, 0, ^-2, ^-5$ (b) $^-7, ^-5, ^-1, ^+4, ^+5, ^+7$
    (c) $0, ^-1, ^-4, ^-5, ^-7$ (d) $^+2, ^+4, ^+7, ^+8$
    (e) $^-3, ^-4, ^-6, ^-7$ (f) $^-2, ^+1, ^+4, ^+5$
    (g) $^+3, 0, ^-3$ (h) $^-4, ^-3, ^-1, 0$

2. Mark the following pairs of numbers on a number line and hence state which is the larger number.

    (a) $^+8, ^+3$ (b) $^+2, ^+5$
    (c) $^+4, ^-1$ (d) $^+6, ^-3$
    (e) $^+1, ^-7$ (f) $^+7, ^-9$
    (g) $^-3, ^+4$ (h) $^-6, ^+2$
    (i) $^-4, ^+6$ (j) $^-3, ^+1$
    (k) $^-7, ^-2$ (l) $^-10, ^-5$
    (m) $^-1, ^-4$ (n) $^-3, 0$

3. Mark the following groups of numbers on a number line and state which is (i) the largest and (ii) the smallest.

    (a) $^+2, ^+6, ^+1$ (b) $^+4, ^+7, ^+2$
    (c) $^-2, ^+2, ^+4$ (d) $^-3, ^+6, ^+1$
    (e) $^+2, ^-3, ^-5, ^+1$ (f) $^+4, ^-3, ^+1, ^-6$
    (g) $^-4, ^-1, ^-5, ^-2$ (h) $^-6, ^-3, ^-5, ^-7$
    (i) $^+1, ^-6, ^-2, 0$ (j) $^-7, ^-2, ^+3, ^+7$
    (k) $^-2, ^-8, ^-9, ^-6$ (l) $^-3, ^-2, ^-5, ^-10$
    (m) $^-6, 0, ^-3, ^+1$ (n) $^-8, ^+2, 0, ^+3$

4. Insert < or > sign between each of the following pairs of numbers.

(a) $^+2$ $^+8$ (b) $^+7$ $^+3$
(c) $^+14$ $^+12$ (d) $^+37$ $^+49$
(e) $^+2$ $^-1$ (f) $^+8$ $^-3$
(g) $^-8$ $^+7$ (h) $^-13$ $^+6$
(i) $^+1$ 0 (j) 0 $^+2$
(k) $^-24$ $^-3$ (l) $^-14$ $^-45$
(m) $^-6$ $^-7$ (n) $^-4$ $^-1$
(o) 0 $^-8$ (p) $^-11$ 0
(q) $^-13$ $^+12$ (r) $^+6$ $^-10$
(s) $^-3$ $^-5$ (t) $^-8$ $^-1$
(u) $^-2$ $^+3$ (v) $^+6$ $^-6$
(w) $^+4$ $^-4$ (x) $^+12$ $^-7$
(y) $^-6$ $^-3$ (z) $^-1$ $^-17$

> When numbers are represented on a number line, the further to the right, the larger the number.

3.3   Addition and Subtraction

When adding and subtracting directed numbers, we find the number line very useful. We can translate arithmetic expressions into instructions for moving along the number line so as to find the finishing point or answer.

3.3.1   Addition

Example 3.4

On a number line find the following finishing points:
(a) start at $^+3$ face east and move 5 units forward;
(b) start at $^+5$ face east and move 2 units forward;
(c) start at $^-8$ face east and move 4 units forward;
(d) start at $^-3$ face east and move 7 units forward;

Solution

(a)

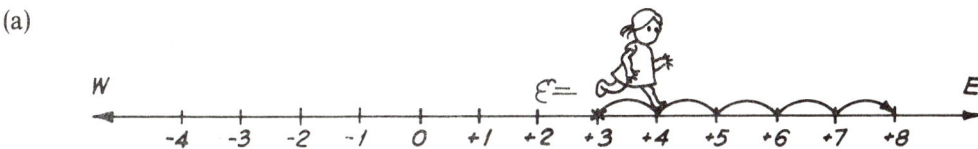

Finishing point $^+8$.

(b)

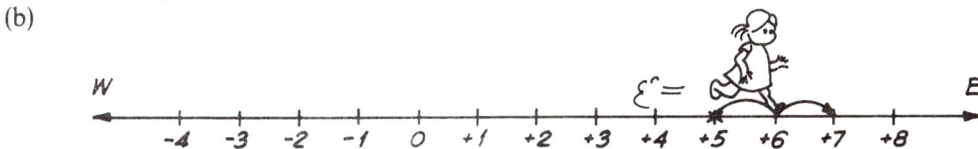

Finishing point $^+7$.

(c)

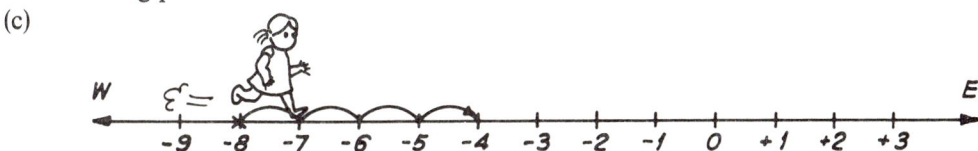

Finishing point $^-4$.

(d)

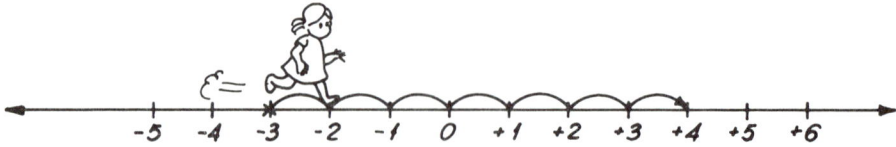

Finishing point $^+4$.

In each of the above examples, the instructions remain the same with only the starting point and the number of units moved being changed. The instruction to 'face east' can be equated to addition and the instruction to 'move forward' can be equated to the number of units to move being positive.

From example 3.4(a) we find

start at $^+3$	face east	move 5 units forward—finish at $^+8$
$^+3$	$+$	$^+5$ $= ^+8$

From example 3.4(b) we find

start at $^+5$	face east	move 2 units forward—finish at $^+7$
$^+5$	$+$	$^+2$ $= ^+7$

From example 3.4(c) we find

start at $^-8$	face east	move 4 units forward—finish at $^-4$
$^-8$	$+$	$^+4$ $= ^-4$

From example 3.4(d) we find

start at $^-3$	face east	move 7 units forward—finish at $^+4$
$^-3$	$+$	$^+7$ $= ^+4$

Hence $^-5$ $+$ $^+2$ can be translated to give,

start at $^-5$ face east   move 2 units forward

Then from the number line we find

$^-5$ $+$ $^+2$ $= ^-3$

So far we have added positive numbers.

Negative numbers may also be added using the number line.

Example 3.5

Evaluate

(a) $^+5 + ^-2$

(b) $^-1 + ^-4$

Solution

(a) $^+5$ $+$ $^-2$ can be translated to give,

start at $^+5$ face east   move 2 units *backwards.*

On the number line,

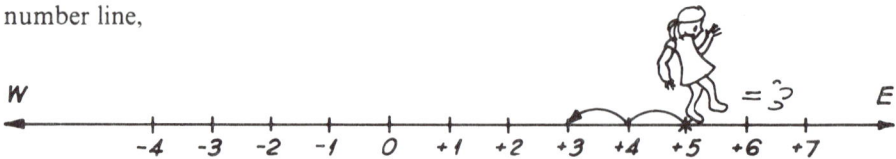

We finish at $^+3$.

Therefore $^+5 + ^-2 = ^+3$

(b) $^-1$ $+$ $^-4$

says start at $^-1$,   face east   move 4 units *backwards.*

On a number line,

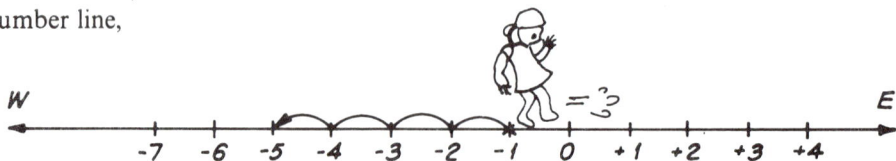

We finish at $^-5$.

Therefore $^-1 + ^-4 = ^-5$

Exercise 3E

1. Rewrite the following, explaining the steps involved:

(a) $^+4 + {}^+2$ (b) $^+6 + {}^+3$
(c) $^+7 + {}^+8$ (d) $^+2 + {}^+5$
(e) $^-4 + {}^+6$ (f) $^-3 + {}^+8$
(g) $^-12 + {}^+9$ (h) $^-10 + {}^+6$
(i) $^-7 + {}^+7$ (j) $^-16 + {}^+16$
(k) $^+12 + {}^-3$ (l) $^+7 + {}^-2$
(m) $^+18 + {}^-17$ (n) $^+22 + {}^-13$
(o) $^+8 + {}^-12$ (p) $^+1 + {}^-4$
(q) $^-13 + {}^-11$ (r) $^-16 + {}^-3$
(s) $^-20 + {}^-15$ (t) $^-6 + {}^-15$

2. Using the number line, find the finishing points (answers) of the examples in question 1.

3.3.2   Subtraction

Similarly, for *subtraction*, we would change the instruction from add ($+$) meaning face east, to subtract ($-$) meaning face west.

Example 3.6

Find the finishing points on a number line of the following.

(a) $^+5 - {}^+3$ (b) $^+1 - {}^+6$
(c) $^-2 - {}^+5$ (d) $^-3 - {}^-4$

Solution

(a) $^+5$ $-$ $^+3$
    says, start at $^+5$ face west   move 3 units forward.
On a number line,

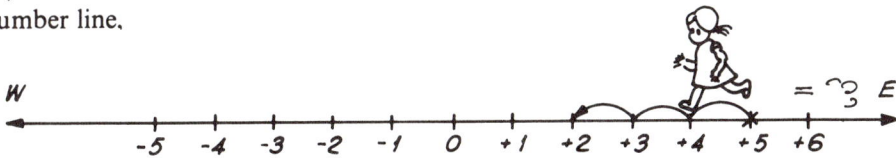

We finish at $^+2$.
Therefore $^+5 - {}^+3 = {}^+2$

(b) $^+1$ $-$ $^+6$
    says, start at $^+1$ face west   move 6 units forward.
On a number line,

We finish at $^-5$.
Therefore $^+1 - {}^+6 = {}^-5$

(c) $^-2$ $-$ $^+5$
    says, start at $^-2$ face west   move 5 units forward.
On a number line,

We finish at $^-7$.
Therefore $^-2 - {}^+5 = {}^-7$

(d) $^-3$ $-$ 4

   says, start at $^-3$ face west   move 4 units backwards.
On a number line,

We finish at $^+1$.
Therefore $^-3 - {}^-4 = {}^+1$

Exercise 3F

1. Write the steps involved in finding the following:

(a) $^+4 - {}^+2$	(b) $^+6 - {}^+3$
(c) $^+7 - {}^+4$	(d) $^+10 - {}^+5$
(e) $^+9 - {}^+12$	(f) $^+3 - {}^+6$
(g) $^-5 - {}^+3$	(h) $^-2 - {}^+5$
(i) $^-1 - {}^+6$	(j) $^-3 - {}^+7$
(k) $^-3 - {}^+1$	(l) $^-8 - {}^+6$
(m) $^+3 - {}^-1$	(n) $^+7 - {}^-5$
(o) $^+9 - {}^-2$	(p) $^+4 - {}^-2$
(q) $^-4 - {}^-8$	(r) $^-6 - {}^-11$
(s) $^-13 - {}^-6$	(t) $^-8 - {}^-3$

2. Using a number line, find the finishing points (answers) of the examples in question 1.

Summary

+ (addition)   —face east
^+N(positive integer) —move forward
– (subtraction)   —face west
^-N(negative integer)—move backward

Referring back to the examples 3.5 and 3.6, we can see that two different sets of instructions lead to movement in the same direction along the number line.
The combination of $N + {}^+M$ says face east and move forwards.
This leads to movement in an easterly/positive direction.
Similarly
The combination of $N - {}^-M$ says face west and move backwards.
This also leads to movement in an easterly/positive direction.
Furthermore,
the combination of $N + {}^-M$ says face east and move backwards.
This leads to movement in a westerly/negative direction.
Similarly,
the combination of $N - {}^+M$ says face west and move forwards.
This also leads to movement in a westerly/negative direction.
Therefore, we should be able to condense two instructions into one, consisting of either move in a positive direction $(+)$ or a negative direction $(-)$.
For example,
$^+6 + {}^+2$ is the same as $^+6 + 2$,
$^+3 - {}^-2$ is the same as $^+3 + 2$,
$^+4 - {}^+6$ is the same as $^+4 - 6$,
$^-2 + {}^-4$ is the same as $^-2 - 4$.

These instructions now become, respectively:

$^+6$ $+$ 2

start at $^+6$ move east/positive 2 units

$^-2$ $-$ 4

start at $^-2$ move west/negative 4 units

$^+4$ $-$ 6

start at $^+4$ move west/negative 6 units

$^+3$ $+$ 2

start at $^+3$ move east/positive 2 units

Note: the director sign attached to the number which is the starting point does *not* alter the addition or subtraction sign between the numbers. The only time that two signs can be simplified is when they lie between two numbers.

Also if a number has *not* got a director sign attached to it, it is assumed to be positive. Thus 6 is read as $^+6$.

Combinations of signs: $+$ $^+$ is $+$
$+$ $^-$ is $-$
$-$ $^+$ is $-$
$-$ $^-$ is $+$

Since $+$ and $-$ are 'binary operations' (that is, each operation can be performed on only two numbers at a time) care must be taken when three or more numbers are to be added or subtracted (Chapter 1). It is safer to calculate the *first pair of numbers first*.

Example 3.7

Evaluate

(a) $3 + 5 + 2$

(c) $-2 - 3 + 4$

(b) $2 - 5 + 6$

(d) $-3 - 5 - 7$

Solution

(a) $3 + 5 + 2 = (3 + 5) + 2$
$= 8 + 2$
$= 10$

(c) $-2 - 3 + 4 = (-2 - 3) + 4$
$= {}^-5 + 4$
$= {}^-1$

(b) $2 - 5 + 6 = (2 - 5) + 6$
$= -3 + 6$
$= 3$

(d) $-3 - 5 - 7 = (-3 - 5) - 7$
$= {}^-8 - 7$
$= {}^-15$

Exercise 3G

1. Simplify the following by expressing them with a single sign between the numbers first and then use a number line to evaluate them.

   (a) $^+6 + {}^+5$
   (c) $^+9 + {}^+12$
   (e) $^-2 + {}^+19$
   (g) $^-14 + {}^+7$
   (i) $^+3 - {}^-4$
   (k) $^+8 - {}^-9$
   (m) $^-7 - {}^-12$
   (o) $^-8 - {}^-2$

   (b) $^+3 + {}^+7$
   (d) $^+10 + {}^+6$
   (f) $^-5 + {}^+14$
   (h) $^-8 + {}^+2$
   (j) $^+7 - {}^-6$
   (l) $^+2 - {}^-5$
   (n) $^-5 - {}^-10$
   (p) $^-15 - {}^-9$

2. Simplify the following by expressing them with a single sign between the numbers first and then use a number line to evaluate them.

   (a) $^+7 + {}^-3$
   (c) $^+2 + {}^-5$
   (e) $^-4 + {}^-3$

   (b) $^+10 + {}^-6$
   (d) $^+6 + {}^-8$
   (f) $^-1 + {}^-7$

(g) $^-9 + {}^-12$ (h) $^-11 + {}^-6$
(i) $^+4 - {}^+1$ (j) $^+7 - {}^+6$
(k) $^+8 - {}^+11$ (l) $^+5 - {}^+15$
(m) $^-3 - {}^+7$ (n) $^-6 - {}^+13$
(o) $^-16 - {}^+5$ (p) $^-14 - {}^+18$

3. Simplify the following by expressing them with a single sign between the numbers and then use a number line to evaluate them.

(a) $^+4 + {}^+9$ (b) $^+3 + {}^+6$
(c) $^+7 + {}^+6$ (d) $^+2 + {}^+10$
(e) $^-3 + {}^+8$ (f) $^-5 + {}^+6$
(g) $^-9 + {}^+4$ (h) $^-14 + {}^+7$
(i) $^+6 + {}^-1$ (j) $^+11 + {}^-8$
(k) $^+8 + {}^-15$ (l) $^+4 + {}^-6$
(m) $^-12 + {}^-3$ (n) $^-7 + {}^-5$
(o) $^+9 - {}^+6$ (p) $^+13 - {}^+10$
(q) $^-7 - {}^+2$ (r) $^-1 - {}^+6$
(s) $^+2 - {}^-12$ (t) $^+6 - {}^-5$
(u) $^+3 - {}^-3$ (v) $^+8 - {}^-9$
(w) $^-12 - {}^-10$ (x) $^-4 - {}^-8$
(y) $^-6 - {}^-6$ (z) $^-3 - {}^-3$

4. Evaluate the following:

(a) $^+2 + 4$ (b) $^+6 + 8$
(c) $^-7 + 2$ (d) $^-5 + 1$
(e) $^-6 + 11$ (f) $^-4 + 13$
(g) $^+2 - 1$ (h) $^+9 - 5$
(i) $^+6 - 8$ (j) $^+12 - 22$
(k) $^-4 - 6$ (l) $^-3 - 9$
(m) $^+13 + 4$ (n) $^+12 + 1$
(o) $^+6 - 10$ (p) $^+8 - 14$
(q) $^-3 + 6$ (r) $^-4 + 7$
(s) $^-2 - 4$ (t) $^-6 - 7$
(u) $^-1 + 1$ (v) $^-4 + 4$
(w) $^+6 - 6$ (x) $^+10 - 10$
(y) $^-2 - 5$ (z) $^-3 - 9$

5. Evaluate the following:

(a) $2 + 4 + 3$ (b) $1 + 6 + 7$
(c) $3 + 5 + 4$ (d) $4 + 1 + 3$
(e) $3 - 2 + 5$ (f) $4 - 1 + 6$
(g) $8 - 11 + 3$ (h) $7 - 13 + 5$
(i) $5 + 7 - 2$ (j) $4 + 1 - 3$
(k) $10 - 2 - 3$ (l) $8 - 4 - 1$
(m) $6 - 5 - 6$ (n) $11 - 8 - 7$

6. Evaluate the following:

(a) $^-5 + 3 + 1$ (b) $^-9 + 2 + 4$
(c) $^-3 + 2 + 8$ (d) $^-7 + 5 + 6$
(e) $^-2 + 4 - 6$ (f) $^-3 + 6 - 8$
(g) $^-7 - 2 + 5$ (h) $^-5 - 10 + 3$
(i) $^-1 - 3 - 11$ (j) $^-2 - 13 - 4$
(k) $4 - 8 + 3$ (l) $6 - 12 + 5$
(m) $^-2 + 10 - 5$ (n) $^-3 + 20 - 16$
(o) $18 - 5 - 2 - 11$ (p) $24 - 3 - 7 - 19$

3.4   Multiplication and Division

3.4.1   Number Patterns

Example 3.8
Complete the following pattern:
⁻6, ⁻3, 0, 3, —, —, —.

Solution
⁻6, ⁻3, 0, 3, 6, 9, 12   as the numbers are increasing 3 at a time.

Exercise 3H
1. Complete the following patterns:
   (a) 2, 4, 6, 8, —, —, —
   (b) 3, 6, 9, —, —, —
   (c) ⁻4, 0, 4, 8, —, —, —
   (d) ⁻10, ⁻5, 0, —, —, —
   (e) 6, 3, 0, —, —, —
   (f) 2, 0, ⁻2, —, —, —
   (g) ⁻7, ⁻17, ⁻27, —, —, —
   (h) ⁻5, ⁻8, ⁻11, —, —, —
   (i) 1, ⁻4, ⁻9, —, —, —
   (j) 6, ⁻6, ⁻18, —, —, —
   (k) 3, ⁻2, ⁻7, —, —, —
   (l) ⁻4, ⁻2, 0, —, —, —

2. Write sequences of 6 numbers by following the directions given.
   (a) Counting by 2's and starting at 3;
   (b) Counting by 3's and starting at ⁻6;
   (c) Counting by 5's and starting at ⁻30;
   (d) Counting by 7's and starting at ⁻14;
   (e) Counting by 4's and starting at ⁻9;
   (f) Counting backwards by 2's starting at 14;
   (g) Counting backwards by 4's starting at 12;
   (h) Counting backwards by 10's starting at 25;
   (i) Counting backwards by 6's starting at ⁻12;
   (j) Counting backwards by 3's starting at 6.

3.4.2   Multiplication
Multiplication can be expressed as a form of addition. For example, 3×4 is the same as 3 lots of 4 $(4 + 4 + 4)$ which is also the same as 4×3, 4 lots of 3 $(3 + 3 + 3 + 3)$.
The same applies to directed numbers and can be seen clearly on a number line. It should be remembered that one should always *start from zero* when using the number line for multiplication.

Example 3.9
Evaluate
(a) ⁺3 × ⁺5
(b) ⁺6 × ⁻3
(c) ⁻2 × ⁺6

Solution
(a) ⁺3 × ⁺5 is 3 lots of ⁺5
         or ⁺5 + ⁺5 + ⁺5
         = 5 + 5 + 5
On a number line, starting at zero, we can see 3 lots of 5

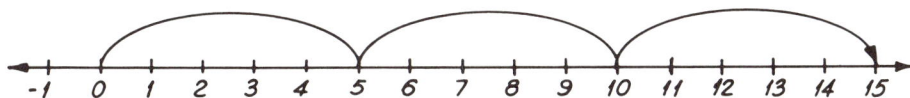

   ∴ ⁺3 × ⁺5 = ⁺15
(b) ⁺6 × ⁻3 is 6 lots of ⁻3
         or ⁻3 + ⁻3 + ⁻3 + ⁻3 + ⁻3 + ⁻3
         = ⁻3 − 3 − 3 − 3 − 3 − 3
On a number line, starting at zero, we can see 6 lots of ⁻3,

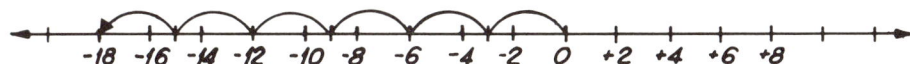

∴ ⁺6 × ⁻3 = ⁻18

(c) $^-2 \times {}^+6$ is $^-2$ lots of 6.

It is an unfamiliar concept to have a negative lot of a number *but* this can be expressed in another way using the commutative law (Chapter 6).

Thus $^-2 \times {}^+6 = {}^+6 \times {}^-2$

So $^+6 \times {}^-2$ is 6 lots of $^-2$

$$\text{or } {}^-2 + {}^-2 + {}^-2 + {}^-2 + {}^-2 + {}^-2$$
$$= {}^-2 - 2 - 2 - 2 - 2 - 2$$

On the number line, starting at zero, we can see 6 lots of $^-2$:

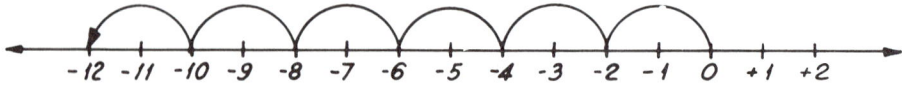

$\therefore \ ^-2 \times {}^+6 = {}^-12.$

Exercise 3I

1. Rewrite the following multiplications as additions and hence use the number line to evaluate.

(a) $^+2 \times {}^+7$

(b) $^+4 \times {}^+8$

(c) $^+3 \times {}^+4$

(d) $^+7 \times {}^+1$

(e) $^+6 \times {}^-2$

(f) $^+9 \times {}^-6$

(g) $^+10 \times {}^-6$

(h) $^+5 \times {}^-3$

(i) $^-8 \times {}^+2$

(j) $^-12 \times {}^+1$

(k) $^-4 \times {}^+4$

(l) $^-3 \times {}^+7$

(m) $^-2 \times {}^+11$

(n) $^-1 \times {}^+6$

(o) $^+4 \times {}^-5$

(p) $^+3 \times {}^-6$

(q) $^+7 \times {}^+4$

(r) $^+6 \times {}^+2$

(s) $^-4 \times {}^+3$

(t) $^-5 \times {}^+5$

(u) $^-1 \times {}^+5$

(v) $^-2 \times {}^+4$

(w) $^-4 \times {}^+9$

(x) $^+3 \times {}^-11$

(y) $^-2 \times {}^+14$

(z) $^-5 \times {}^+6$

2. Copy and complete the following tables using the number line to evaluate:

(a)

$\times$	$^+1$	$^+2$	$^+3$	$^+4$
$^+1$				
$^+2$				
$^+3$				
$^+4$				

(b)

$\times$	$^-1$	$^-2$	$^-3$	$^-4$
$^+1$				
$^+2$				
$^+3$				
$^+4$				

(c)

$\times$	$^-4$	$^-3$	$^-2$	$^-1$	0	$^+1$	$^+2$	$^+3$
$^+1$								
$^+2$								
$^+3$								
$^+4$								

3. Copy and complete the following table using the number line to evaluate:

×	$^{+}4$	$^{+}3$	$^{+}2$	$^{+}1$	0	$^{-}1$	$^{-}2$	$^{-}3$	$^{-}4$
$^{+}4$									
$^{+}3$									
$^{+}2$									
$^{+}1$									
0									
$^{-}1$									
$^{-}2$									
$^{-}3$									
$^{-}4$									

4. Fill in the sections of the table in question 3 which are shaded red by following the pattern of numbers.

5. Copy and complete the following table using the pattern where necessary.

×	4	3	2	1	0	$^{-}1$	$^{-}2$	$^{-}3$	$^{-}4$
$^{-}3$									
$^{-}6$									
$^{-}9$									
$^{-}12$									

Writing multiplications as additions has allowed us to evaluate, for example,

$$^{+}2 \times {}^{+}3 = {}^{+}3 + {}^{+}3 = {}^{+}6$$
$$^{+}2 \times {}^{-}3 = {}^{-}3 + {}^{-}3 = {}^{-}6$$
$$^{-}2 \times {}^{+}3 = {}^{+}3 \times {}^{-}2 = {}^{-}2 + {}^{-}2 + {}^{-}2 = {}^{-}6$$

However in exercise 3I it was necessary to resort to number patterns to evaluate multiplication of two negative numbers. In fact, the above examples lead us to the rules of multiplication which hold true for all directed numbers.

The product of two numbers with *the same signs* will always be *positive*.
E.g. $^{+}2 \times {}^{+}3 = {}^{+}6$
$\quad\;\; ^{-}3 \times {}^{-}2 = {}^{+}6$

The product of two numbers with *different signs* will always be *negative*.
E.g. $^{+}2 + {}^{-}3 = {}^{-}6$
$\quad\;\; ^{-}2 \times {}^{+}3 = {}^{-}6$

Example 3.10
Evaluate
(a) $^{+}3 \times {}^{+}5$ (b) $^{+}3 \times {}^{-}5$
(c) $^{-}3 \times {}^{+}5$ (d) $^{-}3 \times {}^{-}5$
(e) $4 \times {}^{-}6$ (f) $^{-}2 \times 7$

Solution
(a) $^{+}3 \times {}^{+}5 = {}^{+}(3 \times 5)$

      like   positive
     signs

$$= {}^{+}15$$

(b) $^{+}3 \times {}^{-}5 = {}^{-}(3 \times 5)$

      unlike   negative
     signs

$$= {}^{-}15$$

(c) $^-3 \times {}^+5 = {}^-(3 \times 5)$

unlike   negative
signs

$= {}^-15$

(e) $4 \times {}^-6 = {}^+4 \times {}^-6$

unlike
signs

$= {}^-(4 \times 6)$

negative

$= {}^-24$

(d) $^-3 \times {}^-5 = {}^+(3 \times 5)$

like   positive
signs

$= {}^+15$

(f) $^-2 \times 7 = {}^-2 \times {}^+7$

unlike
signs

$= {}^-(2 \times 7)$

negative

$= {}^-14$

When three or more multiplications are to be performed, the same rules of multiplication can be applied, taking two numbers at a time.

Example 3.11

Evaluate

(a) $^-2 \times {}^+4 \times {}^-5$

(b) $3 \times {}^-2 \times 6$

Solution

(a) $^-2 \times {}^+4 \times {}^-5$

$= ({}^-2 \times {}^+4) \times {}^-5$

unlike
signs

$= {}^-(2 \times 4) \times {}^-5$

$= {}^-8 \times {}^-5$

like
signs

$= {}^+(8 \times 5)$

$= {}^+40$

(b) $3 \times {}^-2 \times 6$

$= ({}^+3 \times {}^-2) \times {}^+6$

unlike
signs

$= {}^-(3 \times 2) \times {}^+6$

$= {}^-6 \times {}^+6$

unlike
signs

$= {}^-(6 \times 6)$

$= {}^-36$

Exercise 3J

1. Find the following products:

(a) ${}^+3 \times {}^+5$

(b) ${}^+6 \times {}^+2$

(c) ${}^+7 \times {}^+8$

(d) ${}^+9 \times {}^+6$

(e) ${}^+12 \times {}^+7$

(f) ${}^+14 \times {}^+3$

(g) ${}^+6 \times {}^+9$

(h) ${}^+5 \times {}^+8$

(i) ${}^-4 \times {}^-4$

(j) ${}^-6 \times {}^-7$

(k) ${}^-3 \times {}^-10$

(l) ${}^-2 \times {}^-15$

(m) ${}^-11 \times {}^-9$

(n) ${}^-4 \times {}^-6$

(o) ${}^-7 \times {}^-4$

(p) ${}^-5 \times {}^-10$

2. Find the following products:

(a) ${}^-5 \times {}^+3$

(b) ${}^-6 \times {}^+4$

(c) ${}^-8 \times {}^+8$

(d) ${}^-9 \times {}^+3$

(e) ${}^-2 \times {}^+20$

(f) ${}^-5 \times {}^+6$

(g) ${}^-9 \times {}^+8$

(h) ${}^-12 \times {}^+11$

(i) ${}^+3 \times {}^-14$

(j) ${}^+5 \times {}^-15$

(k) ${}^+10 \times {}^-16$

(l) ${}^+6 \times {}^-8$

(m) ${}^+12 \times {}^-3$

(n) ${}^+7 \times {}^-6$

(o) ${}^+1 \times {}^-24$

(p) ${}^+18 \times {}^-2$

3. Find the following products:

(a) $^+6 \times {}^+12$

(b) $^+3 \times {}^+4$

(c) $^+2 \times {}^+15$

(d) $^+8 \times {}^+11$

(e) $^+7 \times {}^-3$

(f) $^+4 \times {}^-10$

(g) $^+5 \times {}^-5$

(h) $^+6 \times {}^-2$

(i) $^-8 \times {}^+4$

(j) $^-1 \times {}^+6$

(k) $^-4 \times {}^+2$

(l) $^-16 \times {}^+3$

(m) $^-12 \times {}^-2$

(n) $^-10 \times {}^-6$

(o) $^-14 \times {}^-4$

(p) $^-3 \times {}^-20$

(q) $^+3 \times {}^+9$

(r) $^+4 \times {}^+11$

(s) $^-9 \times {}^-8$

(t) $^-6 \times {}^-5$

(u) $^+3 \times {}^-18$

(v) $^+11 \times {}^-11$

(w) $^-6 \times {}^+7$

(x) $^-1 \times {}^+32$

4. Evaluate the following:

(a) 8×3

(b) 4×6

(c) $9 \times {}^-2$

(d) $^-8 \times 7$

(e) $^-3 \times 1$

(f) $20 \times {}^-3$

(g) $^-11 \times {}^-11$

(h) $^-6 \times {}^-12$

(i) $^-2 \times 4$

(j) $8 \times {}^-10$

(k) $12 \times {}^-6$

(l) $7 \times {}^-5$

(m) 10×9

(n) 4×8

(o) $3 \times {}^-3$

(p) $9 \times {}^-6$

(q) $^-7 \times {}^-9$

(r) $^-2 \times {}^-24$

(s) $6 \times 4 \times 5$

(t) $9 \times 3 \times 3$

(u) $2 \times {}^-5 \times 12$

(v) $9 \times 7 \times {}^-2$

(w) $^-3 \times {}^-4 \times 6$

(x) $^-8 \times 6 \times {}^-3$

(y) $^-4 \times {}^-4 \times {}^-4$

(z) $^-5 \times {}^-3 \times {}^-6$

3.4.3   Division

It has been shown that multiplication is more easily performed using rules, rather than repeated addition and the number line. The same rules may be used for division of two directed numbers.

E.g. $^-6 \div {}^+2 = {}^-6 \times {}^+\frac{1}{2}$ (change to multiplication and invert second number)

unlike
signs

$= {}^-(6 \times \frac{1}{2})$

negative

$= {}^-3$

So $^-6 \div {}^+2 = {}^-(6 \div 2)$

unlike   negative
signs

$= {}^-3$

Similarly $^-10 \div {}^-5 = {}^-10 \times {}^-\frac{1}{5}$

like
signs

$= {}^+(10 \times \frac{1}{5})$

positive

$= {}^+2$

So $^-10 \div {}^-5 = {}^+(10 \div 5)$

like   positive
signs

$= {}^+2$

> The quotient of two numbers with *like signs* will always be *positive*.
> E.g. $^+8 \div {}^+4 = {}^+2$
> $^-4 \div {}^-2 = {}^+2$
> The quotient of two numbers with *unlike signs* will always be *negative*.
> E.g. $^+8 \div {}^-4 = {}^-2$
> $^-4 \div {}^+2 = {}^-2$
>
> N.B. This is the same rule as for multiplication.

Example 3.12

Evaluate

(a) $^+12 \div {}^-4$

(b) $^-15 \div 3$

(c) $6 \times {}^-5 \div {}^-2$

(d) $18 \div {}^-2 \times {}^-5$

Solution

(a) $^+12 \div {}^-4 = {}^-(12 \div 4)$

    unlike   negative
    signs

$= {}^-3$

(b) $^-15 \div 3 = {}^-15 \div {}^+3$

    unlike
    signs

$= {}^-(15 \div 3)$

    negative

$= {}^-5$

(c) $6 \times {}^-5 \div {}^-2 = {}^+6 \times {}^-5 \div {}^-2$

    unlike
    signs

$= {}^-30 \div {}^-2$

    like
    signs

$= {}^+15$

(d) $18 \div {}^-2 \times {}^-5 = {}^+18 \div {}^-2 \times {}^-5$

    unlike
    signs

$= {}^-9 \times {}^-5$

    like
    signs

$= {}^+45$

Exercise 3K

1. Find the quotients of the following:

(a) $^+20 \div {}^+5$

(b) $^+40 \div {}^+8$

(c) $^+36 \div {}^+6$

(d) $^+56 \div {}^+7$

(e) $^+100 \div {}^+25$

(f) $^+75 \div {}^+15$

(g) $^+84 \div {}^+12$

(h) $^+16 \div {}^+4$

(i) $^-32 \div {}^-8$

(j) $^-42 \div {}^-6$

(k) $^-80 \div {}^-10$

(l) $^-64 \div {}^-16$

(m) $^-144 \div {}^-12$

(n) $^-132 \div {}^-11$

(o) $^-90 \div {}^-30$

(p) $^-12 \div {}^-1$

2. Find the quotients of the following:

(a) $^+16 \div {}^-4$

(b) $^+48 \div {}^-8$

(c) $^+96 \div {}^-12$

(d) $^+36 \div {}^-12$

(e) $^+56 \div {}^-7$

(f) $^+66 \div {}^-11$

(g) $^+9 \div {}^-3$

(h) $^+14 \div {}^-7$

(i) $^-63 \div {}^+9$

(j) $^-108 \div {}^+12$

(k) $^-46 \div {}^+2$

(l) $^-65 \div {}^+5$

(m) $^-64 \div {}^+16$

(n) $^-100 \div {}^+20$

(o) $^-132 \div {}^+12$

(p) $^-84 \div {}^+7$

3. Find the quotients of the following:
  (a) $^+32 \div ^+8$
  (b) $^+16 \div ^+4$
  (c) $^+100 \div ^-10$
  (d) $^+30 \div ^-5$
  (e) $^-24 \div ^+3$
  (f) $^-12 \div ^+2$
  (g) $^-10 \div ^-5$
  (h) $^-81 \div ^-9$
  (i) $^-15 \div ^+3$
  (j) $^-108 \div ^+12$
  (k) $^+42 \div ^-7$
  (l) $^+40 \div ^-8$
  (m) $^-125 \div ^-25$
  (n) $^-180 \div ^-10$
  (o) $^+56 \div ^+8$
  (p) $^+72 \div ^+8$
  (q) $^-144 \div ^+12$
  (r) $^-26 \div ^+13$

4. Evaluate the following:
  (a) $88 \div 11$
  (b) $49 \div 7$
  (c) $16 \div ^-2$
  (d) $42 \div ^-6$
  (e) $^-18 \div 6$
  (f) $^-72 \div 8$
  (g) $^-20 \div ^-2$
  (h) $^-54 \div ^-9$
  (i) $^-15 \div 5$
  (j) $^-77 \div 7$
  (k) $121 \div 11$
  (l) $350 \div 50$
  (m) $^-72 \div 2$
  (n) $^-36 \div 6$
  (o) $^-50 \div ^-10$
  (p) $^-39 \div ^-13$
  (q) $8 \div ^-4$
  (r) $16 \div ^-8$
  (s) $^-108 \div 12$
  (t) $144 \div ^-3$
  (u) $^-625 \div ^-5$
  (v) $^-328 \div ^-4$
  (w) $^-558 \div 6$
  (x) $^-432 \div 8$
  (y) $760 \div ^-10$
  (z) $924 \div ^-3$

5. Find the quotients or products of the following:
  (a) 5×2
  (b) 7×6
  (c) $8 \div 4$
  (d) $48 \div 6$
  (e) $9 \times ^-5$
  (f) $11 \times ^-3$
  (g) $100 \div ^-10$
  (h) $64 \div ^-8$
  (i) $^-12 \times 8$
  (j) $^-2 \times 32$
  (k) $^-75 \div 5$
  (l) $^-96 \div 4$
  (m) $^-7 \times ^-11$
  (n) $^-6 \times ^-4$
  (o) $^-12 \div ^-3$
  (p) $^-15 \div ^-5$
  (q) $^-72 \div 8$
  (r) $^-16 \div 4$
  (s) $5 \times ^-6$
  (t) $3 \times ^-15$
  (u) $121 \div ^-11$
  (v) $171 \div ^-3$
  (w) $^-56 \div 7$
  (x) $^-34 \div 17$
  (y) $^-9 \times ^-6$
  (z) $^-3 \times ^-37$

3.5   Combined Operations

It should be remembered that the order of operations discussed in chapter 1 can be applied to directed numbers.

Order of Operations:

1. Brackets ( )
2. Of, Division, Multiplication $\div$. $\times$
3. Addition, Subtraction $+$, $-$

Work the problem from left to right.

This is easily remembered using the word BODMAS.

Example 3.13

Evaluate

(a) $^-6 \times 3 - (4 \div {}^-2)$

(b) $15 \div {}^-5 \times {}^-4 + 7$

Solution

(a) $^-6 \times 3 - (4 \div {}^-2)$

$= {}^-6 \times {}^+3 - ({}^+4 \div {}^-2)$ inserting director signs and evaluating brackets;

$= {}^-6 \times {}^+3 - {}^-2$ multiplication;

$= {}^-18 - {}^-2$ condense double signs;

$= {}^-18 + 2$ addition.

$= {}^-16$

(b) $15 \div {}^-5 \times {}^-4 + 7$

$= {}^+15 \div {}^-5 \times {}^-4 + {}^+7$ inserting director signs and evaluating division;

$= {}^-3 \times {}^-4 + {}^+7$ multiplication;

$= {}^+12 + {}^+7$ condense double sign;

$= {}^+12 + 7$ addition.

$= 19$

Exercise 3L

Evaluate the following:

1. $3 + (2 \times 5)$
2. $4 + (72 \div 12)$
3. $6 + (4 \div {}^-2)$
4. $9 + (35 \div {}^-7)$
5. $15 - ({}^-38 \div 2)$
6. $10 - ({}^-42 \div 7)$
7. $8 + ({}^-9 \times 3)$
8. $6 + ({}^-5 \times 8)$
9. $46 - (6 \times {}^-11)$
10. $59 - (4 \times {}^-6)$
11. $(39 \div {}^-3) + 16$
12. $(45 \div {}^-9) + 33$
13. $({}^-7 \times 6) - 13$
14. $({}^-12 \times 4) - 2$
15. $4 \times {}^-3 + (7 \times 2)$
16. $6 \times {}^-2 + (4 \times 9)$
17. $^-5 \times 9 - ({}^-6 \div 3)$
18. $^-4 \times 3 - ({}^-32 \div 8)$
19. $^-6 \times {}^-7 - (81 \div {}^-9)$
20. $^-2 \times {}^-10 - (60 \div {}^-12)$
21. $24 \div {}^-6 + ({}^-14 \div 7)$
22. $32 \div {}^-2 + (30 \div {}^-10)$
23. $^-96 \div 8 - (8 \times {}^-7)$
24. $^-104 \div 2 - (3 \times {}^-16)$
25. $(72 \div {}^-9) + 13 \times {}^-3$
26. $(84 \div {}^-4) + 12 \times {}^-5$
27. $({}^-39 \div {}^-3) - {}^-3 \times 8$
28. $({}^-65 \div 5) - {}^-4 \times 10$
29. $45 \div {}^-9 \times 8 - 2$
30. $132 \div {}^-11 \times 6 - 5$
31. $^-88 \div {}^-11 \times 4 + 15$
32. $^-64 \div {}^-16 \times 10 + 18$
33. $8 \times 7 \div {}^-2 - 14$
34. $6 \times 8 \div 12 - 18$
35. $^-13 \times 6 \div {}^-3 + 4$
36. $^-3 \times 14 \div {}^-6 + 7$
37. $^-42 \div {}^-7 \times {}^-5 - 3$
38. $^-75 \div {}^-5 \times {}^-6 - 10$
39. $36 \div {}^-6 \times {}^-4 + 1$
40. $99 \div {}^-11 \times {}^-5 + 5$
41. $^-9 \times {}^-4 \div {}^-6 - 8$
42. $^-7 \times {}^-6 \div {}^-2 - 15$
43. $16 \div {}^-4 + 12 \div {}^-3$
44. $^-25 \div 5 + 9 \div {}^-3$
45. $^-6 \times 3 - 5 \times {}^-2$
46. $^-12 \times 4 - 7 \times {}^-9$
47. $32 \div {}^-4 + 7 \times 3$
48. $^-108 \div 9 + 4 \times 6$
49. $^-75 \div {}^-15 - {}^-8 \times 3$
50. $^-5 \times 7 - {}^-90 \div 9$
51. $^-35 \div 7 \times {}^-12 \div 6$
52. $^-63 \div 9 \times 6 \div {}^-2$

Symbolic Expression

4

4.1   Introduction

In most businesses, there is a need for at least one secretary whose job involves such tasks as typing and shorthand. Shorthand is a *series of symbols* which are used *to take the place of words*. This enables a good secretary to write as quickly as a person can speak. Similarly, Mathematics has a type of shorthand called *Algebra*. Algebra uses *symbols or single letters in place of numbers*. This enables mathematicians to write general rules in symbols rather than writing them in words or giving many examples.

For example, just as 1 apple + 1 apple = 2 apples

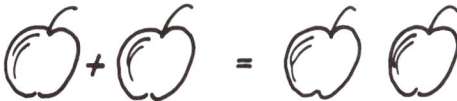

Using the letter '*a*' in place of the words,

$$1a \quad + 1a \quad = 2a$$

1*a* is usually written as *a* (the 1 is left out) so that we have

$$a \quad + a \quad = 2a$$

Similarly   1 banana + 1 banana + 1 banana = 3 bananas

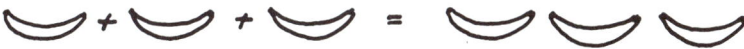

$$1b \quad + 1b \quad + 1b \quad = 3b$$
$$\text{or } b \quad + b \quad + b \quad = 3b$$

Also   2 carrots + 1 carrot = 3 carrots

$$\text{or } 2c \quad + c \quad = 3c$$

Of course   5 dots − 3 dots = 2 dots

$$5d \quad - 3d \quad = 2d$$

Of course the letters (symbols) used above can take the place of anything. Thus, we can add and subtract symbols just as we add and subtract numbers, apples, bananas etc, as long as they are the same.

Example 4.1

Simplify:
(a) $x + x + x$
(b) $x + 3x$
(c) $6x - 2x$
(d) $4x + 6x - 3x$

Solution

(a) $x + x + x = 1x + 1x + 1x$
$$= 3x$$

(b) $x + 3x = 1x + 3x$
$$= 4x$$

(c) $6x - 2x = 4x$

(d) Operations $(+, -, \times, \div)$ can only be performed on two numbers at a time and so we must write $4x + 6x - 3x = (4x + 6x) - 3x$
$$= \quad 10x \quad - 3x$$
$$= 7x$$

Of course, we could have written
$$4x + 6x - 3x = 4x + (6x - 3x)$$
$$= 4x + 3x$$
$$= 7x$$

which gives the same answer in this case.

Care must be taken when subtracting and therefore it is considered safer to work from left to right as in the first method.

Exercise 4A

1. Simplify
    (a) 2 dogs + 5 dogs
    (b) 7 cats + 4 cats
    (c) 8 trees + 7 trees
    (d) 9 hats + 7 hats
    (e) 4 cups + 3 cups
    (f) 10 chairs + 5 chairs
    (g) 6 pears − 2 pears
    (h) 7 apples − 6 apples
    (i) 9 dogs − 3 dogs
    (j) 10 sweets − 5 sweets
    (k) 8 books − 7 books
    (l) 14 seats − 6 seats

2. Simplify
    (a) $3 + 3 + 3 + 3$
    (b) $4 + 4 + 4 + 4 + 4$
    (c) $a + a + a$
    (d) $m + m + m + m$
    (e) $p + p + p + p + p$
    (f) $s + s + s + s + s$
    (g) $y + y + y + y + y + y + y + y$
    (h) $h + h + h + h + h + h + h + h + h$
    (i) $h + h + h + h + h$
    (j) $l + l + l + l + l$

3. Simplify
    (a) $3a + 2a$
    (b) $5m + 7m$
    (c) $8p + 6p$
    (d) $3s + 4s$
    (e) $7g + 11g$
    (f) $4m + 8m$
    (g) $12p + 6p$
    (h) $5y + 6y$
    (i) $5n + n$
    (j) $7m + m$
    (k) $3g + g$
    (l) $8y + y$
    (m) $m + 4m + 2m$
    (n) $s + 3s + 6s$
    (o) $5y + 3y + 2y$
    (p) $l + 4l + 5l$
    (q) $t + 3t + 8t$
    (r) $9m + 8m + 2m$

4. Simplify
    (a) $8s - 3s$
    (b) $5p - 2p$
    (c) $7m - 4m$
    (d) $12a - 3a$
    (e) $11p - 5p$
    (f) $16g - 9g$
    (g) $4m - m$
    (h) $3p - p$
    (i) $18h - 11h$
    (j) $10m - 7m$
    (k) $13y - 5y$
    (l) $15n - 8n$

5. Simplify
    (a) $3a + 5a - 2a$
    (b) $7l + 3l - 2l$
    (c) $11h + 3h - 2h$
    (d) $4p + 8p - 9p$
    (e) $9m + 5m - 6m$
    (f) $15t + 6t - 8t$
    (g) $4s - s + 2s$
    (h) $8y - 3y + y$
    (i) $6l - 3l - 2l$
    (j) $4m - m - m$
    (k) $15t - 6t - 7t$
    (l) $11x - 5x - 4x$

Remember that we use letters or symbols in Algebra. These are called *pronumerals* because they take the place of *numerals* (numbers). Sometimes we are given the value that the pronumeral takes and so we are able to *substitute* this value back in place of that pronumeral.

Thus, if $x = 5$ then $2x = x + x = 5 + 5 = 10$

$$6x = x + x + x + x + x + x = 5 + 5 + 5 + 5 + 5 + 5 = 30$$

Of course $2x = x + x$ can be thought of as $2 \times x$, so if

$$x = 5 \quad 2x = 2 \times x = 2 \times 5 = 10$$

Similarly $6x = x + x + x + x + x + x$ can be thought of as $6 \times x$, so if

$$x = 5 \quad 6x = 6 \times x = 6 \times 5 = 30.$$

In the example above, x was given a value 5. This does not mean that x is always given this value; it can be given any value. However, within a single problem x will have the same value throughout. Therefore, we can use pronumerals to write general rules about any numbers.

E.g. $\qquad\qquad\qquad\qquad x + x + x + x = 4x = 4 \times x.$

Example 4.2

Find the value of each of the following when $b = 2$:

(a) $3b$ (b) $6b$

(c) $5b + b$ (d) $3b - b$

Solution

(a) $3b = 3 \times b$

$\qquad = 3 \times 2 \qquad$ letting $b = 2$

$\qquad = 6$

(b) $6b = 6 \times b$

$\qquad = 6 \times 2 \qquad$ letting $b = 2$

$\qquad = 12$

(c) $5b + b = 6b \qquad$ adding pronumeral
$\qquad\qquad\qquad\qquad$ terms first

$\qquad\qquad = 6 \times 2 \quad$ letting $b = 2$

$\qquad\qquad = 12$

(d) $3b - b = 2b \qquad$ subtracting pronumeral
$\qquad\qquad\qquad\qquad$ terms first

$\qquad\qquad = 2 \times 2 \quad$ letting $b = 2$

$\qquad\qquad = 4$

In the last two examples it is easier to simplify expressions first before substituting $b = 2$.

Exercise 4B

1. Find the value of each of the following if $a = 3$:

   (a) $5a$ (b) $2a$

   (c) $7a$ (d) $9a$

   (e) $6a$ (f) $3a$

2. Find the value of each term in question 1 if $a = 9$.

3. Simplify and then find the value of each of the following if $x = 7$:

   (a) $2x + x$ (b) $5x + 3x$

   (c) $7x + 11x$ (d) $6x + 10x$

   (e) $5x - 2x$ (f) $7x - 4x$

   (g) $11x - 5x$ (h) $6x - 2x$

4. Simplify and then find the value of each of the following when $y = 3$:

   (a) $3y + 2y + y$ (b) $5y + 7y + 2y$

   (c) $6y + 3y - 2y$ (d) $8y + 6y - 5y$

   (e) $7y - 2y + 3y$ (f) $3y - y + 2y$

4.2   Terms and Expressions

A *term* is the product or quotient of either numeral and pronumerals or different pronumerals. E.g. $2x = 2 \times x$ is a term formed by the product of the number 2 and the pronumeral x. The number, 2, is called the *coefficient* of the pronumeral, x.

An *expression* is the sum or difference of terms.

Thus, $2x + 3x$ is an expression.

$2x$ and $3x$ are called *terms* of the *expression* $2x + 3x$.

Consider now

1 apple +   3 bananas   +   2 apples   + 1 banana

=   1 apple +   2 apples   +   3 bananas   + 1 banana

(rearranging the order of the terms)

=   3 apples   +   4 bananas

Similarly $a + 3b + 2a + b$

$= a + 2a + 3b + b$

$=\quad 3a\quad +\quad 4b$

Notice that the 2a is *added* to the a because the 2a is preceded by a *plus* sign.

Just as we cannot add 3 apples and 4 bananas and obtain anything simpler, $3a$ and $4b$ cannot be simplified any further.

Also,

2 apples   +   3 bananas   − 1 apple − 1 banana

=   2 apples   − 1 apple +   3 bananas   − 1 banana

(rearranging the order of the terms)

=   1 apple   +   2 bananas

Similarly $2a + 3b - a - b$

$= 2a - a + 3b - b$

$=\quad a\quad +\quad 2b$

Notice that the a is *subtracted* from 2a because the a is preceded by a *minus* sign. Similarly the b has been subtracted from the 3b.

In algebra, only terms with *identical* pronumeral parts can be added or subtracted. These are called *like terms*.

Example 4.3

Simplify

(a) $2x + 3y - x + y$

(b) $7a - 3 - 4a - 6$

Solution

(a) $2x + 3y - x + y$ grouping like terms

$= 2x - x + 3y + y$ rearranging

$= x + 4y$

(b) $7a - 3 - 4a - 6$

$= 7a - 4a - 3 - 6$

$= 3a - 9$

Of course, a number and a pronumeral term cannot be added or subtracted since they are not like terms.

3 apples + 1   cannot be simplified any further

+ 1

Similarly $3a + 1$ cannot be simplified any further.

Example 4.4
Find the value of each of the following expressions if $a = 3$ and $b = 6$:
(a) $5a + b$
(b) $2b - 3$
(c) $5a + 3b - 4a - b$
(d) $2a + 3 - a - 1$

Solution
(a) $5a + b$
 $= 5 \times a + b$
 $= 5 \times 3 + 6$ letting $a = 3$ and $b = 6$
 $= 15 + 6$ (Using BODMAS)
 $= 21$

(b) $2b - 3$
 $= 2 \times b - 3$
 $= 2 \times 6 - 3$ letting $b = 6$
 $= 12 - 3$
 $= 9$

(c) $5a + 3b - 4a - b$
 $= 5a - 4a + 3b - b$ (simplifying
 $= a + 2b$ before
 $= 3 + 2 \times 6$ substituting)
 $= 3 + 12$
 $= 15$

(d) $2a + 3 - a - 1$
 $= 2a - a + 3 - 1$ (simplifying before
 $= a + 2$ substituting)
 $= 3 + 2$
 $= 5$

Exercise 4C
1. In each of the following state whether or not the pairs of terms are like or unlike:
   (a) a and b
   (b) p and q
   (c) $5a$ and $4a$
   (d) $6b$ and $11b$
   (e) g and $9g$
   (f) y and $7y$
   (g) $2a$ and $2b$
   (h) $3x$ and $3y$
   (i) $3a$ and $2b$
   (j) $4b$ and $3c$

2. Simplify
   (a) 2 apples + 3 oranges + 1 apple + 5 oranges
   (b) 6 books + 5 records + 4 books + 2 records
   (c) 3 chocolate frogs + 4 humbugs + 7 chocolate frogs + 5 humbugs
   (d) 6 witches + 2 devils + 3 devils + 7 witches
   (e) 5 pears + 2 bananas − 2 pears − 1 banana
   (f) 10 ice creams + 3 drinks − 2 drinks − 8 ice creams

3. Simplify
   (a) $3a + 2b + 2a$
   (b) $4x + 3y + 2x$
   (c) $4c + 3d - 2c$
   (d) $5p + 2q - 3p$
   (e) $15l - 3m - 10l$
   (f) $12s - 5t - 10s$
   (g) $10g - 2h + 5g$
   (h) $17h - 5k + 3h$

4. Simplify
   (a) $3a + 2b + 5a + 4b$
   (b) $6s + 3y + 2s + 5y$
   (c) $7m + 5n + 2n + 6m$
   (d) $3l + 5t + 2t + 6l$
   (e) $8g + 4h + 2h + g$
   (f) $7p + 5q + 3q + 2p$
   (g) $6s + 3t + 2s + 5t$
   (h) $8y + 7x + 5x + 4y$
   (i) $4m + 2n + 3n + 5m$
   (j) $8c + 6d + 5d + 3c$
   (k) $2a + 3 + 5a + 7$
   (l) $6b + 2 + b + 8$
   (m) $3m + 2 + 6m - 1$
   (n) $4m + 8 + m - 3$
   (o) $5s + 2 + 6s + 3$
   (p) $7p + 4 + 2p + 3$
   (q) $8g - 2 + 4g - 5$
   (r) $8m - 3 + 6m - 7$
   (s) $3g - 4 - 2g + 5$
   (t) $6s - 5 - 5s - 6$

5. Simplify
   (a) $4m + 2n - 3m + 6n$ (b) $8a + 3b - 5a + b$
   (c) $6s + 3t - 2s + 8t$ (d) $11h + 2k - 8h + 5k$
   (e) $16a + 5b - 11a + 3b$ (f) $9x + 4y - 7x - 3y$
   (g) $15l + 5m - 14l - 3m$ (h) $18p + 9q - 13p - 6q$
   (i) $8h + 3g - 2g - 7h$ (j) $14a + 7b - 3b + 8a$
   (k) $g + h + 2 + g + h + 5$ (l) $p + q + 7 + 2p + 2q + 3$
   (m) $2y + 3 + z + 4 + z + y$ (n) $h + 4g + 3 + g - 2h - 3$

6. Find the value of the following expressions if $a = 7$ and $b = 3$, simplifying first where possible.
   (a) $a + b$ (b) $a - b$
   (c) $b + a$ (d) $b - a$
   (e) $2a$ (f) $3b$
   (g) $2a + 3b$ (h) $3a + 4b$
   (i) $7a - 2b$ (j) $3a - 4b$
   (k) $2b - 7a$ (l) $4b - 3a$
   (m) $4a + 3b - a - b$ (n) $2a + 3b - b - a$
   (o) $5a + 4b - b + 2a$ (p) $6a - 3b - b - 5a$

4.3   Multiplication by an Integer

If I have 3 bags, each containing 2 eggs, how many eggs do I have?

> $= 2$ eggs $+ 2$ eggs $+ 2$ eggs
> $= 3 \times 2$ eggs $= 6$ eggs
> or $3 \times 2e = 6e$

If I have 2 bags, each containing 4 sweets and I am given 3 more sweets, how many sweets do I have?

> $=$ 4 sweets $+$ 4 sweets $+$ 3 sweets
> $=$ 2×4 sweets $+$ 3 sweets
> $=$ 8 sweets $+$ 3 sweets $= 11$ sweets
> or $2 \times 4s + 3s$
> $=$ $8s + 3s = 11s$

Notice the use of BODMAS.

Example 4.5

Simplify
(a) $2 \times 3y$ (b) $3 \times 5z + 2z$
(c) $4 \times 3p - 2p$ (d) $3 \times 5q - 2 \times 2q$

Solution

(a) $2 \times 3y = 6y$ (b) $3 \times 5z + 2z$ multiplication
                                         $= 15z + 2z$ addition
                                         $= 17z$

(c) $4 \times 3p - 2p$ multiplication   (d) $3 \times 5q - 2 \times 2q$ multiplication
    $= 12p - 2p$ subtraction $= 15q - 4q$ subtraction
    $= 10p$ $= 11q$

Exercise 4D

1. (a) If I have 2 bags each containing 6 sweets, how many sweets do I have?
   (b) If there are 3 sets of 12 pencils, how many pencils are there?
   (c) If there are 5 classes of 20 students, how many students are there?
   (d) If there are 4 records each containing 5 songs, how many songs are there?
   (e) If there are 3 teams each consisting of 4 players, how many players are there?
   (f) If there are 6 trays each loaded with 8 glasses, how many glasses are there?

2. Simplify

   (a) $4 \times 2a$ (b) $6 \times 3a$
   (c) $7 \times 9m$ (d) $8 \times 9g$
   (e) $9 \times 5s$ (f) $11 \times 3t$
   (g) $5 \times 7x$ (h) $10 \times 4y$
   (i) $6 \times 3p$ (j) $2 \times 9m$
   (k) $2p \times 2$ (l) $6 \times 7g$
   (m) $4m \times 6$ (n) $7y \times 7$
   (o) $3y \times 2$ (p) $3s \times 5$

3. (a) $8 \times 2a + a$ (b) $3 \times 4g + 2g$
   (c) $6 \times 4m + 3m$ (d) $2 \times 7y + 3y$
   (e) $5 \times 3a - 7a$ (f) $6 \times 4d - 8d$
   (g) $11 \times 5s - 16s$ (h) $5 \times 8m - 7m$
   (i) $2 \times 5a + 3 \times 4a$ (j) $6 \times 3a + 4 \times 3a$
   (k) $3 \times 2a + 7 \times 3a$ (l) $9 \times 4g + 5 \times 3g$
   (m) $6 \times 7m - 3 \times 2m$ (n) $4 \times 6t - 3 \times 5t$
   (o) $5 \times 3s - 2 \times 7s$ (p) $6 \times 5y - 3 \times 7y$

4. Simplify and then find the value of each of the following when $a = 5$:

   (a) $4 \times 3a$ (b) $6 \times 3a$
   (c) $3 \times 4a$ (d) $3 \times 6a$
   (e) $5 \times 2a$ (f) $4 \times 7a$
   (g) $2 \times 5a$ (h) $7 \times 4a$
   (i) $2 \times 3a + a$ (j) $5 \times 2a + 2a$
   (k) $6 \times 3a - 8a$ (l) $9 \times 4a - 6a$
   (m) $2 \times 5a - 3 \times 2a$ (n) $7 \times 5a - 5 \times 3a$
   (o) $6 \times 4a - 2 \times 2a$ (p) $5 \times 5a - 3 \times 5a$

4.4 Multiplication of Pronumerals

Remember that $2 \times x$ is usually written as $2x$
         and $5 \times y$ is usually written as $5y$
Similarly, $a \times b$ is usually written as ab
What does ab mean?

$$2x = x + x \qquad \text{(2 x's added together)}$$
$$5y = y + y + y + y + y \qquad \text{(5 y's added together)}$$
$$ab = \underbrace{b + b + b + \ldots + b}_{a \text{ of these terms}} \qquad \text{(a b's added together)}$$

Notice that this shorthand notation for multiplication ($a \times b = ab$) only applies when there is at least one pronumeral involved. It *cannot* be used for purely numeral terms.
E.g. $2 \times 3 \neq 23$

Also $3 \times 2 = 2 + 2 + 2 = 6$
and $2 \times 3 = 3 + 3 = 6$
so $3 \times 2 = 2 \times 3$
Also $2x = 2 \times x = x \times 2$ and $5y = 5 \times y = y \times 5$

$$ab = a \times b = b \times a = ba$$

Of course, with three or more multiplications of pronumerals we use the same shorthand notation.

E.g. $a \times b \times c \times d = abcd$

Example 4.6

Find the value of each of the following expressions when $a = 2$ and $b = 5$:

(a) ab (b) ba

(c) $ab + ba$ (d) $ab - ba$

Solution

(a) $ab = a \times b$ (b) $ba = b \times a$

$\quad\quad = 2 \times 5$ $\quad\quad\quad = 5 \times 2$

$\quad\quad = 10$ $\quad\quad\quad = 10$

(c) $ab + ba = a \times b + b \times a$ (d) $ab - ba = 0$ since $ab = ba$

$\quad\quad\quad = 2 \times 5 + 5 \times 2$

$\quad\quad\quad = 10 + 10$

$\quad\quad\quad = 20$

(Using BODMAS.)

Exercise 4E

1. Simplify

   (a) $a \times b$ (b) $b \times c$

   (c) $p \times q$ (d) $r \times s$

   (e) $g \times h$ (f) $a \times r$

   (g) $d \times b$ (h) $m \times p$

2. Find the value of each of the following expressions when $m = 6, p = 3, q = 2, r = 4$:

   (a) mp (b) pm

   (c) $mp + pm$ (d) $mp - mp$

   (e) pq (f) qp

   (g) rq (h) qr

   (i) $mp + pq$ (j) $pq + qr$

   (k) $mr + pq$ (l) $pr + mq$

   (m) $2mp$ (n) $3pq$

   (o) $5qr$ (p) $2pq + mr$

   (q) $5mp + pq$ (r) $3pr + mq$

   (s) $4mr - 3pq$ (t) $6pq - 2mr$

   (u) $2mq - 2pr$ (v) $4pq - mr$

Even more complicated terms with the identical pronumeral parts can be added or subtracted.

E.g. $2mn + 7mn = 9mn$

$\quad\quad 5xy + 2xy - 3xy = 4xy$

$\quad\quad 4abc - 2abc + 3abc = 5abc.$

However $3abc + 2ab$ cannot be simplified because the pronumeral parts are not identical.

Example 4.7

Simplify

(a) $2ab + 3cd - ab + 4cd$ (b) $5ab - 3ba + 2ab$

Solution

(a) $2ab + 3cd - ab + 4cd$ (b) $5ab - 3ba + 2ab$

$\quad = 2ab - ab + 3cd + 4cd$ $\quad$ but $ba = ab$

$\quad = \quad\quad ab + 7cd$ $\quad \therefore\ 5ab - 3ba + 2ab$

$\quad\quad\quad\quad\quad\quad\quad\quad\quad\quad\quad\quad$ $\quad = 5ab - 3ab + 2ab$

$\quad\quad\quad\quad\quad\quad\quad\quad\quad\quad\quad\quad$ $\quad = \quad\quad 2ab + 2ab$ (calculating first two

$\quad\quad\quad\quad\quad\quad\quad\quad\quad\quad\quad\quad$ $\quad = 4ab$ $\quad\quad\quad\quad\quad$ terms first)

Exercise 4F

1. Which of the following pairs of terms are like terms?

   (a) ab and $4ab$

   (b) $3pq$ and $5pq$

   (c) st and t

   (d) xy and y

   (e) $3mp$ and cp

   (f) $5ad$ and $5a$

   (g) $7pq$ and $8pq$

   (h) $8mn$ and $7mn$

   (i) $6st$ and $3ts$

   (j) $7mn$ and $3nm$

2. Simplify each of the following:

   (a) $ab + 3ab$

   (b) $5mn + 7mn$

   (c) $8st + 3st$

   (d) $4pq + 5pq$

   (e) $4xy + 7yx$

   (f) $8gh + 4hg$

   (g) $13pq - 5pq$

   (h) $11st - 3st$

   (i) $8rs - 3rs$

   (j) $6lm - 2lm$

   (k) $5pq - qp$

   (l) $5yz - zy$

   (m) $7abc - 4abc$

   (n) $6rst - 3trs$

3. Simplify each of the following:

   (a) $ab + 4pq + 2ab + 3pq$

   (b) $4mn + 5st + 3st + 2mn$

   (c) $8hg + 2rs + 3rs + 4hg$

   (d) $6ab + 4cd + 2cd + 3ab$

   (e) $5xy + 5st - 3xy + 4st$

   (f) $7mn + 12pq - 8pq - 5mn$

   (g) $6lm + 2mn - mn - 6lm$

   (h) $5yz + 8ab - 3ba - 2yz$

   (i) $16abc + 2a - 3abc - a$

   (j) $5rs + 3r - 4rs + 2r$

   (k) $6lm + 2l + 3l - 4ml$

   (l) $8yz + 2z - 3zy - z$

4.5   Division of Pronumerals

$12 \div 4 = ?$ asks 'how many 4's in 12?' or 'how many sets of 4 in a set of 12?' or 'how many times can 4 be subtracted from 12 before there is no remainder?'.

Thus $12 \div 4 = 3$

If 6 apples are shared between 3 people, how many apples does each person receive?

Fig 4.1

It can be seen from fig 4.1 that each person received

(6 apples) $\div 3 = 2$ apples.

Similarly $6a \div 3 = 2a$

This could be written as $6a \div 3 = \dfrac{6a}{3}$

and cancelled down to give $\dfrac{\overset{2}{\cancel{6}} \times a}{\underset{1}{\cancel{3}}} = \dfrac{2a}{1} = 2a.$

> Remember that simplification by cancelling can only take place between parts which are joined by multiplication or division.

Example 4.8

Simplify

(a) $10a \div 2$

(b) $\dfrac{15a}{5}$

(c) $\dfrac{2a}{10}$

Solution

(a) $10a \div 2$

  (aaaaa) (aaaaa) dividing 10*a*'s into 2 groups
   each containing 5 *a*'s

  $10a \div 2 = 5a$

(b) $\dfrac{{}^{3}\cancel{15}a}{\cancel{5}_{1}} = \dfrac{3a}{1}$ cancelling down

   $= 3a$

(c) $\dfrac{2a}{10} = \dfrac{{}^{1}\cancel{2} \times a}{{}_{1}\cancel{2} \times 5} = \dfrac{a}{5}$ cancelling down

Exercise 4G

1. Simplify each of the following:
   (a) $\frac{12}{4}$

   (c) $\frac{18}{12}$

   (e) $\frac{6}{27}$

   (g) $\dfrac{3a}{3}$

   (i) $\dfrac{6c}{3}$

   (k) $\dfrac{54l}{9}$

   (m) $\dfrac{15a}{5}$

   (o) $\dfrac{32h}{8}$

   (q) $\dfrac{144p}{12}$

   (s) $\dfrac{84t}{7}$

   (b) $\frac{24}{6}$

   (d) $\frac{28}{21}$

   (f) $\frac{9}{33}$

   (h) $\dfrac{5b}{5}$

   (j) $\dfrac{27d}{9}$

   (l) $\dfrac{63z}{7}$

   (n) $\dfrac{24b}{4}$

   (p) $\dfrac{25c}{5}$

   (r) $\dfrac{36g}{9}$

   (t) $\dfrac{96d}{12}$

2. Simplify each of the following:
   (a) $\dfrac{4a}{2}$

   (c) $\dfrac{9y}{27}$

   (e) $\dfrac{15l}{10}$

   (g) $\dfrac{64y}{40}$

   (i) $\dfrac{18p}{12}$

   (k) $\dfrac{9xy}{30}$

   (m) $\dfrac{90lm}{100}$

   (o) $\dfrac{16ab}{12}$

   (q) $\dfrac{27xy}{18}$

   (s) $\dfrac{14pq}{21}$

   (u) $\dfrac{5lmn}{20}$

   (w) $\dfrac{8ab}{16}$

   (y) $\dfrac{24lm}{48}$

   (b) $\dfrac{15x}{5}$

   (d) $\dfrac{6b}{18}$

   (f) $\dfrac{24d}{20}$

   (h) $\dfrac{15h}{35}$

   (j) $\dfrac{16a}{40}$

   (l) $\dfrac{12pq}{16}$

   (n) $\dfrac{25st}{10}$

   (p) $\dfrac{15pq}{10}$

   (r) $\dfrac{42mn}{35}$

   (t) $\dfrac{16yz}{40}$

   (v) $\dfrac{15abc}{30}$

   (x) $\dfrac{9mn}{18}$

   (z) $\dfrac{16pq}{48}$

3. Simplify each of the following:

(a) $8mnp \div 4$ (b) $9lmn \div 6$

(c) $8x \div 16$ (d) $4y \div 24$

(e) $35lm \div 21$ (f) $9yz \div 27$

(g) $15pq \div 30$ (h) $25ab \div 30$

Example 4.9

Find the value of each of the following when $a = 3$:

(a) $\dfrac{8a}{4}$ (b) $\dfrac{7a}{3}$

Solution

(a) $\dfrac{^2 8a}{4_1} = 2a$ (simplifying first)

      $= 2 \times 3$ (substituting $a = 3$)

      $= 6$

(b) $\dfrac{7a}{3}$ cannot be simplified.

      Thus substituting $a = 3$

      $\dfrac{7a}{3} = \dfrac{7 \times a}{3}$

           $= \dfrac{7 \times 3}{3}$

           $= 7$

Exercise 4H

Find the value of each of the following when $m = 5$ and $n = 2$:

1. $\dfrac{6m}{3}$ 2. $\dfrac{8m}{4}$

3. $\dfrac{4m}{5}$ 4. $\dfrac{3m}{5}$

5. $\dfrac{4m}{10}$ 6. $\dfrac{8m}{10}$

7. $\dfrac{10m}{15}$ 8. $\dfrac{3m}{15}$

9. $\dfrac{12mn}{3n}$ 10. $\dfrac{8mn}{4n}$

11. $\dfrac{15mn}{3m}$ 12. $\dfrac{24mn}{12m}$

13. $\dfrac{15m}{10mn}$ 14. $\dfrac{20m}{15mn}$

Exercise 4I   Miscellaneous Exercise on Symbolic Expression

Simplify each of the following:

1. $d + d + d + d$ 2. $4m + 2m - m$

3. $2p + 3q + 4p$ 4. $7t + 2s + 5t + s$

5. $2b + 5c - b + 2c$ 6. $6g + 8k - 5g - 4k$

7. $4x + 3 + 2x + 6$ 8. $7y + 7 - 3y - 5$

9. $6m - 6 - 2m - 1$ 10. $5 \times 3a$

11. $4 \times 2n + 4n$ 12. $3 \times 6q - 2 \times 7q$

13. $4rt + 3lm + rt - lm$ 14. $12pqr + 3abc - 7pqr - abc$

15. $\dfrac{16ab}{12}$ 16. $\dfrac{54xy}{108}$

4.6   Indices

In 4.4 we saw that instead of $a \times b$ we wrote ab and instead of $5 \times x$ we wrote $5x$. Of course, we could write $a \times a = aa$ but we have an even quicker way of writing it:

$$2 \times 2 = 2^2$$

and $3 \times 3 = 3^2$

so $a \times a = a^2$

The small 2 is called the *index* or *power* and for a^2 we say '*a* squared' or '*a* to the power of 2'. This tells us that a has been multiplied by itself 2 times.

Similarly $2 \times 2 \times 2 = 2^3$ (2 multiplied by itself 3 times)

so $a \times a \times a = a^3$ (a multiplied by itself 3 times)

The index 3 means 'cubed' or 'to the power of 3' and for a^3 we say '*a* cubed' or '*a* to the power of 3'.

Of course, an index is even more useful when we have a large number of like numerals or pronumerals multiplied together.

$$7 \times 7 \times 7 \times 7 \times 7 \times 7 \times 7 \times 7 = 7^8 \text{ or 'seven to the power of 8'}$$

8 factors

$$a \times a \times a \times a \times a = a^5 \text{ or 'a to the power of 5'}$$

5 factors

Example 4.10

Write each of the following in index form:

(a) $5 \times 5 \times 5$

(b) $m \times m \times m \times m \times m \times m$

Solution

(a) $5 \times 5 \times 5 = 5^3$

3 factors

(b) $m \times m \times m \times m \times m \times m = m^6$

6 factors

Exercise 4J

Write each of the following in index form

1. $7 \times 7 \times 7 \times 7$

2. 3×3

3. $8 \times 8 \times 8 \times 8 \times 8$

4. $11 \times 11 \times 11 \times 11 \times 11 \times 11$

5. $s \times s \times s \times s \times s \times s$

6. $p \times p \times p \times p \times p$

7. $m \times m$

8. $h \times h \times h$

9. $y \times y \times y \times y \times y \times y \times y \times y$

10. $g \times g \times g \times g \times g \times g \times g \times g \times g$

11. $b \times b \times b \times b \times b$

12. $a \times a \times a \times a \times a$

To find the value of a number written in index form, it must first be expanded as a product of its factors:

$$2^3 = 2 \times 2 \times 2 = (2 \times 2) \times 2 = 4 \times 2 = 8$$

and $3^4 = 3 \times 3 \times 3 \times 3 = (3 \times 3) \times (3 \times 3) = 9 \times 9 = 81.$

We can also expand

$$a^5 = a \times a \times a \times a \times a$$

but it cannot be evaluated unless we know the value of a.

Example 4.11
Write the following as a product of their prime factors:
(a) 3^5 (b) p^7

Solution
(a) $3^5 = 3 \times 3 \times 3 \times 3 \times 3$ (b) $p^7 = p \times p \times p \times p \times p \times p \times p$

Example 4.12
Evaluate, when $a = 2$:
(a) a^3 (b) $3a^3$

Solution

(a) $a^3 = a \times a \times a$
$= 2 \times 2 \times 2$ since $a = 2$
$= 8$

(b) $3a^3 = 3 \times a \times a \times a$
$= 3 \times 2 \times 2 \times 2$
$= 24$

or, since $a^3 = 8$,
$3 \times a^3 = 3 \times 8$
$= 24$

Exercise 4 K
1. Write each of the following as the product of prime factors:
   (a) 3^2 (b) 2^2
   (c) 5^3 (d) 7^3
   (e) a^5 (f) m^6
   (g) b^4 (h) p^7
   (i) x^6 (j) y^2
   (k) p^8 (l) h^5

2. Evaluate:
   (a) 3^2 (b) 5^2
   (c) 7^2 (d) 10^2
   (e) 30^2 (f) 10^3
   (g) 6^3 (h) 2^4
   (i) 10^4 (j) 1^7
   (k) 0^2 (l) 10^6

3. If $x = 4$ find the value of :
   (a) x^3 (b) x^4
   (c) x^1 (d) x^2
   (e) $2x^2$ (f) $3x^2$

4. If $y = 3$ find the value of :
   (a) y^1 (b) y^4
   (c) y^5 (d) y^2
   (e) $5y^2$ (f) $2y^4$

5 Matrices

5.1 Introduction

In this section we shall look at *matrices*. We speak of one *matrix* but several *matrices*. A matrix is a set of numbers arranged in a rectangular pattern like this $\begin{pmatrix} 15 & 6 & 17 \\ 9 & -4 & 0 \end{pmatrix}$.

Matrices can be used in many different ways in Maths but in this chapter we shall look at how to use them for making up coded messages and storing information.

5.2 Addition and Subtraction

If you wanted to send a message to someone which only he could read you would use a code. The following message is written in code:

NCSVJP   UTBKOU   IQSUFU

Can you tell what the message says? It uses a code made up using matrices and it is very hard to crack such a code. The message is 'MARTIN TRAINS HORSES'. Here is how it was coded.

First of all the letters of the alphabet are numbered:

A	B	C	D	E	F	G	H	I	J	K	L	M
1	2	3	4	5	6	7	8	9	10	11	12	13

N	O	P	Q	R	S	T	U	V	W	X	Y	Z
14	15	16	17	18	19	20	21	22	23	24	25	26

Next, each word of the message is turned into numbers:

$$
\begin{aligned}
\text{MARTIN} &= 13,1,18,20,9,14 \\
\text{TRAINS} &= 20,18,1,9,14,19 \\
\text{HORSES} &= 8,15,18,19,5,19
\end{aligned}
$$

Then a matrix is made up for each word:

$$\text{MARTIN} = \begin{pmatrix} 13 & 1 & 18 \\ 20 & 9 & 14 \end{pmatrix} \quad \text{TRAINS} = \begin{pmatrix} 20 & 18 & 1 \\ 9 & 14 & 19 \end{pmatrix} \quad \text{HORSES} = \begin{pmatrix} 8 & 15 & 18 \\ 19 & 5 & 19 \end{pmatrix}$$

Now, each matrix is changed using the matrix $\begin{pmatrix} 1 & 2 & 1 \\ 2 & 1 & 2 \end{pmatrix}$

$$\begin{pmatrix} 13 & 1 & 18 \\ 20 & 9 & 14 \end{pmatrix} + \begin{pmatrix} 1 & 2 & 1 \\ 2 & 1 & 2 \end{pmatrix} = \begin{pmatrix} 14 & 3 & 19 \\ 22 & 10 & 16 \end{pmatrix}$$

Looking back at the alphabet we see that 14 is N, 3 is C, 19 is S, 22 is V, 10 is J and 16 is P, so 'MARTIN' becomes 'NCSVJP'.

The method is used for the other two words in the message:

Matrix for TRAINS   Code matrix   Code for TRAINS

$$\begin{pmatrix} 20 & 18 & 1 \\ 9 & 14 & 19 \end{pmatrix} \quad + \quad \begin{pmatrix} 1 & 2 & 1 \\ 2 & 1 & 2 \end{pmatrix} \quad = \quad \begin{pmatrix} 21 & 20 & 2 \\ 11 & 15 & 21 \end{pmatrix}$$

TRAINS   becomes   UTBKOU

Matrix for HORSES   Code matrix   Code for HORSES

$$\begin{pmatrix} 8 & 15 & 18 \\ 19 & 5 & 19 \end{pmatrix} \quad + \quad \begin{pmatrix} 1 & 2 & 1 \\ 2 & 1 & 2 \end{pmatrix} \quad = \quad \begin{pmatrix} 9 & 17 & 19 \\ 21 & 6 & 21 \end{pmatrix}$$

HORSES   becomes   IQSUFU

The person receiving the coded message only needs to know the code matrix and he can decode your message by reversing the above processes.

For example, IQSUFU has the matrix $\begin{pmatrix} 9 & 17 & 19 \\ 21 & 6 & 21 \end{pmatrix}$

To decode we subtract the code matrix $\begin{pmatrix} 9 & 17 & 19 \\ 21 & 6 & 21 \end{pmatrix} - \begin{pmatrix} 1 & 2 & 1 \\ 2 & 1 & 2 \end{pmatrix} = \begin{pmatrix} 8 & 15 & 18 \\ 19 & 5 & 19 \end{pmatrix}$

$\begin{pmatrix} 8 & 15 & 18 \\ 19 & 5 & 19 \end{pmatrix}$ is the matrix for HORSES.

Exercise 5A

1. Using the code matrix $\begin{pmatrix} 1 & 1 & 1 \\ 2 & 2 & 2 \end{pmatrix}$ make up codes for the following words:

(a) MATRIX   (b) SCHOOL
(c) TENNIS   (d) EATING
(e) VIOLIN   (f) SAUCER
(g) FOLLOW   (h) BEHIND

2. Use the matrix $\begin{pmatrix} 1 & 1 & 1 \\ 2 & 2 & 2 \end{pmatrix}$ to decide the following message:

MPPMKP   UIFDKI   MPVPIG   GPSVJG   QBDMGV   GVMNQH   TFDTGV   QBQGTU

5.3 Order of a Matrix

All of our matrices so far have had 2 rows and 3 columns.

$$\begin{pmatrix} 1 & 2 & 1 \\ 2 & 1 & 2 \end{pmatrix} \begin{matrix} \leftarrow \\ \leftarrow \end{matrix} \quad \text{2 rows}$$
$$\uparrow \quad \uparrow \quad \uparrow$$
3 columns

However a matrix can be any size you wish.

(1 3) has 1 row and 2 columns

$$\begin{pmatrix} 4 & 3 & 9 \\ 1 & 2 & 8 \\ 7 & 6 & 11 \end{pmatrix}$$ has 3 rows and 3 columns. It is a square matrix.

$$\begin{pmatrix} 1 \\ 7 \\ -2 \end{pmatrix} \text{ has 3 rows and 1 column}$$

$$\begin{pmatrix} 4 & 3 & 9 & 17 \\ 21 & 3 & 4 & 11 \\ 6 & 2 & 12 & 13 \\ 4 & 9 & 16 & 2 \\ 8 & 6 & 14 & 1 \end{pmatrix} \text{ has 5 rows and 4 columns}$$

The number of rows and columns is called the *order* of a matrix.

$\begin{pmatrix} 2 & 1 & 3 & 8 \\ 7 & 6 & 4 & 6 \end{pmatrix}$ has 2 rows and 4 columns. Its order is 2×4.

This is pronounced 'two by four'. Notice that the number of rows is first.

ORDER = ROWS × COLUMNS

We used a matrix of order 2×3 for our codes because it was convenient for six letter words. However you can use a matrix of any order. Put a zero in the matrix for any spaces you want to leave in the message and fill up your matrices with zeroes if there are any gaps.

Example 5.1

Use the matrix $\begin{pmatrix} 1 & 2 \\ 2 & 1 \end{pmatrix}$, of order 2×2, to make up a code for the message:

'MY CAT HAD KITTENS'

Solution

A 2×2 matrix has 4 numbers in it so split the message up into groups of four letters or spaces:
MY C/AT H/AD K/ITTE/NS

Group one: MY C $= 13,25,0,3$. The matrix is $\begin{pmatrix} 13 & 25 \\ 0 & 3 \end{pmatrix}$

Group two: AT H $= 1,20,0,8$. The matrix is $\begin{pmatrix} 1 & 20 \\ 0 & 8 \end{pmatrix}$

Group three: AD K $= 1,4,0,11$. The matrix is $\begin{pmatrix} 1 & 4 \\ 0 & 11 \end{pmatrix}$

Group four: ITTE $= 9,20,20,5$. The matrix is $\begin{pmatrix} 9 & 20 \\ 20 & 5 \end{pmatrix}$

Group five: NS $= 14,19,0,0$. The matrix is $\begin{pmatrix} 14 & 19 \\ 0 & 0 \end{pmatrix}$

In matrices one, two and three, the zeroes stand for spaces; in group five the zeroes are there to fill up the matrix so it has the order 2×2.

$$\begin{pmatrix} 13 & 25 \\ 0 & 3 \end{pmatrix} + \begin{pmatrix} 1 & 2 \\ 2 & 1 \end{pmatrix} = \begin{pmatrix} 14 & 27 \\ 2 & 4 \end{pmatrix} \qquad N \quad A \quad B \quad D$$

(There is not a 27th letter
so start again at A)

$$\begin{pmatrix} 1 & 20 \\ 0 & 8 \end{pmatrix} + \begin{pmatrix} 1 & 2 \\ 2 & 1 \end{pmatrix} = \begin{pmatrix} 2 & 22 \\ 2 & 9 \end{pmatrix} \qquad B \quad V \quad B \quad I$$

$$\begin{pmatrix} 1 & 4 \\ 0 & 11 \end{pmatrix} + \begin{pmatrix} 1 & 2 \\ 2 & 1 \end{pmatrix} = \begin{pmatrix} 2 & 6 \\ 2 & 12 \end{pmatrix} \qquad B \quad F \quad B \quad L$$

$$\begin{pmatrix} 9 & 20 \\ 20 & 5 \end{pmatrix} + \begin{pmatrix} 1 & 2 \\ 2 & 1 \end{pmatrix} = \begin{pmatrix} 10 & 22 \\ 22 & 6 \end{pmatrix} \qquad J \quad V \quad V \quad F$$

$$\begin{pmatrix} 14 & 19 \\ 0 & 0 \end{pmatrix} + \begin{pmatrix} 1 & 2 \\ 2 & 1 \end{pmatrix} = \begin{pmatrix} 15 & 21 \\ 2 & 1 \end{pmatrix} \qquad O \quad U \quad B \quad A$$

'MY CAT HAD KITTENS'   becomes   'NABD BVBI BFBL JVVF OUBA'

Exercise 5 B

1. State the order of the following matrices:

(a) $\begin{pmatrix} 1 & 4 \\ 2 & 3 \end{pmatrix}$ _____ (b) $(-6 \quad 7 \quad 1)$

(c) $\begin{pmatrix} 5 & 2 & 1 \\ 1 & 4 & 2 \\ 0 & -9 & 17 \\ 3 & 6 & 11 \end{pmatrix}$ (d) $\begin{pmatrix} 23 & 14 & -17 & 13 \\ 49 & 1 & 0 & 21 \end{pmatrix}$

(e) $\begin{pmatrix} 1 & 0 & 0 \\ 0 & 1 & 0 \\ 0 & 0 & 1 \end{pmatrix}$ (f) $\begin{pmatrix} 1069 & -21 \\ -4 & 17 \\ 1 & 596 \end{pmatrix}$

2. Use the matrix $\begin{pmatrix} 1 & 2 & 3 \\ 3 & 2 & 1 \end{pmatrix}$ to decode the following message:

UJHCJJ   EFGQBI   BTERWS   AKVCVI   SGHCOJ   MGVCPP   SVKCQG
AVKHBM   JIKWBI   PWVHBA

3. Evaluate:

(a) $\begin{pmatrix} 3 & 14 & 1 \\ 2 & 9 & 16 \end{pmatrix} + \begin{pmatrix} 17 & 3 & 46 \\ 21 & 19 & 18 \end{pmatrix}$ (b) $\begin{pmatrix} 21 & 16 \\ 13 & 14 \\ 11 & 10 \end{pmatrix} - \begin{pmatrix} 0 & 5 \\ 6 & 13 \\ 0 & 5 \end{pmatrix}$

(c) $\begin{pmatrix} 13 & -2 & 5 \\ 14 & 11 & -12 \\ 21 & 3 & 15 \end{pmatrix} + \begin{pmatrix} -21 & 5 & 16 \\ 13 & 29 & 20 \\ -16 & 91 & 14 \end{pmatrix}$ (d) $\begin{pmatrix} -3 & 2 \\ 15 & -16 \end{pmatrix} - \begin{pmatrix} 4 & 7 \\ 13 & 11 \end{pmatrix}$

(e) $\begin{pmatrix} 100 & 50 \\ -99 & 80 \end{pmatrix} + \begin{pmatrix} -301 & 29 \\ 56 & -79 \end{pmatrix}$ (f) $\begin{pmatrix} 18 & 46 \\ -97 & -3 \\ 30 & 9 \end{pmatrix} - \begin{pmatrix} -17 & 23 \\ 89 & -99 \\ -15 & 56 \end{pmatrix}$

5.4   Multiplication by a scalar

There are other ways of using matrices to code messages. One way is like this.

HEAD has the matrix $\mathbf{M} = \begin{pmatrix} 7 & 5 \\ 1 & 4 \end{pmatrix}$

We can multiply this matrix by, say, 3 to produce the coded word.

$3 \times \mathbf{M} = 3 \times \begin{pmatrix} 7 & 5 \\ 1 & 4 \end{pmatrix} = \begin{pmatrix} 21 & 15 \\ 3 & 12 \end{pmatrix}$

HEAD becomes VOCL

$\mathbf{M}$ is a matrix.
3 is called a *scalar* to distinguish it from the matrix.
We have multiplied a *matrix* by a *scalar* to get a *matrix*.
Later on we shall learn how to multiply one matrix by another.

Example 5.2

Multiply the matrix $\mathbf{P}$ by 4, where $\mathbf{P} = \begin{pmatrix} 2 & -3 \\ 4 & -1 \end{pmatrix}$

Solution

Each number in $\mathbf{P}$ has to be multiplied by 4

$4 \times \mathbf{P} = 4 \times \begin{pmatrix} 2 & -3 \\ 4 & -1 \end{pmatrix} = \begin{pmatrix} 8 & -12 \\ 16 & -4 \end{pmatrix}$

Exercise 5C

1. Multiply each of the following matrices by the given scalar:

(a) $\begin{pmatrix} -1 & 2 \\ 0 & 17 \end{pmatrix}$ by 3

(b) $\begin{pmatrix} 15 & 4 \\ 6 & 34 \end{pmatrix}$ by 2

(c) $\begin{pmatrix} -6 & 13 \\ -2 & 27 \\ 3 & 16 \\ 7 & 11 \end{pmatrix}$ by 5

(d) $(6 \quad 3 \quad 9)$ by 11

(e) $\begin{pmatrix} -47 & 13 & 12 \\ 91 & -15 & 6 \end{pmatrix}$ by 7

(f) $\begin{pmatrix} -103 & 84 & -22 \\ 35 & -27 & 16 \\ 46 & 3 & 0 \end{pmatrix}$ by -2

(g) $\begin{pmatrix} 9 & 12 \\ 3 & -9 \\ 6 & 18 \end{pmatrix}$ by $\frac{1}{3}$

(h) $\begin{pmatrix} 68 & 32 & 312 \\ 94 & 106 & 48 \\ 64 & 16 & 18 \\ 78 & 38 & 82 \end{pmatrix}$ by $\frac{1}{2}$

2. Use the scalar 2 to code this sentence:
   FEED ALL CAMELS

3. Decode your answer to question 2 by multiplying by the scalar $\frac{1}{2}$.

4. Given that $\mathbf{A} = \begin{pmatrix} 5 & 9 \\ 3 & 1 \end{pmatrix}$ $\mathbf{B} = \begin{pmatrix} -1 & 4 & 20 \\ 12 & 9 & 11 \end{pmatrix}$ $\mathbf{C} = \begin{pmatrix} 33 & 42 \\ -15 & -9 \end{pmatrix}$

Evaluate:
(a) $3\mathbf{A}$ (b) $4\mathbf{B}$
(c) $2\mathbf{C}$ (d) $\frac{1}{3}\mathbf{C}$
(e) $\mathbf{A} - \mathbf{C}$ (f) $3\mathbf{A} - \frac{1}{3}\mathbf{C}$
(g) $\frac{1}{2}\mathbf{B}$ (h) $\frac{1}{4}\mathbf{A}$

5.5   Storing information

Matrices can also be used for storing information.

A toy factory produces three types of toy. These are cars, soldiers and guns. Each toy can be made of metal or of plastic. In the first week of the year toys are made as shown in Table 5.1

	CARS	SOLDIERS	GUNS
METAL	100	300	29
PLASTIC	50	170	51

Table 5.1

This information can be put in a 2×3 matrix: $\begin{pmatrix} 100 & 300 & 29 \\ 50 & 170 & 51 \end{pmatrix}$

The figures for the second week are given by: $\begin{pmatrix} 203 & 250 & 35 \\ 59 & 109 & 29 \end{pmatrix}$

The total for the first two weeks of the year is given by:

$$\begin{pmatrix} 100 & 300 & 29 \\ 50 & 170 & 51 \end{pmatrix} + \begin{pmatrix} 203 & 250 & 35 \\ 59 & 109 & 29 \end{pmatrix} = \begin{pmatrix} 303 & 550 & 64 \\ 109 & 279 & 80 \end{pmatrix}$$

Example 5.3

In the third week of the year the numbers of toys produced in the above factory are given by:

$$\begin{pmatrix} 50 & 200 & 21 \\ 150 & 57 & 70 \end{pmatrix}$$

Work out the matrix which gives the total production figures for the first three weeks of the year. Use the matrix to say how many plastic toy cars were made up to the end of the third week.

Solution
Using the matrix for the first two weeks and the matrix for the third week:

$$\begin{pmatrix} 303 & 550 & 64 \\ 109 & 279 & 80 \end{pmatrix} + \begin{pmatrix} 50 & 200 & 21 \\ 150 & 57 & 70 \end{pmatrix} = \begin{pmatrix} 353 & 750 & 85 \\ 259 & 336 & 150 \end{pmatrix}$$

The matrix for the first three weeks is: $\begin{pmatrix} 353 & 750 & 85 \\ 259 & 336 & 150 \end{pmatrix}$

By looking back at Table 5.1 we can see that the number of plastic cars made in the first three weeks of the year is 259 because this is the number in the second row and first column of the matrix and the table.

Exercise 5 D

A restaurant is open for 5 days each week and serves breakfast, lunch and afternoon tea. The numbers of each type of meal served are given below:

MEALS SERVED AT SUNNY VIEW RESTAURANT
from 1st to 28th February

Day	Date	Breakfast	Lunch	Tea
Mon	1st	25	34	26
Tues	2nd	CLOSED ALL DAY		
Wed	3rd	19	29	30
Thur	4th	CLOSED ALL DAY		
Fri	5th	10	24	16
Sat	6th	13	16	25
Sun	7th	5	45	56
Mon	8th	27	33	21
Tues	9th	CLOSED ALL DAY		
Wed	10th	14	30	24
Thur	11th	CLOSED ALL DAY		
Fri	12th	9	20	15
Sat	13th	15	20	30
Sun	14th	4	41	50
Mon	15th	25	31	33
Tues	16th	CLOSED ALL DAY		
Wed	17th	16	32	35
Thur	18th	CLOSED ALL DAY		
Fri	19th	12	23	35
Sat	20th	14	39	42
Sun	21st	10	55	30
Mon	22nd	15	35	19
Tues	23rd	CLOSED ALL DAY		
Wed	24th	11	21	33
Thur	25th	CLOSED ALL DAY		
Fri	26th	13	25	41
Sat	27th	16	21	38
Sun	28th	8	51	29

1. The restaurant is always closed on Tuesday and Thursday. Use a 3×5 matrix to show the figures for the first week.

2. Use the same order of matrix as you used in question 1 to show the figures for the second, third and fourth weeks.

3. Use the matrices that you wrote down for questions 1 and 2 to work out the figures for the whole of February.

4. On which day of the week is breakfast   i) most popular
                                               ii) least popular.

5. Which meal is least popular on Wednesday?

Revision Exercises for Chapters 1–5

1. Evaluate
   (a) $214 + 106 + 71 + 8$
   (b) $305 + 962 + 43 + 7$
   (c) $1010 + 101 + 1 + 1101$
   (d) $6909 + 69 + 9 + 960$
   (e) $7659 - 3842$
   (f) $8218 - 2561$
   (g) $8021 - 785$
   (h) $3106 - 429$
   (i) 416×37
   (j) 205×18
   (k) $32 \times 46 \times 91$
   (l) $14 \times 27 \times 38$
   (m) $856 \div 4$
   (n) $208 \div 6$
   (o) $3025 \div 5$
   (p) $6904 \div 8$

2. Evaluate
   (a) $\frac{4}{5} + \frac{8}{15}$
   (b) $\frac{5}{6} + \frac{5}{18}$
   (c) $2\frac{3}{8} + 1\frac{5}{12} + \frac{1}{6}$
   (d) $\frac{3}{7} + 1\frac{2}{3} + 2\frac{6}{21}$
   (e) $\frac{7}{8} - \frac{2}{3}$
   (f) $\frac{5}{12} - \frac{4}{15}$
   (g) $1\frac{2}{9} - \frac{5}{6}$
   (h) $2\frac{7}{10} - 1\frac{8}{15}$
   (i) $\frac{6}{7} \times \frac{5}{18} \times \frac{3}{10}$
   (j) $\frac{2}{3} \times \frac{9}{10} \times \frac{7}{12}$
   (k) $1\frac{2}{3} \times 2\frac{5}{8} \times 1\frac{1}{3}$
   (l) $2\frac{1}{4} \times 1\frac{4}{5} \times 2\frac{1}{2}$
   (m) $\frac{4}{7} \div \frac{2}{3}$
   (n) $\frac{5}{8} \div \frac{5}{6}$
   (o) $1\frac{3}{5} \div 2\frac{4}{14}$
   (p) $3\frac{3}{4} \div 2\frac{2}{9}$

3. Evaluate
   (a) $1.43 + 2.61 + 0.3$
   (b) $0.65 + 12.86 + 1.5$
   (c) $210 + 0.56 + 14.021 + 3.9$
   (d) $8.6 + 107 + 23.52 + 0.815$
   (e) $85.276 - 21.38$
   (f) $72.041 - 8.15$
   (g) $2.78 - 0.825$
   (h) $3.07 - 0.723$
   (i) 1.6×0.4
   (j) 2.7×0.05
   (k) 3.02×0.61
   (l) 0.45×1.063
   (m) $272 \div 0.4$
   (n) $385 \div 0.8$
   (o) $392.5 \div 10$
   (p) $730.8 \div 1000$

4. Express the following as decimals correct to 3 decimal places:
   (a) $\frac{7}{10}$
   (b) $\frac{3}{10}$
   (c) $\frac{76}{100}$
   (d) $\frac{12}{100}$
   (e) $\frac{3}{5}$
   (f) $\frac{7}{20}$
   (g) $\frac{18}{25}$
   (h) $\frac{43}{50}$
   (i) $\frac{1}{2}$
   (j) $\frac{1}{4}$
   (k) $\frac{2}{3}$
   (l) $\frac{5}{6}$
   (m) $\frac{3}{8}$
   (n) $\frac{4}{9}$
   (o) $\frac{11}{12}$
   (p) $\frac{7}{5}$

5. Express the following as fractions in simplest form:
   (a) 0.2
   (b) 0.4
   (c) 0.25
   (d) 0.75
   (e) 0.08
   (f) 0.05
   (g) 1.16
   (h) 1.18
   (i) 0.375
   (j) 0.445
   (k) 0.3
   (l) 0.9

6. Use the following information to complete the statements below.
   $\mathscr{E} = \{$whole number from 1 to 30$\}$
   $T = \{2, 6, 10, 14, 18, 22, 26, 30\}$
   $X = \{5, 10, 15, 20, 25, 30\}$
   $D = \{9, 12, 15, 18, 21, 24, 27, 30\}$
   (a) $6 \in$
   (b) $27 \in$
   (c) $15 \in$
   (d) $10 \in$
   (e) $30 \in$
   (f) $18 \in$
   (g) $n(T) =$
   (h) $n(X) =$
   (i) $D \leftrightarrow$
   (j) $n(D) =$
   (k) $\{10, 18, 26\} \subset$
   (l) $\{10, 25, 30\} \subset$
   (m) $\{15, 18, 27\} \subset$
   (n) $\{1, 8, 16\} \subset$

(o) $X \subset$ (p) $T \subset$

(q) $T' =$ (r) $D \subset$

(s) $X' =$ (t) $D' =$

(u) $\mathscr{E}' =$ (v) $n(T') =$

(w) $n(D') =$ (x) $n(X') =$

7. Draw a Venn diagram of sets T, X, D and $\mathscr{E}$ from question 6 and complete the statements below.

(a) $T \cap X =$ (b) $T \cap D =$

(c) $X \cap T \cap D =$ (d) $X \cap D =$

(e) $n(X \cap D) =$ (f) $n(T \cap X) =$

(g) $n(T \cap D) =$ (h) $n(T \cap X \cap D) =$

(i) $(T \cap X)' =$ (j) $(X \cap D)' =$

(k) $T' \cap D =$ (l) $D' \cap X =$

(m) $X' \cap T' =$ (n) $T' \cap D' =$

(o) $T \cup D$ (p) $D \cup X$

(q) $T \cup D \cup X$ (r) $T \cup X$

(s) $n(X \cup D)$ (t) $n(T \cup D)$

(u) $(T \cup X)'$ (v) $(T \cup D)'$

(w) $X' \cup D$ (x) $T \cup D'$

(y) $D' \cup T'$ (z) $X' \cup T'$

8. Draw a Venn diagram for each part in question 7 (except cardinal numbers), shading the region stated.

9.

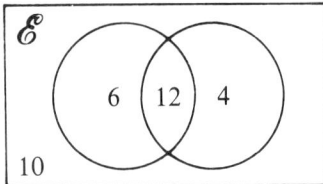

The Venn diagram represents a second year class of whom some have blond hair (H) and some blue eyes (E). The numbers represent the number of children in each group.

(i) How many pupils in the class?

(ii) How many blue-eyed blonds in the class?

(iii) How many blonds do not have blue eyes?

(iv) How many have neither blue eyes nor blond hair?

(v) How many children are not blue-eyed?

(vi) How many have blond hair?

(vii) Write the number of blue-eyed children as a fraction of the whole class.

10. Using $<$ or $>$ signs complete the statements below.

(a) $+2$ $+4$ (b) $+7$ $+1$

(c) -6 $+2$ (d) -3 $+4$

(e) $+5$ -8 (f) $+2$ -6

(g) -4 0 (h) 0 -5

(i) -3 -7 (j) -6 -1

(k) -12 -10 (l) -14 -15

11. Evaluate

(a) $6 + 3$ (b) $8 + 7$

(c) $-2 + 5$ (d) $-3 + 10$

(e) $-8 + 2$ (f) $-14 + 11$

(g) $14 - 9$ (h) $12 - 7$

(i) $5 - 8$ (j) $4 - 15$

(k) $-2 - 3$ (l) $-6 - 2$

(m) $-12 - 16$ (n) $-20 - 13$

(o) $8 + 4 - 10$ (p) $6 + 3 - 8$

(q) $6 - 8 + 4$ (r) $8 - 13 + 7$

(s) $2 - 5 - 8$ (t) $4 - 6 - 10$

(u) $-7 - 4 + 5$ (v) $-15 - 6 + 11$

(w) $-19 + 25 - 14$ (x) $-12 + 23 - 10$

(y) $-4 - 23 - 15$ (z) $-16 - 3 - 18$

12. Evaluate
    (a) 5×5
    (b) 6×8
    (c) -3×9
    (d) -7×4
    (e) 6×-11
    (f) 10×-12
    (g) -3×-14
    (h) -15×-2
    (i) $45 \div 9$
    (j) $64 \div 8$
    (k) $-32 \div 4$
    (l) $-108 \div 9$
    (m) $100 \div -25$
    (n) $28 \div -7$
    (o) $-56 \div -7$
    (p) $-81 \div -9$
    (q) $-6 \times 5 \div 15$
    (r) $-3 \times 12 \div 9$
    (s) $60 \div -12 \times 4$
    (t) $84 \div -7 \times 11$
    (u) $33 \div -11 \times -15$
    (v) $54 \div -6 \times -7$
    (w) $-8 \times -6 \div 12$
    (x) $-5 \times -20 \div 25$
    (y) $-18 \div -3 \times -9$
    (z) $-42 \div -6 \times -10$

13. Simplify
    (a) $p + p + p$
    (b) $m + m + m + m + m$
    (c) $2a + 3a$
    (d) $5s + 9s$
    (e) $6g - 3g$
    (f) $8h - 10h$
    (g) $2p + 4p - 3p$
    (h) $8h + 3h - 4h$

14. Find the value of each of the following if $a = 2$:
    (a) $6a$
    (b) $-3a$
    (c) $5a - a$
    (d) $14a - 8a$

15. Simplify
    (a) $4 \times 5p$
    (b) $6 \times 7g$
    (c) $2a \times 5$
    (d) $4a \times 7$
    (e) $2 \times 3a + 3 \times 2a$
    (f) $7 \times 5a - 3 \times 4a$

16. Simplify
    (a) $2a + 3b + 6a$
    (b) $7m + 4n + 3m$
    (c) $8y + 4x + 2x - 3y$
    (d) $p + 2 - 4 - 3p$

17. Simplify
    (a) $2ab + 4ba$
    (b) $3pq - 2qp$
    (c) $5st + 3s + 2s + ts$
    (d) $ab - 3a + 4b - 2ba$

18. Simplify
    (a) $\dfrac{6a}{3}$
    (b) $\dfrac{18b}{12}$
    (c) $\dfrac{16s}{12}$
    (d) $\dfrac{8y}{16}$
    (e) $\dfrac{3a}{5a}$
    (f) $\dfrac{15h}{55h}$
    (g) $\dfrac{4pqr}{pr}$
    (h) $\dfrac{5abc}{3b}$

19. Write in factor form
    (a) a^3
    (b) b^5
    (c) c^4
    (d) d^7
    (e) e^8
    (f) f^6

20. Write in index form
    (a) $3 \times 3 \times 3 \times 3$
    (b) $p \times p \times p \times p \times p \times p$
    (c) $g \times g \times g \times g \times g \times g \times g$
    (d) $h \times h \times h \times h$
    (e) $y \times y \times y \times y \times y$
    (f) $s \times s \times s \times s \times s \times s \times s \times s \times s$

21. Evaluate
    (a) 2^3
    (b) 3^2
    (c) 4^4
    (d) 5^3

22. Perform the following matrix additions and subtractions:

(a) $\begin{pmatrix} 11 & 17 & 31 \\ 23 & 0 & 10 \end{pmatrix} + \begin{pmatrix} 21 & 13 & 9 \\ 11 & 22 & 11 \end{pmatrix}$

(b) $\begin{pmatrix} 12 & -14 \\ 25 & -2 \end{pmatrix} + \begin{pmatrix} 13 & 22 \\ -17 & 3 \end{pmatrix}$

(c) $\begin{pmatrix} 15 & -16 \\ -9 & -2 \\ 14 & 3 \end{pmatrix} + \begin{pmatrix} 0 & 9 \\ 12 & -1 \\ 7 & 4 \end{pmatrix}$

(d) $\begin{pmatrix} 4 & 9 \\ 16 & 3 \end{pmatrix} - \begin{pmatrix} 1 & 5 \\ 5 & 1 \end{pmatrix}$

(e) $\begin{pmatrix} 0 & 3 & 9 \\ 27 & 15 & 2 \end{pmatrix} - \begin{pmatrix} 3 & 2 & 8 \\ 15 & 16 & 1 \end{pmatrix}$

(f) $\begin{pmatrix} 14 & 1 \\ -3 & 2 \\ 0 & 15 \end{pmatrix} - \begin{pmatrix} 9 & -3 \\ 8 & 17 \\ 15 & 16 \end{pmatrix}$

23. Given $\mathbf{A} = \begin{pmatrix} 17 & 51 & 2 \\ 0 & -13 & 5 \end{pmatrix}$, $\mathbf{B} = \begin{pmatrix} -7 & 3 \\ 6 & 0 \\ 8 & 1 \end{pmatrix}$, $\mathbf{C} = \begin{pmatrix} 14 & 5 \\ 16 & 2 \end{pmatrix}$, $\mathbf{D} = \begin{pmatrix} 12 & 27 \\ -13 & 14 \end{pmatrix}$,

$\mathbf{E} = \begin{pmatrix} 51 & 30 \\ -3 & 12 \end{pmatrix}$, $\mathbf{F} = \begin{pmatrix} 14 & 10 & 0 \\ -2 & -12 & 1 \end{pmatrix}$

find:

(a) $3\mathbf{A}$

(b) $11\mathbf{B}$

(c) $\frac{1}{2}\mathbf{C}$

(d) $-2\mathbf{D}$

(e) $\frac{1}{3}\mathbf{E}$

(f) $\frac{1}{4}\mathbf{F}$

(g) $\mathbf{A} - \mathbf{F}$

(h) $\mathbf{F} - \mathbf{A}$

Number Laws and Algebra

6

6.1 Introduction

In arithmetic there are certain rules which must be followed. It is necessary to know whether these same rules apply in algebra where pronumerals take the place of numbers.

6.2 Commutative Law

6.2.1 Addition

Consider two snails, Lazy and Creepy, who are standing side by side. If Lazy crawls 2 cms, has a rest and then crawls 6 cm, he will finish at the same place as Creepy who crawls 6 cm, has a rest and then crawls 2 cm.

I.e. $\qquad 6 + 2 = 2 + 6$

Similarly $\quad 7 + 6 = 6 + 7$

$\qquad 10 + 15 = 15 + 10$ etc.

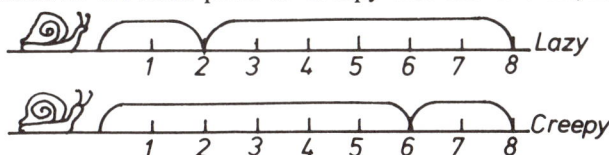

Thus, in arithmetic, the order in which two numbers are added can be changed without altering the result. This is called the *commutative law* for addition.

Now, does the commutative law apply to the addition of two terms in algebra?

If I have one apple and my friend has three apples, we have 4 apples altogether.

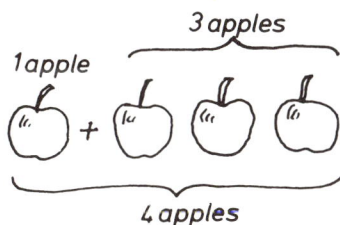

I.e. $\quad 1a + 3a = 4a$

Similarly, if I have 3 apples and my friend has 1 apple, we have 4 apples altogether.

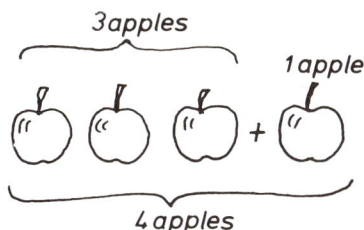

I.e. $\quad 3a + 1a = 4a$

Thus $\quad 1a + 3a = 3a + 1a = 4a$.

Again, the order of addition does not change the result.

Similarly, if I had 2 bananas and my friend had 3 apples, we would have exactly the same amount of each fruit as if I had 3 apples and my friend the 2 bananas.

I.e. $\quad 2b + 3a = 3a + 2b$

Note that in this case the expression cannot be further simplified as the terms are unlike.

In general, changing the order in which two algebraic terms are added does not alter the result:

$$a + b = b + a$$

Thus the commutative law holds for addition in algebra.

6.2.2   Subtraction

I walk 6 paces to the right and then 2 paces to the left.

Where do I finish? Of course I am 4 paces to the right of my starting point. I.e. $6 - 2 = 4$

If, instead, I had walked 2 paces right and then 6 paces left would I have finished at the same place? No. In this case I would have finished 4 paces to the left of where I started. I.e. $2 - 6 = -4$

Thus $6 - 2 \neq 2 - 6$
Similarly $8 - 4 \neq 4 - 8$
                 $12 - 8 \neq 8 - 12$

Thus in arithmetic, changing the order in which two numbers are subtracted *does* alter the result. Hence the commutative law does not hold for subtraction in arithmetic.
If I have 20p and spend 15p, I will have 5p left.
I.e. $20p - 15p = 5p$
However, if I have 15p and spend 20p, I will *owe* 5p
I.e. $15p - 20p = -5p$
Thus $20p - 15p \neq 15p - 20p$
Similarly $2a - 3b$ is not the same as $3b - 2a$.
I.e. $2a - 3b \neq 3b - 2a$

Thus in algebra, changing the order in which two numbers are subtracted *does* alter the result.
In general, $a - b \neq b - a$
Thus the commutative law does *not* hold for subtraction in algebra.

Exercise 6A

1. Evaluate each of the following given $a = 6$ and $b = 3$:

   (a) $a + b$ (b) $2a + b$
   (c) $b + a$ (d) $b + 2a$
   (e) $a - b$ (f) $2a - b$
   (g) $b - a$ (h) $b - 2a$
   (i) $2a + 3b$ (j) $3a + b$
   (k) $3a + 2b$ (l) $b + 3a$
   (m) $b + 2a$ (n) $b - 3a$
   (o) $2b + a$ (p) $2a - b$
   (q) $3a - 2b$ (r) $a + 2b$
   (s) $2b - 3a$ (t) $3b + a$

2. State whether each of the following is True (T) or False (F):

(a) $a + b = b + a$ (b) $2a + b = b + 2a$

(c) $a - b = b - a$ (d) $2a - b = b - 2a$

(e) $2a + b = a + 2b$ (f) $a + 2b = b + 2a$

(g) $a - 2b = b - 2a$ (h) $b - a = a + b$

6.2.3   Multiplication

If I have 2 lots of three apples, I will have 6 apples

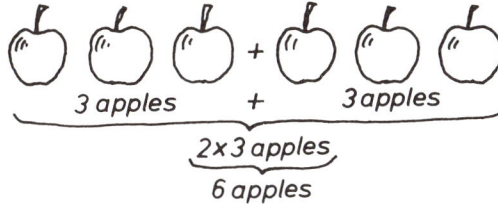

3 apples   +   3 apples

2 × 3 apples

6 apples

But if I have 3 lots of 2 apples, I will still have 6 apples.

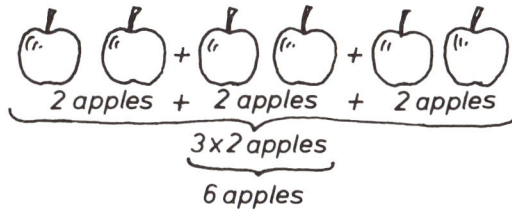

2 apples   +   2 apples   +   2 apples

3 × 2 apples

6 apples

Thus 2 lots of 3 is the same as 3 lots of 2.

I.e. $2 \times 3 = 3 \times 2$

Thus in arithmetic, changing the order in which numbers are multiplied does not alter the result.

Now consider the algebraic terms ab and ba.

If $a = 6$ and $b = 4$

$$ab = a \times b = 6 \times 4 = 24$$

Similarly $ba = b \times a = 4 \times 6 = 24$

Thus, for $a = 6$ and $b = 4$, ab has the same value as ba.

This would be true for any numeric values of a and b.

> In general, changing the order in which two algebraic terms are multiplied does not alter the result:
>
> $$a \times b = b \times a$$
>
> Thus the commutative law holds for multiplication in algebra.

6.2.4   Division

If I have 6 chocolates to divide equally between 3 people, how many should I give each person? Of course, each person should receive 2 chocolates.

$$\frac{6}{3} = \frac{{}^{1}\!\cancel{3} \times 2}{\cancel{3}_{1}} = 2$$

However, if I have 3 chocolates to divide between 6 people, they will each receive $\frac{1}{2}$ chocolate.

$$\frac{3}{6} = \frac{\cancel{3}^{1}}{{}_{1}\cancel{3} \times 2} = \frac{1}{2}$$

Thus $\dfrac{6}{3} \neq \dfrac{3}{6}$

Hence changing the order of division of numbers will alter the result.

Also, $\dfrac{20}{5} \neq \dfrac{5}{20}$ $\dfrac{12}{18} \neq \dfrac{18}{12}$

Consider, now, the algebraic terms $\dfrac{a}{b}$ and $\dfrac{b}{a}$.

If $a = 12$ and $b = 2$,

$$\dfrac{a}{b} = \dfrac{12}{2} = \dfrac{6 \times 2^1}{2_1} = 6$$

and $$\dfrac{b}{a} = \dfrac{2}{12} = \dfrac{2^1}{6 \times 2_1} = \dfrac{1}{6}$$

Thus, evaluating $\dfrac{a}{b}$ and $\dfrac{b}{a}$ for $a = 12$ and $b = 2$ shows us that $\dfrac{a}{b}$ does not have the same value as $\dfrac{b}{a}$.

In general, changing the order in which two algebraic terms are divided *does* alter the result:

$$\dfrac{a}{b} \neq \dfrac{b}{a}$$

Thus the commutative law does not hold for division in algebra.

Exercise 6B

1. Evaluate (i) $a \times b$ and (ii) $b \times a$ if
   (a) $a = 6, b = 3$ (b) $a = 8, b = 2$
   (c) $a = 3, b = 12$ (d) $a = 4, b = 12$
   (e) $a = 2, b = 9$ (f) $a = 9, b = 7$
   (g) $a = 5, b = 12$ (h) $a = 12, b = 9$

2. Evaluate (i) $\dfrac{a}{b}$ and (ii) $\dfrac{b}{a}$ for each pair of values of a and b in question 1.

3. State whether each of the following is true (T) or false (F)
   (a) $m \times n = n \times m$ (b) $s \times t = t \times s$
   (c) $\dfrac{p}{q} = \dfrac{q}{p}$ (d) $\dfrac{r}{t} = \dfrac{t}{r}$
   (e) $a + b = b + a$ (f) $2p + q = q + 2p$
   (g) $y - s = s - y$ (h) $g - b = b - g$
   (i) $2a + b = 2b + a$ (j) $s + 2t = 2s + t$
   (k) $3g - h = 3h - g$ (l) $4y - z = 4z - y$

6.3   Associative Law

For addition, subtraction, multiplication and division of numbers, only two numbers can be dealt with at one time.

Thus, for example, $2 + 4 + 7 = (2 + 4) + 7$

$$= \quad 6 + 7$$

$$= \quad\quad 13$$

and $$3 \times 2 \times 5 = 3 \times (2 \times 5)$$

$$= \quad 3 \times 10$$

$$= \quad\quad 30$$

The associative law states that when three terms are being considered either the first two *or* the second two may be grouped together and dealt with first.

Does this law hold for each of addition, subtraction, multiplication and division?
(a) Addition:
e.g. $7 + 3 + 2 = (7 + 3) + 2 = 7 + (3 + 2)$
Yes, the associative law holds for addition.
(b) Subtraction:
e.g. $7 - 3 - 2 = (7 - 3) - 2 \neq 7 - (3 - 2)$
No, the associative law does *not* hold for subtraction.
(c) Multiplication:
e.g. $12 \times 6 \times 2 = (12 \times 6) \times 2 = 12 \times (6 \times 2)$
Yes, the associative law holds for multiplication.
(d) Division:
e.g. $12 \div 6 \div 2 = (12 \div 6) \div 2 \neq 12 \div (6 \div 2)$
No, the associative law does *not* hold for division.

> Similarly, in algebra the associative law holds for addition and multiplication, but does not hold for subtraction and division.

Example 6.1
Simplify
(a) $(2m + 3m) + 7m$
(b) $2m + (3m + 7m)$
(c) $(8b - 5b) - 2b$
(d) $8b - (5b - 2b)$
(e) $(3x \times 5x) \times 2x$
(f) $3x \times (5x \times 2x)$
(g) $(18y \div 6y) \div 3y$
(h) $18y \div (6y \div 3y)$

Solution
(a) $(2m + 3m) + 7m$

$= 5m + 7m$
$= 12m$

(b) $2m + (3m + 7m)$

$= 2m + 10m$
$= 12m$

(c) $(8b - 5b) - 2b$

$= 3b - 2b$
$= b$

(d) $8b - (5b - 2b)$

$= 8b - 3b$
$= 5b$

(e) $(3x \times 5x) \times 2x$

$= 15x^2 \times 2x$
$= 30x^3$

(f) $3x \times (5x \times 2x)$

$= 3x \times 10x^2$
$= 30x^3$

(g) $(18y \div 6y) \div 3y$

$= \dfrac{^3 18y}{_1 6y} \div 3y$

$= \dfrac{3^1}{_1 3y}$

$= \dfrac{1}{y}$

(h) $18y \div (6y \div 3y)$

$= 18y \div \dfrac{^2 6y}{_1 3y}$

$= \dfrac{^9 18y}{_1 2}$

$= 9y$

Example 6.2
Simplify
(a) $7 \times 5a$
(b) $6n \times n$
(c) $3a \times 5b$

Solution

(a) $7 \times 5a = (7 \times 5) \times a$
$= 35a$

(b) $6n \times n$
$= 6 \times (n \times n)$
$= 6n^2$

(c) $3a \times 5b = 3 \times a \times 5 \times b$
$= (3 \times 5) \times (a \times b)$
$= 15 \times ab$
$= 15ab$

Exercise 6C

1. Which of the following are true (T) and which are false (F):
   (a) $8 + (3 + 5) = (8 + 3) + 5$
   (b) $7 - (5 - 2) = (7 - 5) - 2$
   (c) $16 \div (8 \div 2) = (16 \div 8) \div 2$
   (d) $8 \times (3 \times 2) = (8 \times 3) \times 2$
   (e) $(2a + 3a) + a = 2a + (3a + a)$
   (f) $(5b - 2b) - b = 5b - (2b - b)$
   (g) $(3p \times 2p) \times p = 3p \times (2p \times p)$
   (h) $(4x \div 2x) \div 2 = 4x \div (2x \div 2)$
   (i) $6m - (m - 2m) = (6m - m) - 2m$
   (j) $2g + (5g + 7g) = (2g + 5g) + 7g$
   (k) $(3a \div 6a) \div q = 3a \div (6a \div q)$
   (l) $4x \times (x \times 3x) = (4x \times x) \times 3x$

2. Simplify:
   (a) $3x + 2x + 4x$
   (b) $5a + 6a + 7a$
   (c) $2m + 7m + 3m$
   (d) $9a + 2a + 6a$
   (e) $12a - 3a - 2a$
   (f) $7p - p - p$
   (g) $4m + 12m - 9m$
   (h) $6a + 7a - 3a$
   (i) $2a - 4a + 5a$
   (j) $a - 5a + 6a$
   (k) $g - 3g + 6g$
   (l) $b - 3b + 2b$

3. Simplify each of the following:
   (a) $2 \times 2a$
   (b) $5 \times 5a$
   (c) $6x \times 6$
   (d) $7p \times 3$
   (e) $9g \times 3$
   (f) $8y \times 5$
   (g) $7p \times q$
   (h) $4p \times h$
   (i) $2x \times 4y$
   (j) $6a \times 2b$
   (k) $7m \times 2n$
   (l) $8y \times 4z$
   (m) $6l \times 3m$
   (n) $7p \times 5q$
   (o) $2s \times 3t$
   (p) $5r \times 2t$
   (q) $11g \times 4h$
   (r) $7c \times 8d$
   (s) $9b \times 5c$
   (t) $8x \times 9y$
   (u) $3p \times 5q$
   (v) $7h \times 9k$

4. Simplify each of the following:
   (a) $2a \times 4a$
   (b) $5m \times 4m$
   (c) $6n^2 \times 7n$
   (d) $8y \times 2y^2$

6.3.1   Addition and Subtraction

In fact, the commutative and associative laws enable us to simplify many algebraic expressions such as seen in Chapter 4.

For example,

$4b + a + 2b + 2a$ reordering the terms using the commutative law
$= 4b + 2b + a + 2a$
$= (4b + 2b) + (a + 2a)$ grouping like terms using associative law
$= 6b + 3a$

Other, more complicated algebraic expressions can similarly be simplified using addition and subtraction so long as there are 'like' terms. Remember that 'like' terms have identical pronumeral parts.

For example,

5a and 2a are 'like' terms and so are $2mn$ and $7mn$.

However, 5a and 2b are *not* 'like' terms and neither are $2mn$ and $7m$. Consider now the terms $2a^2$ and $3a^2$. The pronumeral part of both these terms is a^2 (i.e. $a \times a$) and thus they are 'like' terms. However, the terms $5a^2$ and $4a^3$ are *not* 'like' terms because the pronumeral part of the first term is a^2 (i.e. $a \times a$) while the pronumeral part of the second term is a^3 ($a \times a \times a$).

Example 6.2

Simplify where possible:

(a) $5a^3 - 2a^3$

(b) $a^2 + a^2$

(c) $2a^2 + 7a^4$

(d) $8m^2n - 7mn^2$

(e) $3x^2 + 5x - x^2$

(f) $10a^2b + 4ab - 6ba^2$

Solution

(a) $5a^3 - 2a^3$
$= 3a^3$

(b) $a^2 + a^2$
$= 1a^2 + 1a^2$
$= 2a^2$

(c) $2a^2 + 7a^4$ cannot be simplified because the two terms do not have identical pronumeral parts.

(d) $8m^2n - 7mn^2$ cannot be simplified because the first pronumeral part is $m \times m \times n$ while the second is $m \times n \times n$.

(e) $3x^2 + 5x - x^2$
$= 3x^2 - x^2 + 5x$ regrouping terms
$= (3x^2 - x^2) + 5x$
$= \quad 2x^2 + 5x$ and this cannot be simplified further

(f) $10a^2b + 4ab - 6ba^2$
$= 10a^2b + 4ab - 6a^2b$ because $a^2b = ba^2$ using the commutative law
$= 10a^2b - 6a^2b + 4ab$ regrouping terms
$= (10a^2b - 6a^2b) + 4ab$
$= \quad 4a^2b + 4ab$ and this cannot be simplified further.

Exercise 6D

1. Which of the following pairs of terms are like terms:
   (a) a^2 and $2a^2$
   (b) c^3 and $5c^3$
   (c) g^2 and g
   (d) e^2 and e^3
   (e) $3a^3$ and $5a^3$
   (f) $4m^2$ and $3m^3$
   (g) a^2b and ab^2
   (h) lm^2n and l^2mn
   (i) xy^2 and y^2x
   (j) pq^2r and q^2pr

2. Simplify each of the following:
   (a) $2a^2 + 3a^2$
   (b) $5p^2 + 7p^2$
   (c) $6p^3 + 5p^3$
   (d) $7x^2 + 8x^2$
   (e) $5y^2 - 2y^2$
   (f) $11h^3 - 3h^3$
   (g) $12x^3 - 5x^3$
   (h) $7x^2 - 6x^2$
   (i) $5x^2 - 5x^2$
   (j) $3a^3 - 3a^3$

3. Simplify where possible:
   (a) $3a^2 + 5a^2$
   (b) $4b^3 - 3b^3$
   (c) $2a^4 - a^4$
   (d) $3b^5 - b^5$
   (e) $4x^2 + 3x^3 - 2x^2$
   (f) $5y^3 - 3y^2 + 2y^3$
   (g) $2a^2 + 3a + 5a$
   (h) $7m^2 + 4m^2 + 2m$
   (i) $3b^2 - 4b^2 + 5b$
   (j) $6c^2 - 7c^2 + 7c$
   (k) $2ab + 4ba - 3b^3$
   (l) $6gh + 5hg - 7g^2$
   (m) $3st^2 + 4st^2 - 4s^2$
   (n) $5gp^2 + 3gp^2 - 2g^2$

    (o) $6s^2t + 4s + 2s^2t$ (p) $6p^2q + 4p - 2p^2q$

    (q) $2gh + 3hg - 4h^2g$ (r) $5st + 4st - 3s^2t$

    (s) $6lmn + 4lm + 6lm$ (t) $5pq + 6pqr - 7pq$

4. Simplify:

    (a) $a^2 + 4a^3 + 2a^2 + a^3$ (b) $6p^2 + 4p + 3p + 2p^2$

    (c) $7m^2 + 5m + 4m^2 + 3m$ (d) $7g^2 + 5g^3 + 2g^3 + 3g^2$

    (e) $2p + 6p^2 + 5p - 2p^2$ (f) $6m^3 + 3m - 5m - 2m^3$

    (g) $h + 5h^2 - 3h^2 + 10h$ (h) $2gh + 4g^2 - 2g^2 - gh$

    (i) $7m^2p - 3p + 4m^2p + 7p$ (j) $3s^2t + 4s - s^2t + 5s$

    (k) $6g^2h + 5gh^2 - 3gh^2 - 5g^2h$ (l) $6pq^2 + 2p^2q + 5p^2q - 4pq^2$

    (m) $12a^2b + 4ab^2 - 3ab^2 - 7a^2b$ (n) $9g^2h + 4gh^2 - 2gh^2 + 4g^2h$

6.4   Distributive Law

We have seen in Chapter 1 that in arithmetic, brackets must always be calculated first.

E.g. $2 \times (5 + 3) \div 4$

     $= 2 \times 8 \div 4$ (brackets)

     $=\quad 16 \div 4$ (multiplication)

     $= 4$ (division)

However, in algebra, the brackets cannot always be simplified.

E.g. in $2 \times (x + y)$, $x + y$ cannot be simplified but can anything be done with an expression such as this?

Consider, if we have 2 bags each containing 3 apples and 4 bananas, how many apples and bananas do we have?

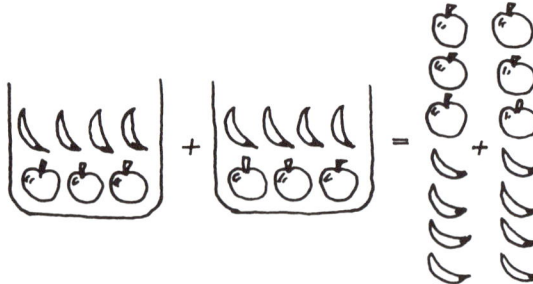

$$2 \times (3 \text{ apples} + 4 \text{ bananas}) = 2 \times 3 \text{ apples} + 2 \times 4 \text{ bananas}$$
$$= 6 \text{ apples} + 8 \text{ bananas}$$

Using pronumerals

$$2 \times (3a + 4b) = 2 \times 3a + 2 \times 4b$$
$$= 6a + 8b$$

This is called multiplying out or expanding brackets using the distributive law.

The distributive law states that *each* term in the bracket is to be multiplied by the term outside the brackets.

E.g. $4(a + 3) = 4 \times a + 4 \times 3 = 4a + 12$

Example 6.3

Use the distributive law to expand the following brackets and simplify where possible:

  (a) $3(x + y)$ (b) $2(a - b)$

  (c) $a(a - 2c)$ (d) $3x(2y + 1)$

  (e) $(x + y) \times 3$ (f) $3(x + y) + 2(x - y)$

Solution

(a) $3 \times (x + y) = 3 \times x + 3 \times y$ distributive law
$\qquad\qquad\quad = 3x + 3y$

(b) $2(a - b) = 2 \times (a - b)$
$\qquad\qquad = 2 \times a - 2 \times b$ distributive law
$\qquad\qquad = 2a - 2b$

(c) $a(a - 2c) = a \times a - a \times 2c$
$\qquad\qquad\;\; = a^2 - a \times 2 \times c$
$\qquad\qquad\;\; = a^2 - 2ac$

(d) $3x(2y + 1) = 3x \times 2y + 3x \times 1$ distributive law
$\qquad\qquad\quad = 6xy + 3x$

(e) $(x + y) \times 3 = 3 \times (x + y)$ using commutative law
$\qquad\qquad\quad\;\; = 3x + 3y$ using distributive law

(f) $3(x + y) + 2(x - y) = 3x + 3y + 2x - 2y$ distributive law
$\qquad\qquad\qquad\qquad\; = 5x + y$ adding 'like' terms.

Exercise 6E

1. Expand each of the following:

   (a) $2(a + 1)$
   (b) $3(b + 4)$
   (c) $5(c + 7)$
   (d) $9(d + 12)$
   (e) $3(b + 4)$
   (f) $2(m + 6)$
   (g) $5(g - 2)$
   (h) $8(p - 5)$
   (i) $9(y - 4)$
   (j) $7(y - 11)$
   (k) $6(s - 3)$
   (l) $12(t - 8)$
   (m) $4(3 - a)$
   (n) $5(9 - h)$
   (o) $2(4 - l)$
   (p) $3(6 - z)$
   (q) $8(7 - t)$
   (r) $5(12 - f)$
   (s) $11(2 + p)$
   (t) $6(3 + h)$
   (u) $9(5 + y)$
   (v) $11(7 + s)$
   (w) $8(8 + a)$
   (x) $12(2 + b)$
   (y) $5(2 + m)$
   (z) $9(12 + c)$

2. Expand each of the following:

   (a) $a(b + 7)$
   (b) $b(c + 3)$
   (c) $d(e + 5)$
   (d) $l(m + 4)$
   (e) $h(y - 3)$
   (f) $k(a - 4)$
   (g) $z(x - 8)$
   (h) $g(h - 2)$
   (i) $m(p + q)$
   (j) $l(s + t)$
   (k) $y(x + z)$
   (l) $a(b + c)$
   (m) $x(p + h)$
   (n) $c(d + e)$
   (o) $h(k + l)$
   (p) $e(f + g)$
   (q) $b(c - d)$
   (r) $s(t - u)$
   (s) $h(i - j)$
   (t) $g(h - k)$

3. Expand each of the following:

   (a) $4(2a + 3)$
   (b) $6(3b + 5)$
   (c) $7(4m + 8)$
   (d) $3(8 + 5p)$
   (e) $5(6 + 7y)$
   (f) $7(2 + 3c)$
   (g) $4(2m - 3)$
   (h) $3(5z - 7)$
   (i) $7(5p - 4)$
   (j) $12(2a - 3)$
   (k) $6(2b + 5c)$
   (l) $8(3l + 4m)$
   (m) $4(2x - 3y + 4z)$
   (n) $3(7m - 8n + 5p)$
   (o) $5(2b + 5c + 7d)$
   (p) $9(5c + 3d + 5e)$
   (q) $7(4y + 7z - 3p)$
   (r) $8(4p + 3q - 7r)$

4. Expand each of the following:

(a) $a(a + 2)$
(b) $b(b + 3)$
(c) $c(c - 5)$
(d) $d(d - 2)$
(e) $m(m + 8)$
(f) $q(q - 5)$
(g) $d(d + 1)$
(h) $e(e + 1)$
(i) $a(a + b)$
(j) $c(c + d)$
(k) $m(m + n)$
(l) $p(p + q)$
(m) $s(s - t)$
(n) $x(x - y)$
(o) $s(2s + t)$
(p) $y(3y - z)$
(q) $a(4a - b)$
(r) $m(2m - 3n)$
(s) $l(2m + 3n + 4p)$
(t) $a(2a + 3b - 4c)$
(u) $2a(3b - 4c)$
(v) $5x(2y - 3z)$
(w) $3p(4r + 7t)$
(x) $8l(3m - 4n)$
(y) $5a(2b - c + 3d)$
(z) $6s(5t + 3u - 6v)$

5. Expand each of the following:

(a) $(a + b)g$
(b) $(b + c)d$
(c) $(e + f)h$
(d) $(m + n)p$
(e) $(2a + b)b$
(f) $(x + y)y$
(g) $(3a - 2b)a$
(h) $(4a - 2b)a$
(i) $(l - 3m)l$
(j) $(5y - 6z)z$

6. Expand and simplify each of the following:

(a) $2(2a + 3) + 3(a + 5)$
(b) $7(4b + 3) + 2(b - 5)$
(c) $6(d + 5) + 4(2d - 3)$
(d) $3(3h + 5) + 5(7 - h)$
(e) $4(2x + 1) + 2(x + 2)$
(f) $7(3x + 4) + 5(4x + 1)$
(g) $5(5x + 6) + 3(6x - 7)$
(h) $8(4 + a) + 5(3 + 4a)$
(i) $a(a + b) + a(2a + 3b)$
(j) $h(h + k) + 3h(h - k)$
(k) $z(2z + y) + z(3z - y)$
(l) $a(a + b) + a(2a + 5b)$
(m) $l(2l - m) + l(l + m)$
(n) $2s(s - t) + s(s + 4t)$
(o) $2p(3p + q) + p(4p - q)$
(p) $2h(h + t) + h(t - 2h)$
(q) $4(3a + b) + 1(a - 4b)$
(r) $7(5b + 3c) + 1(2b - c)$
(s) $6(l + 4m) + (m + 5l)$
(t) $6(y + 3z) + (2y - 5z)$
(u) $3(p + 2q) + (p + 6q)$
(v) $7(a + 3b) + (2a - 4b)$
(w) $(s + 3t) + 2(s + 4t)$
(x) $(5g + 4h) + 3(2g - 2h)$
(y) $(2p + 4q) + 3(q - p)$
(z) $(7d + 5e) + 2(2e - 3d)$

Example 6.4

Find the value of each of the following when $p = 5$ and $q = 3$.

(a) $2(p + q)$
(b) $p(q - 2)$

Solution

The brackets may be first expanded and then values substituted.

(a) $2(p + q) = 2p + 2q$
$$= 2 \times 5 + 2 \times 3$$
$$= 10 + 6$$
$$= 16$$

Alternatively, values may be substituted and then bracket simplified.

(b) $p(q - 2) = 5(3 - 2)$
$$= 5 \times 1$$
$$= 5$$

It is often easier to substitute first.

Exercise 6F

Find the value of each of the following when $a = 6$ and $b = 2$.

1. $2(a + b)$
2. $3(a + b)$
3. $4(a - b)$
4. $2(a - b)$

5. $3(2a + b)$

6. $5(2a - b)$

7. $8(b + a)$

8. $4(4b - a)$

9. $a(b + 1)$

10. $a(b - 1)$

11. $b(a + 3)$

12. $b(a + 5)$

13. $b(a - 4)$

14. $b(a - 1)$

15. $(a + b)a$

16. $(a - b)a$

17. $(a - b)b$

18. $(a + b)b$

6.5   Negative Numbers in Algebra

All the rules we have derived earlier in this chapter have used positive numbers and pronumerals which have positive values. However, we can apply the same rules when negative numbers are involved.

Example 6.5

Evaluate each of the following when $a = 2$ and $b = {}^-3$.

(a) $5a$

(b) $2b$

(c) $a + b$

(d) $a - b$

(e) $2ba + b^2$

(f) $a^2 - b^2$

(g) $2b \div a$

(h) $a \times b - b$

Solution

(a) $5a = 5 \times a$
$\qquad = 5 \times 2$
$\qquad = 10$

(b) $2b = 2 \times b$
$\qquad = 2 \times {}^-3$
$\qquad = {}^-6$

(c) $a + b = 2 + {}^-3$
$\qquad\quad = 2 - 3$
$\qquad\quad = {}^-1$

(d) $a - b = 2 - {}^-3$
$\qquad\quad = 2 + 3$
$\qquad\quad = 5$

(e) $2ba + b = 2 \times {}^-3 \times 2 + {}^-3$
$\qquad\quad\;\; = {}^-12 - 3$
$\qquad\quad\;\; = {}^-15$

(f) $a^2 - b^2 = 2^2 - ({}^-3)^2$
$\qquad\qquad = 4 - 9$
$\qquad\qquad = {}^-5$

(g) $2b \div a = 2 \times {}^-3 \div 2$
$\qquad\quad\; = \dfrac{{}^12 \times {}^-3}{2_1}$
$\qquad\quad\; = {}^-3$

(h) $a \times b - b = 2 \times {}^-3 - {}^-3$
$\qquad\qquad\; = {}^-6 + 3$
$\qquad\qquad\; = {}^-3$

Exercise 6G

Find the value of the following when $m = {}^-2$, $n = 2$ and $p = {}^-3$.

1. $6m$

2. $3p$

3. ${}^-5n$

4. ${}^-2m$

5. ${}^-4p$

6. ${}^-6n$

7. mn

8. mp

9. pn

10. np

11. $\dfrac{m}{n}$

12. $\dfrac{p}{n}$

13. $\dfrac{n}{p}$

14. $\dfrac{m}{p}$

15. $m + n$

16. $m + p$

17. $m - p$

18. $n - p$

19. $p - m$

20. $p - n$

21. $2m + 4$

22. $3n + 5$

23. $6p + 10$

24. $7n + 3$

25. $3n - m$ 26. $2m - p$
27. $4m - 2p$ 28. $5n - 3p$
29. m^2 30. n^2
31. p^2 32. $p^2 + 4$
33. $m^2 - p^2$ 34. $n^2 - p^2$
35. $m^2 + np$ 36. $n^2 + mp$
37. $p^2 + np$ 38. $p^2 + nm$
39. $p(m + n)$ 40. $n(m + p)$
41. $m(p + n)$ 42. $m(m + p)$
43. $m(m - n)$ 44. $p(p - m)$
45. $m(m + 1) + p(p + 1)$ 46. $n(n + 2) + p(p + 4)$
47. $m(n + p) - n(3 - m)$ 48. $n(m + p) - n(2 - p)$
49. $n(2 - m) - p(2 - n)$ 50. $p(3 - n) - m(2 - m)$

6.5.1   Addition and Subtraction
Like terms with negative coefficients may be added or subtracted.

Example 6.6
Simplify
(a) $5 - 9$ (b) $-7 - 12$
(c) $5a - 9a$ (d) $-7b - 12b$
(e) $4m - 2n - 6m - 8n$ (f) $2p - 4q - 3p + q$

Solution
(a) $5 - 9 = -4$ (see 3.3)   (b) $-7 - 12 = -19$ (see 3.3)
(c) $5a - 9a = -4a$ (d) $-7b - 12b = -19b$
(e) $4m - 2n - 6m - 8n = 4m - 6m - 2n - 8n$

$$= -2m \quad -10n$$

(f) $2p - 4q - 3p + q$
$$= 2p - 3p - 4q + q$$
$$= -p \quad -3q$$

Exercise 6H
1. Simplify
   (a) $3 - 5$ (b) $6 - 9$
   (c) $-8 + 7$ (d) $-12 + 5$
   (e) $3a - 6a$ (f) $7m - 15m$
   (g) $2b - 12b$ (h) $6b - 13b$
   (i) $3m - 9m$ (j) $12m - 18m$
   (k) $-6p + 4p$ (l) $-9p + 8p$
   (m) $-15c + 7c$ (n) $-8c + 4c$
   (o) $-6gh + 5gh$ (p) $-3lm + lm$
   (q) $-8st + st$ (r) $-15s^2 + s^2$

2. Simplify
   (a) $-6 - 7$ (b) $-3 - 8$
   (c) $-4 - 9$ (d) $-8 - 6$
   (e) $-2a - 4a$ (f) $-3a - 5a$
   (g) $-6a - 9a$ (h) $-2g - 3g$
   (i) $-4pq - 5pq$ (j) $-7pq - 8pq$
   (k) $-6a^2 - 9a^2$ (l) $-11g^2 - 7g^2$
   (m) $-3mp - 5mp$ (n) $-9ab - 3ab$

3. Simplify

(a) $2a - 3a - 3a$

(b) $7p - 8p - 3p$

(c) $5g - 12g - 3g$

(d) $8c - 9c - 4c$

(e) $-2p + p - 3p$

(f) $-6d + d - 4d$

(g) $-8y + 3y - 6y$

(h) $-4y + y - 3y$

(i) $-2x - 3x - 4x$

(j) $-5p - 6p - 2p$

(k) $-9h - 2h + h$

(l) $-3h - h + 3h$

(m) $-4p - 6p + 10p$

(n) $-3c - 7c + 10c$

(o) $-2c - 4c + 6c$

(p) $-8x + 2x + 6x$

4. Simplify

(a) $2a + 4b - 6a - 8b$

(b) $3s + 2y - 9s - 12y$

(c) $6l - 3p - 9l - 7p$

(d) $-5c - 7d - 3c + 2d$

(e) $-8x - 5y + y - x$

(f) $-3h - 2g + g - 4h$

(g) $-2xy + 3x - 7x + xy$

(h) $-9pq + 4q - 12q + 2pq$

(i) $7hk - 4p - 9p - 8hk$

(j) $l^2 - 5m - 7m - 8l^2$

(k) $6p^2 - 3q^2 - 5q^2 - 8p^2$

(l) $4st + 3x - 6st - 7x$

(m) $-5z^2 - 3xy + 2z^2 - xy$

(n) $-9t^2 - 8p + 7t^2 + 3p$

6.5.2   Multiplication and Division

In Chapter 3 we found that

$$+3 \times +2 = +6$$
$$+3 \times -2 = -6$$
$$-3 \times +2 = -6$$
$$-3 \times -2 = +6$$

and

$$+6 \div +2 = +3$$
$$+6 \div -2 = -3$$
$$-6 \div +2 = -3$$
$$-6 \div -2 = -3$$

i.e.   The product or quotient of two numbers with the same signs will always be positive.
The product or quotient of two numbers with different signs will always be negative.

Example 6.7

Simplify

(a) $2 \times -3a$

(b) $-3 \times 2a$

(c) $-2a \times -3a$

(d) $-4a \div 2$

(e) $-4a \div -2$

(f) $-2a \div a$

(g) $-4pq \div 2q$

(h) $-8a^2 \div -4a$

Solution

(a) $2 \times -3a$

$= 2 \times -3 \times a$

$= -6a$

(b) $-3 \times 2a$

$= -3 \times 2 \times a$

$= -6a$

(c) $-2a \times -3a$

$= -2 \times a \times -3 \times a$

$= -2 \times -3 \times a \times a$

$= 6a^2$

(d) $-4a \div 2$

$= \dfrac{{-4}^2 \times a}{2_1}$

$= -2a$

(e) $-4a \div -2$

$= \dfrac{{}^{-2}4 \times a}{-2_1}$

$= 2a$

(f) $-2a \div a$

$= \dfrac{-2 \times a^1}{a_1}$

$= -2$

(g) $-4pq \div -2p$

$$= \frac{{}^{2}\cancel{-4pq}{}^{1}}{-2\cancel{q}_{1}}$$

$$= 2p$$

(h) $-8a^2 \div -4a$

$$= \frac{-8a^2}{-4a}$$

$$= \frac{{}^{2}\cancel{-8} \times a \times \cancel{a}}{\cancel{-}4\cancel{a}}$$

$$= 2a$$

Exercise 6I

1. Simplify each of the following:

   (a) $2 \times -4m$

   (c) $6 \times -7q$

   (e) $7 \times -3h$

   (g) $-2 \times 4n$

   (i) $-5 \times 3a$

   (k) $-8 \times 5c$

   (m) $-2a \times 2b$

   (o) $5b \times -3c$

   (q) $-8g \times -3p$

   (s) $-3a \times 2a$

   (u) $-5c \times -3c$

   (w) $-4p \times -2p$

   (b) $5 \times -3p$

   (d) $8 \times -5g$

   (f) $7 \times -4m$

   (h) $-3 \times 6p$

   (j) $-6 \times 2q$

   (l) $-4 \times 9d$

   (n) $-3p \times 4q$

   (p) $6m \times -5n$

   (r) $-2s \times -4t$

   (t) $-4b \times 6b$

   (v) $-8h \times -3h$

   (x) $-c \times -3c$

2. Simplify each of the following:

   (a) $-4a \div 4$

   (c) $-12a \div 4$

   (e) $-6p \div 3$

   (g) $-40p \div -4$

   (i) $-55a \div -11$

   (k) $-3a \div a$

   (m) $-12s \div s$

   (o) $-18q \div q$

   (q) $-3a \div -a$

   (s) $-7c \div -c$

   (u) $-6b \div 2b$

   (w) $-15c \div 5c$

   (y) $-24p \div -12p$

   (b) $-6a \div 6$

   (d) $-8m \div 4$

   (f) $-48a \div 12$

   (h) $-36p \div -9$

   (j) $-18q \div -6$

   (l) $-7g \div g$

   (n) $-15t \div t$

   (p) $-14p \div p$

   (r) $-4b \div -b$

   (t) $-5z \div -z$

   (v) $-10p \div 2p$

   (x) $-18g \div -6g$

   (z) $-30q \div -10q$

3. Simplify each of the following:

   (a) $-2ab \div 2a$

   (c) $-4pq \div 4q$

   (e) $-3mn \div mn$

   (g) $-8st \div -st$

   (i) $-a^2 \div a$

   (k) $-3q^2 \div 3q$

   (m) $-6x^2 \div -2x$

   (c) $-9b^2 \div 3b$

   (q) $2x^2 \div -4x$

   (b) $-5mn \div 5n$

   (d) $-6st \div 6s$

   (f) $-4pq \div pq$

   (h) $-6xy \div -xy$

   (j) $-m^2 \div m$

   (l) $-15x^2 \div 5x$

   (n) $-18p^2 \div 6p$

   (p) $-8s^2 \div 16s$

   (r) $3m^2 \div -9m$

6.5.3   Distributive Law

In arithmetic

$$-3 \times 5 + -3 \times 4 = -15 + -12$$
$$= -15 - 12$$
$$= -27$$

and $\qquad -3(5 + 4) = -3 \times 9 = -27$

thus $\qquad -3(5 + 4) = -3 \times 5 + -3 \times 4$

$$\therefore \ -3(5 + 4) = -3 \times 5 - 3 \times 4$$

Also

$$-2 \times 4 - -2 \times 3 = -8 - -6$$
$$= -8 + 6$$
$$= -2$$

and
$$-2(4 - 3) = -2 \times 1 = -2$$

thus
$$-2(4 - 3) = -2 \times 4 - -2 \times 3$$
$$\therefore -2(4 - 3) = -2 \times 4 + 2 \times 3$$

Similarly in algebra

$$-2(x + 5) = -2 \times x + -2 \times 5$$
$$= -2 \times x - 2 \times 5$$
$$\therefore -2(x + 5) = -2x - 10$$

and

$$-3(y - 7) = -3 \times y - -3 \times 7$$
$$=: -3 \times y + 3 \times 7$$
$$\therefore -3(y - 7) = -3y + 21$$

Notice that when the number in front of the bracket is negative, expansion results in the sign of *every* term in the bracket being changed.

In general, when expanding brackets, a simple rule can be applied.

E.g. $3 \times x = 3x$

(a) $3 \ (\quad x \quad + \quad 2 \ )$

$3 \times +2 = +6$
$$= 3x + 6$$

$3 \times x = 3x$

(b) $3 \ (\quad x \quad - \quad 2 \ )$

$3 \times -2 = -6$
$$= 3x - 6$$

$-3 \times x = -3x$

(c) $-3 \ (\quad x \quad + \quad 2 \ )$

$-3 \times +2 = -6$
$$= -3x - 6$$

$-3 \times x = -3x$

(d) $-3 \ (\quad x \quad - \quad 2 \ )$

$-3 \times -2 = +6$
$$= -3x + 6$$

Example 6.8

Expand:

(a) $-3(x + 9)$ (b) $-5(y - 7)$

(c) $-a(b + c)$ (d) $-5y(2y - z)$

(e) $-(m + n)$ (f) $-(a - b)$

Solution

(a) $-3(x + 9) = -3x - 27$

(b) $-5(y - 7) = -5y + 35$

(c) $-a(b + c) = -a \times b - a \times c$
$$= -ab - ac$$

(d) $-5y(2y - z) = -5y \times 2y + 5y \times z$
$$= -10y^2 + 5yz$$

(e) $-(m + n) = -1(m + n)$
$= -1 \times m - 1 \times n = -m - n$

(f) $-(a - b) = -1(a - b)$
$= -1 \times a + 1 \times b$
$= -a + b$

In general

$$a(b + c) = ab + ac$$
$$a(b - c) = ab - ac$$
$$-a(b + c) = -ab - ac$$
$$-a(b - c) = -ab + ac$$

Exercise 6J

1. Expand each of the following:
   (a) $-2(a + 2)$ (b) $-3(b + 4)$
   (c) $-4(c + 5)$ (d) $-8(d + 7)$
   (e) $-3(x + 9)$ (f) $-5(y + 1)$
   (g) $-6(f + 4)$ (h) $-8(g + 5)$
   (i) $-9(m + 10)$ (j) $-6(n + 7)$
   (k) $-3(a - 2)$ (l) $-4(b - 4)$
   (m) $-4(h - 5)$ (n) $-6(s - 9)$
   (o) $-9(p - 4)$ (p) $-10(h - 5)$
   (q) $-6(s - 11)$ (r) $-2(t - 1)$
   (s) $-7(4 - y)$ (t) $-3(2 - x)$
   (u) $-6(7 - p)$ (v) $-8(1 - y)$
   (w) $-4(4 + z)$ (x) $-3(3 + a)$
   (y) $-7(2 + y)$ (z) $-3(9 + m)$

2. Expand each of the following:
   (a) $-a(b + c)$ (b) $-c(m + n)$
   (c) $-4(e + f)$ (d) $-g(h + k)$
   (e) $-l(m + n)$ (f) $-p(q + r)$
   (g) $-x(y + z)$ (h) $-g(s + t)$
   (i) $-a(b - c)$ (j) $-c(d - e)$
   (k) $-e(f - g)$ (l) $-h(l - m)$
   (m) $-m(n - p)$ (n) $-s(r - t)$
   (o) $-g(h + l)$ (p) $-p(q + r)$
   (q) $-s(v + w)$ (r) $-l(r + s)$
   (s) $-y(x - z)$ (t) $-w(u - v)$

3. Expand each of the following:
   (a) $-2(3a + 5)$ (b) $-3(4b + 7)$
   (c) $-5(2m + 3)$ (d) $-7(4l + 5)$
   (e) $-2(3x - 4)$ (f) $-5(2y - 7)$
   (g) $-6(8a - 9)$ (h) $-9(3l - 9)$
   (i) $-4(2a + 5b)$ (j) $-5(3x + 7y)$
   (k) $-10(4l - 5m)$ (l) $-6(5a - 9b)$
   (m) $-4(2a - 7b)$ (n) $-9(8s - 4b)$
   (o) $-3(2a - 5b + 6c)$ (p) $-7(3x - 4y - 5z)$

4. Expand each of the following:
   (a) $-a(a + 3)$ (b) $-b(b + 7)$
   (c) $-m(m + 2)$ (d) $-n(n + 9)$
   (e) $-g(g - 5)$ (f) $-s(s - 9)$
   (g) $-y(y - 4)$ (h) $-n(n - 12)$
   (i) $-x(x + y)$ (j) $-s(y + z)$

(k) $-a(b + a)$

(l) $-c(c + d)$

(m) $-e(f - e)$

(n) $-g(y - g)$

(o) $-a(3a - 2)$

(p) $-y(2y + 5)$

(q) $-b(4b + 7)$

(r) $-c(9c - 11)$

(s) $-2a(3a - 9)$

(t) $-7b(2b + 5)$

(u) $-3c(8 + 4c)$

(v) $-5y(2y - 9)$

(w) $-3c(8 + 4c + 3d)$

(x) $-8h(4h - 3k + 6j)$

(y) $-6b(4a - 2b + 3c)$

(z) $-2z(3x - 4y - 5z)$

5. Expand

(a) $-(a + 2)$

(b) $-(l + 5)$

(c) $-(c + 3)$

(d) $-(p + 7)$

(e) $-(h + 4)$

(f) $-(c + 6)$

(g) $-(m - 3)$

(h) $-(a - 4)$

(i) $-(c - 3)$

(j) $-(d - 5)$

(k) $-(x - 7)$

(l) $-(y - 10)$

(m) $-(a + b)$

(n) $-(c + d)$

(o) $-(e + f)$

(p) $-(g + h)$

(q) $-(2a - 3b)$

(r) $-(6c - 7d)$

(s) $-(8e - 9f)$

(t) $-(2a - 4b)$

(u) $-(6c - 3d)$

(v) $-(5e - 4f)$

(w) $-(2l + 3m - 4n)$

(x) $-(5p - 4q + 7r)$

(y) $-(6a + 3b + 4c)$

(z) $-(2x - 3y - 4z)$

Example 6.9

Expand and simplify:

(a) $8x - 3(a + 2x)$

(b) $2(a - 4) - 3(a - 2)$

(c) $a(a - 3) - a(2a + 5)$

Solution

(a) $8x - 3(a + 2x)$

$= 8x - 3(a + 2x)$

$= 8x - 3a - 6x$

$= 2x - 3a$

(b) $2(a - 4) - 3(a - 2)$

$= 2(a - 4) - 3(a - 2)$

$= 2a - 8 - 3a + 6$

$= -a - 2$

(c) $a(a - 3) - a(2a + 5)$

$= a^2 - 3a - 2a^2 - 5a$

$= -a^2 - 8a$

Exercise 6K

1. Expand and simplify:

(a) $7(2a + 1) - 3a$

(b) $4(3m + 5) - 6m$

(c) $2(3g + 4) - 5g$

(d) $3(6m + 1) - 8m$

(e) $4(1 + 2p) - 9p$

(f) $3(1 + 2g) - 6g$

(g) $3(2g - 1) - 8g$

(h) $6(g - 3) - 7g$

(i) $5(g - 4) - 3g$

(j) $2(d - 5) - 6d$

2. Expand and simplify:

(a) $7a - 2(a + 4)$

(b) $6b - 3(b + 5)$

(c) $9c - 4(c + 5)$

(d) $7g - 3(g + 2)$

(e) $11f - 5(2f - 3)$

(f) $12d - 3(4d - 7)$

(g) $6a - 2(3 - 2a)$

(h) $4b - 5(1 - 3b)$

(i) $7c - 4(2 - 3c)$

(j) $6g - 3(4 - 3g)$

3. Expand and simplify:

(a) $3(a - 4) - 2(a + 2)$

(b) $7(b - 3) - 4(2b + 1)$

(c) $6(2s + 3) - 2(5s + 8)$

(d) $6(4 - s) - 3(10 + s)$

(e) $5(2a - b) - 3(2a - b)$

(f) $2(m - n) - 3(4m - 2n)$

(g) $3(4g - h) - 2(g - h)$

(h) $5(a - b) - 3(b - a)$

(i) $5(2b - c) - 3(b - 4c)$

(k) $-3(a - b) - 4(a - b)$

(m) $-2(a - b) - 3(2a - b)$

(o) $-5(b - c) - 3(c - b)$

(q) $-4(p + q) - (p - q)$

(s) $-6(2p - q) - (p + q)$

(u) $-3(2m - n) - (m - n)$

(w) $-2(a - b) - (b + a)$

(y) $-4(a - 3b) - (b - 2a)$

(j) $7(m - n) - 2(2n - m)$

(l) $-5(a + 3b) - 2(a - 2b)$

(n) $-3(a + b) - 2(b - 3a)$

(p) $-3(s - 2t) - 2(s + 3t)$

(r) $-3(a + b) - (a - b)$

(t) $-2(s - t) - (s + 3t)$

(v) $-4(2c - 3d) - (3d + 2c)$

(x) $-3(p - q) - (2p - 3q)$

(z) $-5(g - 4h) - (h - 4g)$

4. Expand and simplify:

(a) $a(a - 1) - a(2a + 1)$

(c) $s(2s - 3) - 2s(3s + 4)$

(e) $t(3t - 4) - t(1 - t)$

(g) $-p(5p - 7) - p(1 - p)$

(i) $-q(q - 3) - q(2q - 5)$

(k) $-a(4a - 3) - a(3 - 2a)$

(b) $b(2b - 3) - b(b - 2)$

(d) $y(2 - 3y) - y(2y + 5)$

(f) $h(4 + h) - h(h + 6)$

(h) $-y(2y + 3) - 2y(1 + 3y)$

(j) $-g(2g - 1) - g(3 - g)$

(l) $-d(2d - 4) - d(3 - d)$

6.6   Factors

It has been seen in 1.2 that the factors of a certain number are those numbers which multiply together to give that certain number.

E.g. $4 \times 5 = 20$ $\therefore$ 4 and 5 are factors of 20.

Similarly $2 \times 10 = 20$ $\therefore$ 2 and 10 are factors of 20.

and $1 \times 20 = 20$ $\therefore$ 1 and 20 are factors of 20.

Thus, the factors of 20 are 1, 2, 4, 5, 10 and 20.

Example 6.10

Find the factors of 42.

Solution

$42 = 7 \times 6$

$ = 14 \times 3$

$ = 21 \times 2$

$ = 42 \times 1$

Thus, the factors of 42 are 1, 2, 3, 6, 7, 14, 21 and 42.

It has also been seen in Chapter 4 that

$2 \times x$ may be written as $2x$

$2 \times x = 2x$ $\therefore$ 2 and x are the factors of $2x$.

also $1 \times 2x = 2x$ $\therefore$ 1 and $2x$ are the factors of $2x$.

Thus, the factors of $2x$ are 1, 2, x, $2x$.

Similarly,

$8 \times g = 8g$ $\therefore$ 8 and g are factors of $8g$.

also $4 \times 2g = 8g$ $\therefore$ 4 and $2g$ are factors of $8g$.

and $2 \times 4g = 8g$ $\therefore$ 2 and $4g$ are factors of $8g$.

and $1 \times 8g = 8g$ $\therefore$ 1 and $8g$ are factors of $8g$.

Thus, the factors of $8g$ are 1, 2, 4, 8, g, $2g$, $4g$ and $8g$.

Similarly,

$x \times y = xy$ $\therefore$ x and y are factors of xy.

and $1 \times xy = xy$ $\therefore$ 1 and xy are factors of xy.

Thus, the factors of xy are 1, x, y, xy.

Example 6.11

Find the factors of:

(a) $7m$

(b) $12g$

(c) $5xy$

Solution

(a) $7m = 7 \times m$
$\qquad = 7m \times 1$

Thus, the factors of $7m$ are 1, 7, m, $7m$.

(b) $12g = 12 \times g$
$\qquad = 6 \times 2g$
$\qquad = 3 \times 4g$
$\qquad = 4 \times 3g$
$\qquad = 2 \times 6g$
$\qquad = 1 \times 12g$

Thus, the factors of $12g$ are 1, 2, 3, 4, 6, 12, g, $2g$, $3g$, $4g$, $6g$, $12g$.

(c) $5xy = 5 \times xy$
$\qquad = 5x \times y$
$\qquad = 5y \times x$
$\qquad = 5xy \times 1$

Thus, the factors of $5xy$ are 1, 5, x, y, $5x$, $5y$, xy, $5xy$.

Exercise 6L

Find the factors of:

1. 10
2. 15
3. 24
4. 56
5. $2a$
6. $5z$
7. $6m$
8. $14p$
9. $9s$
10. $18t$
11. $20g$
12. $30y$
13. gh
14. mp
15. db
16. kz
17. $3ar$
18. $3aw$
19. $5xy$
20. $7pq$
21. $11gh$
22. $5st$
23. $6cd$
24. $4yz$
25. $8mn$
26. $9pq$
27. $12gh$
28. $15yz$

6.6.1   Common Factors

What are the factors of 12?

$\qquad 12 = 12 \times 1$
$\qquad\quad = 6 \times 2$
$\qquad\quad = 4 \times 3$

Thus, the factors of 12 are 12, 6, 4, 3, 2, 1.

What are the factors of 18?

$\qquad 18 = 18 \times 1$
$\qquad\quad = 9 \times 2$
$\qquad\quad = 6 \times 3$

Thus, the factors of 18 are 18, 9, 6, 3, 2, 1.

What then are the *common factors* of 12 and 18? That is, what numbers are factors of both 12 and 18?

No.	Factors
12	12, 6, 4, 3, 2, 1
18	18, 9, 6, 3, 2, 1

It can be seen from the table above that 6, 3, 2 and 1 are the factors common to both 12 and 18.

Consider now the factors of 4 and $2x$

Term	Factors
4	4, ②, ①
$2x$	②, ①, x, $2x$

The factors which are common to 4 and $2x$ are 2 and 1.

What are the common factors of $2xy$ and yz?

Term	Factors
$2xy$	$2xy$, xy, $2x$, $2y$, x, ⓨ, ①
yz	yz, ⓨ, z, ①

Thus, the common factors are y and 1.

Consider now the common factors of $3p$ and $6p^2$.

Term	Factors
$3p$	③p ⓟ ③ ①
$6p^2$	$6p^2$, $3p^2$, $2p^2$, p^2, $6p$, ③p, $2p$, ⓟ, 6, ③, 2, ①

Thus, the common factors are $3p$, p, 3 and 1.

Exercise 6M

1. Find the common factors of the following pairs of terms:
   (a) 10, 15
   (b) 8, 12
   (c) 24, 32
   (d) 14, 21
   (e) $3, 3a$
   (f) $5, 5m$
   (e) $6, 2x$
   (h) $12, 6x$
   (i) $8, 4p$
   (j) $9, 6g$
   (k) $10, 8h$
   (l) $12, 8m$
   (m) $15, 10m$
   (n) $20, 12s$

2. Find the common factors of the following pairs of terms:
   (a) $4a, 2b$
   (b) $6m, 3p$
   (c) $8c, 12d$
   (d) $9y, 6z$
   (e) $2a, a$
   (f) $4a, a$
   (g) $3b, b$
   (h) $c, 7c$
   (i) $2g, 5g$
   (j) $4d, 3d$
   (k) $7p, 3p$
   (l) $9p, 2p$
   (m) $2m, 10m$
   (n) $3g, 6g$
   (o) $8y, 10y$
   (p) $7y, 14y$
   (q) $9s, 12s$
   (r) $10c, 12c$

3. Find the common factors of the following pairs of terms:
   (a) mn and mp
   (b) db and bc
   (c) yz and az
   (d) gh and gp
   (e) $2ab$ and bc
   (f) $3mp$ and pb
   (g) $4pq$ and pm
   (h) $5ad$ and de
   (i) $3db$ and $3bc$
   (j) $4mp$ and $4pq$
   (k) $5ad$ and $5ab$
   (l) $3st$ and $3tv$

Example 6.12

Factorise each of the following by taking out the highest common factor:
(a) $5p + 10$
(b) $3x - 9$
(c) $14y + 7z$
(d) $6x - 9y$

Solution

(a) $5p + 10 = 5 \times p + 5 \times 2$ HCF = 5
$= 5(p + 2)$

Checking, $5(p + 2) = 5 \times p + 5 \times 2$
$= 5p + 10$

(b) $3x - 9 = 3 \times x - 3 \times 3$ HCF = 3
$= 3(x - 3)$

Checking, $3(x - 3) = 3x - 9$

(c) $14y + 7z = 7 \times 2y + 7 \times z$ HCF = 7
$= 7(2y + z)$

Checking, $7(2y + z) = 14y + 7z$

(d) $6x - 9y = 3 \times 2x - 3 \times 3y$ HCF = 3
$= 3(2x - 3y)$

Checking, $3(2x - 3y) = 6x - 9y$

Exercise 6 N

1. Factorise each of the following by taking out the highest common factor:
    (a) $2a + 2$
    (b) $4m + 4$
    (c) $6p + 6$
    (d) $9g + 9$
    (e) $7y + 7$
    (f) $8s + 8$
    (g) $3p + 3$
    (h) $10t + 10$
    (i) $4 + 4g$
    (j) $7 + 7h$
    (k) $8 + 8p$
    (l) $9 + 9y$
    (m) $5m - 5$
    (n) $6m - 6$
    (o) $9t - 9$
    (p) $5z - 5$
    (q) $12y - 12$
    (r) $7a - 7$
    (s) $11p - 11$
    (t) $6x - 6$
    (u) $3 - 3b$
    (v) $7 - 7c$
    (w) $5 - 5k$
    (x) $10 - 10s$
    (y) $9 - 9l$
    (z) $8 - 8p$

2. Factorise each of the following by taking out the highest common factor:
    (a) $2a + 4$
    (b) $4m + 8$
    (c) $6p + 12$
    (d) $3p + 12$
    (e) $7p + 21$
    (f) $9q + 18$
    (g) $5f + 15$
    (h) $7y + 49$
    (i) $12g - 3$
    (j) $6s - 2$
    (k) $9y - 6$
    (l) $8y - 2$
    (m) $15g - 10$
    (n) $20s - 15$
    (o) $24p - 18$
    (p) $36t - 24$
    (q) $14h - 21$
    (r) $18a - 22$
    (s) $24 - 15t$
    (t) $14 - 35a$
    (u) $20 - 25m$
    (v) $16 - 24h$

3. Factorise each of the following by taking out the highest common factor:
    (a) $2a + 2b$
    (b) $4m + 4n$
    (c) $6x - 6y$
    (d) $7g - 7h$
    (e) $5m + 5n$
    (f) $6m + 6n$
    (g) $11h - 11g$
    (h) $12s - 12t$
    (i) $5a + 10g$
    (j) $12p + 6q$

7 Vectors

7.1 Vector and Scalar Quantities

Imagine a man and his camel at an oasis in the desert. He is about to set off for the next oasis but, before he leaves, he needs to know particular pieces of information; the distance to the next oasis and, equally important, the direction in which he has to travel.

Why does he need to know both the distance and the direction?

If he does not know whether the length of his journey is 20 km or 100 km then he will not be sure of the quantity of supplies he will need for the journey.

If he knows the distance is 30 km but does not know the direction to take, he is likely to miss the next oasis entirely by setting off in the wrong direction.

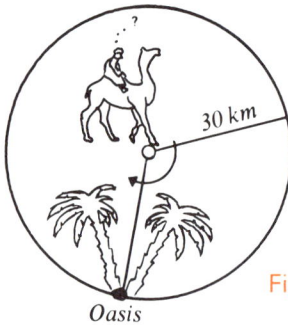

Oasis

Fig 7.1

Thus both the length and the direction of his journey are important to him.

Quantities which have both size (magnitude) and direction are known as *vector quantities.*

Quantities which have no direction, such as the number of sweets in a packet or the cost of the sweets, are known as *scalar quantities.*

As direction is part of a vector, when we show vectors on a diagram it is important to show this direction. We do this by using an arrow.

Example 7.1

Represent the vector 30 km per hour due West by a diagram.

Solution

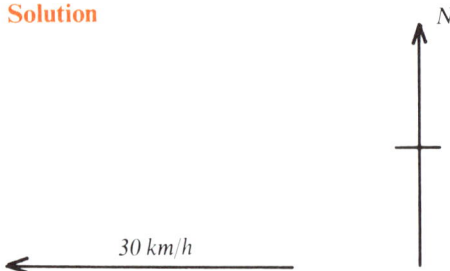

A speed in a fixed direction such as this is known as a velocity.

Exercise 7A

1. State whether the following are vector or scalar quantities:
   (a) 30 km due West.
   (b) 30 cows.
   (c) The year 1984.
   (d) A plane travelling at 400 km per hour in a northerly direction.
   (e) Moving a chess pawn one space.
   (f) 15 km.
   (g) The church is 5 km SE of the town centre.
   (h) The shopkeeper ordered 86 pints of milk.
   (i) A speed of 70 km per hour.
   (j) A speed of 70 km per hour in a westerly direction.

2. Represent each of the vector quantities in question 1 by a diagram.

3. (a) Give two examples of your own of scalar quantities.
   (b) Give two examples of your own of vector quantities.

4. Write down the magnitude and direction of each of the vectors in these diagrams:

(a)

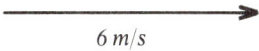

6 m/s

(b)

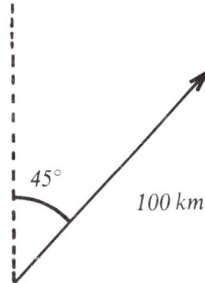

45°   100 km

(c)

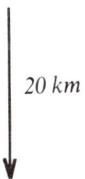

20 km

(d)

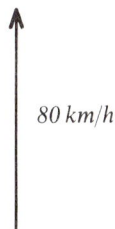

80 km/h

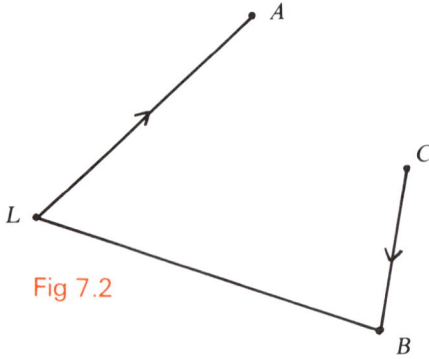

Fig 7.2

An airline flies the routes shown in fig 7.2. The journey from *L* to *A* is a vector quantity, (magnitude and direction) and is written $\vec{LA}$ with the arrow showing that the direction is from *L* to *A*. Similarly, the journey from *C* to *B* would be written as $\vec{CB}$.

Exercise 7B

1.

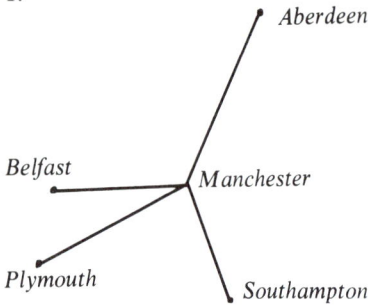

Using the initial letters of the towns on this route map, write down the vectors representing journeys from:
(a) Manchester to Belfast
(b) Manchester to Plymouth
(c) Manchester to Aberdeen
(d) Aberdeen to Manchester
(e) Southampton to Manchester
(f) Manchester to Southampton

2. Mark three points, *P*, *Q* and *R* on your paper and show the following vectors: $\vec{PR}$, $\vec{RQ}$ and $\vec{PQ}$

7.2   Vector Addition

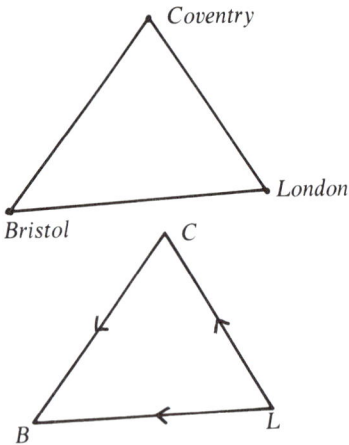

Fig 7.3

If a driver goes from London to Coventry, $\vec{LC}$, and then from Coventry to Bristol, $\vec{CB}$, adding these journeys together gives the same result as if he had travelled from London to Bristol, $\vec{LB}$.

$$\vec{LC} + \vec{CB} = \vec{LB}$$

Example 7.2

Draw a triangle ABC and show on it $\vec{AB} + \vec{BC} = \vec{AC}$

Solution

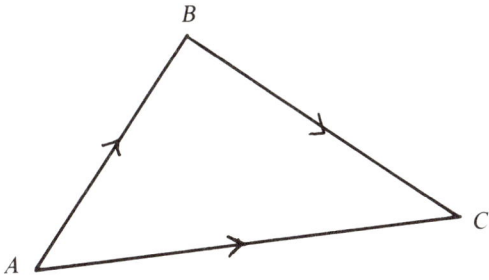

If you go along AB and then along BC it has the same effect as going directly from A to C.

Exercise 7C

1. Draw a triangle RST and put arrows on the sides to show $\vec{TS} + \vec{SR} = \vec{TR}$

2. Mark three points D, E and F, on your paper. Draw in the vectors $\vec{EF}$ and $\vec{FD}$. What vector is equal to $\vec{EF} + \vec{FD}$?

3. Draw a quadrilateral $ABCD$ and show the vectors $\vec{BC}$, $\vec{CD}$ and $\vec{DA}$. What vector is equal to $\vec{BC} + \vec{CD} + \vec{DA}$?

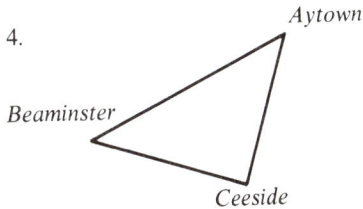

4.

A family go for a Sunday drive from their home in Aytown. They drive to Ceeside and then to Beaminster and back home. What vector does $\vec{AC} + \vec{CB} + \vec{BA}$ equal?

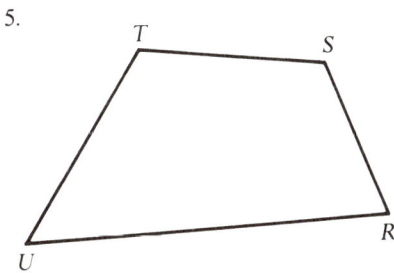

5.

Using this diagram write down the vectors equal to:

(a) $\vec{ST} + \vec{TU}$

(b) $\vec{RS} + \vec{ST}$

(c) $\vec{TS} + \vec{SR}$

(d) $\vec{UR} + \vec{RS}$

(e) $\vec{RS} + \vec{ST} + \vec{TU}$

(f) $\vec{UT} + \vec{TS} + \vec{SR}$

(g) $\vec{UT} + \vec{TR}$

(h) $\vec{UR} + \vec{RT}$

(i) $\vec{UT} + \vec{TS} + \vec{SR} + \vec{RU}$

(j) $\vec{ST} + \vec{TU} + \vec{UR} + \vec{RS}$

(k) $\vec{TS} + \vec{ST}$

(l) $\vec{UR} + \vec{RU}$

7.3   Multiplication of a Vector by a Scalar

7.3.1   Multiplication by a Positive Scalar

Example 7.3

A boat is moving with a velocity of 6 km per hour due South. Its velocity then doubles. What is its new velocity?

Solution

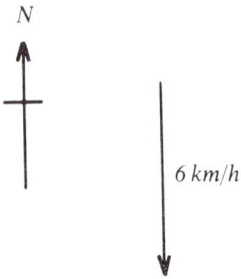

$2 \times 6 = 12$, so the new velocity is 12 km per hour due South.
The direction is unaltered.

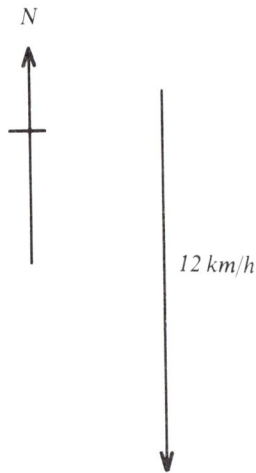

This vector has a magnitude twice that of the first vector, so multiplying a vector by a positive scalar does not change the direction of the vector.

Example 7.4

Three towns, A, B and C, are on a straight road which runs due North from A. A to B is 6 km and B to C is 12 km. Express $\overrightarrow{BC}$ and $\overrightarrow{AC}$ in terms of $\overrightarrow{AB}$.

Solution

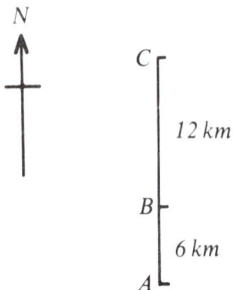

$\overrightarrow{AB}$ is 6 km due North
$\overrightarrow{BC}$ is 12 km due North — therefore $\overrightarrow{BC} = 2\overrightarrow{AB}$
$\overrightarrow{AC}$ is 18 km due North — therefore $\overrightarrow{AC} = 3\overrightarrow{AB}$

7.3.2 Multiplication by a Negative Scalar

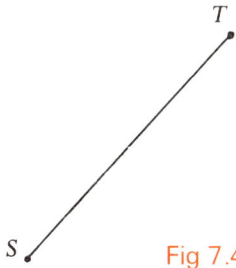

The vectors representing a driver travelling from S to T and back to S would be:

$$\overrightarrow{ST} + \overrightarrow{TS}$$

Since the effect of his journey is to put him back where he started we could write:

$$\overrightarrow{ST} + \overrightarrow{TS} = 0$$

or $\quad \overrightarrow{ST} = -\overrightarrow{TS}$

Fig 7.4

So we can represent the journey from S to T by $\overrightarrow{ST}$ or by $-\overrightarrow{TS}$.
Similarly $-\overrightarrow{ST} = \overrightarrow{TS}$.

Multiplying by a negative scalar reverses the direction.

Exercise 7D

1. Represent each of the following by a diagram:
    (a) A plane flying at 400 km per hour due West halves its velocity.
    (b) A car travelling along a straight road at 60 km per hour increases its velocity by a quarter.
    (c) A tank travelling in a straight line at 20 km per hour reverses its direction and halves its speed.

2. Four stations L, M, E and R are on a straight railway line as shown in the diagram.

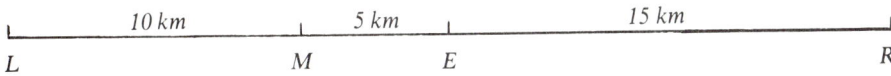

(a) Express the following in terms of $\overrightarrow{ME}$:
    (i) $\overrightarrow{LM}$ (ii) $\overrightarrow{ER}$
    (iii) $\overrightarrow{LE}$ (iv) $\overrightarrow{MR}$
    (v) $2\overrightarrow{ER}$ (vi) $3\overrightarrow{LM}$

(b) Express the following in terms of $\overrightarrow{LM}$:
    (i) $\overrightarrow{MR}$ (ii) $\overrightarrow{LR}$
    (iii) $\overrightarrow{LE}$ (iv) $\overrightarrow{ME}$
    (v) $\overrightarrow{ML}$ (vi) $\overrightarrow{RM}$
    (vii) $\frac{1}{2}\overrightarrow{ME}$ (viii) $\frac{1}{3}\overrightarrow{ER}$

(c) Express the following in terms of $\overrightarrow{RE}$:
    (i) $\overrightarrow{EM}$ (ii) $\overrightarrow{ML}$
    (iii) $\overrightarrow{ME}$ (iv) $\overrightarrow{LM}$
    (v) $\overrightarrow{LE}$ (vi) $\overrightarrow{EL}$
    (vii) $\overrightarrow{RL}$ (viii) $\overrightarrow{LR}$

(d) Express the following in terms of $\overrightarrow{ME}$:
    (i) $\overrightarrow{LM} + \overrightarrow{EM}$ (ii) $\overrightarrow{LM} + \overrightarrow{RE}$

8 Equations and Inequalities

8.1 Equations

8.1.1 Revision

Consider the following mathematical statements:

$$7 + 5 = 12$$
$$5 - 3 = 2$$

Each of the above is called an *equation* because each statement is true and contains two sides (a left-hand side and a right-hand side) joined by an equal sign ($=$).

As pronumerals take the place of numbers, an equation can contain pronumerals as well as numbers.

For example $x + 3 = 10$ and $x - 2 = 7$

The equation $7 + 5 = 12$ states that 'the number 5 added to the number 7 gives a result of 12', but the equation $x + 3 = 10$ asks us to answer the question 'what number, when added to 3, gives 10?' From our number laws we know that $7 + 3 = 10$. Thus the value of x is 7. In fact this is the *only* value of x which will make the equation $x + 3 = 10$ a true statement. Finding the value of the pronumeral in an equation is called *solving the equation*.

> An equation is a true statement containing a left-hand side and a right-hand side joined by an equal sign.

We have seen in Book 1 that to solve equations we use inverse operations.

> The opposite or inverse operation to $+$ is $-$.
> Subtracting 'undoes' adding.

Example 8.1

Find the value of the pronumeral in $a + 5 = 11$, making sure that both sides balance at all times.

Solution

$$a + 5 = 11$$
$$a + 5 - 5 = 11 - 5 \qquad -5 \text{ 'undoes' } +5 \text{ so } -5 \text{ from both sides}$$
$$a + 0 = 6$$
$$a = 6$$

We can check that this is correct by substituting $a = 6$ in the original equation:

$$a + 5 = 11$$
$$\therefore 6 + 5 = 11 \text{ which is a true statement.}$$

> The opposite or inverse operation to $-$ is $+$.
> Adding 'undoes' subtracting.

Example 8.2

Find the value of the pronumeral in $y - 5 = 4$, making sure that both sides of the equation are balanced at all times.

Solution

$$y - 5 = 4$$
$$y - 5 + 5 = 4 + 5 \qquad +5 \text{ 'undoes' } -5 \text{ so } +5 \text{ to both sides}$$
$$y + 0 = 9$$
$$y = 9$$

Checking the solution: $y - 5 = 4$ and if $y = 9$ then $9 - 5 = 4$ which is a true statement.

Exercise 8A

1. Solve each of the following equations, checking your answer:

   (a) $a + 4 = 5$ (b) $b + 3 = 9$
   (c) $c + 7 = 11$ (d) $x + 5 = 19$
   (e) $f + 4 = 2$ (f) $z + 8 = 5$
   (g) $y + 5 = 4$ (h) $p + 7 = 3$
   (i) $a + 2 = -9$ (j) $q + 2 = -4$
   (k) $l + 3 = -5$ (l) $g + 4 = -9$
   (m) $m + 0.5 = 1.7$ (n) $h + 0.9 = 3.9$
   (o) $s + 9.6 = 10.8$ (p) $q + 3.4 = 6.1$
   (q) $a + \frac{1}{4} = \frac{3}{4}$ (r) $b + \frac{1}{8} = \frac{5}{8}$
   (s) $m + \frac{4}{7} = -\frac{2}{7}$ (t) $y + \frac{1}{5} = -\frac{4}{5}$
   (u) $5 + x = 8$ (v) $8 + m = 12$
   (w) $7 + p = 12$ (x) $6 + m = 15$
   (y) $4 + z = 1$ (z) $8 + p = 2$

2. Solve

   (a) $a - 2 = 3$ (b) $b - 5 = 7$
   (c) $h - 5 = 8$ (d) $y - 3 = 4$
   (e) $p - 2 = 7$ (f) $p - 5 = 1$
   (g) $l - 4 = -1$ (h) $m - 2 = -5$
   (i) $z - 5 = -5$ (j) $p - 9 = -9$
   (k) $m - 2 = -1$ (l) $h - 3 = -2$
   (m) $y - 0.6 = 0.3$ (n) $l - 0.1 = 4.7$
   (o) $t - 1.4 = 2.7$ (p) $s - 4.6 = -2.3$
   (q) $h - \frac{1}{4} = \frac{1}{4}$ (r) $p - \frac{1}{6} = \frac{1}{6}$
   (s) $s - \frac{3}{7} = \frac{2}{7}$ (t) $y - \frac{4}{5} = -\frac{1}{5}$

> The opposite or inverse operation of $\times$ is $\div$.
> Dividing 'undoes' multiplying.

Example 8.3

Find the value of the pronumeral in $4a = 12$ making sure that both sides balance at all times.

Solution

$$4a = 12$$
$$4a \div 4 = 12 \div 4 \qquad \div 4 \text{ 'undoes' } \times 4$$
$$\frac{4a}{4} = \frac{12}{4}$$
$$a = 3$$

Checking the solution: $4a = 12$ and if $a = 3$ then $4 \times 3 = 12$ which is a true statement.

> The opposite or inverse operation of ÷ is ×.
> Multiplying 'undoes' dividing.

Example 8.4

Find the value of the pronumeral in $\dfrac{m}{4} = -4$ making sure that both sides balance at all times.

Solution

$$\frac{m}{4} = -4$$

$$\frac{m}{4} \times 4 = -4 \times 4 \qquad \times 4 \text{ 'undoes' } \div 4$$

$$m = -16$$

Checking the solution: $\dfrac{m}{4} = -4$ and if $m = -16$ then $-\dfrac{16}{4} = -4$ which is a true statement.

Exercise 8B

1. Solve

(a) $2a = 6$ (b) $4b = 16$

(c) $6m = 42$ (d) $2y = 18$

(e) $2p = -10$ (f) $5y = -25$

(g) $6l = -18$ (h) $7g = -49$

(i) $3a = 24$ (j) $4p = 32$

(k) $7d = 147$ (l) $5y = 125$

(m) $-2b = 12$ (n) $-7m = 21$

(o) $-3h = 15$ (p) $-9d = 81$

(q) $-8k = -64$ (r) $-5c = -25$

(s) $7x = 4$ (t) $8y = 3$

(u) $6p = 3$ (v) $4y = 2$

(w) $12x = 18$ (x) $12y = 4$

(y) $8a = 12$ (z) $6z = 9$

2. Solve

(a) $\dfrac{a}{2} = 3$ (b) $\dfrac{b}{5} = 7$

(c) $\dfrac{c}{3} = 9$ (d) $\dfrac{d}{8} = 4$

(e) $\dfrac{p}{7} = 2$ (f) $\dfrac{y}{8} = 1$

(g) $\dfrac{g}{9} = 3$ (h) $\dfrac{c}{4} = 5$

(i) $\dfrac{h}{2} = -3$ (j) $\dfrac{a}{9} = -6$

(k) $\dfrac{m}{4} = -5$ (l) $\dfrac{l}{3} = -6$

(m) $\dfrac{p}{9} = -5$ (n) $\dfrac{s}{7} = -9$

(o) $\dfrac{l}{3} = -4$ (p) $\dfrac{g}{5} = -7$

(q) $\dfrac{a}{-2} = 4$ (r) $\dfrac{b}{-9} = 7$

(s) $\dfrac{h}{-5} = 7$ (t) $\dfrac{y}{-4} = 1$

(u) $\dfrac{s}{-4} = -2$ (v) $\dfrac{h}{-3} = -5$

(w) $\dfrac{m}{-2} = -6$ (x) $\dfrac{y}{-5} = -9$

(y) $\dfrac{p}{4} = \dfrac{1}{2}$ (z) $\dfrac{q}{5} = \dfrac{1}{8}$

8.1.2   Equations Involving More Than One Operation.

Sometimes it is necessary to use more than one inverse operation in order to obtain the pro-numeral on its own on one side of the equation.

Consider the expression $2 \times 4 + 5$. Rules for the order of operations state that multiplication is carried out before addition or subtraction and thus, to evaluate this expression, 2 is *first* multiplied by 4 and *then* 5 is added.

I.e.
$$2 \times 4 + 5$$
$$= 8 + 5$$
$$= 13$$

Similarly, in the algebraic expression $2y + 5$, the unknown, y, is *first* multiplied by 2, and *then* 5 is added.

Now consider the equation $2y + 5 = 7$.

The unknown value, y, is multiplied by 2 and then 5 is added to obtain the result of 7. In order to find the value of y, the expression $2y + 5$ must be 'undone' in the reverse order to that in which it was built up.

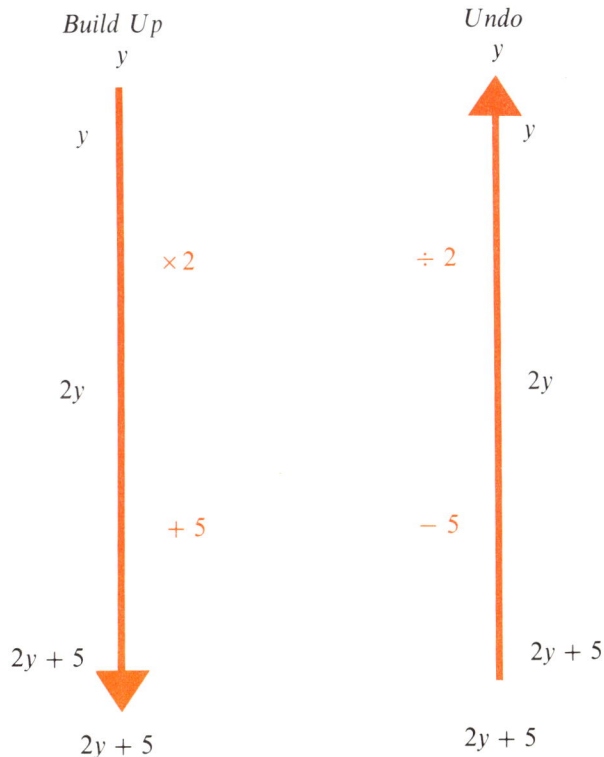

Fig 8.1

$$\therefore 2y + 5 - 5 = 7 - 5 \qquad -5 \text{ 'undoes' } +5$$
$$2y = 2$$
$$\frac{2y}{2} = \frac{2}{2} \qquad \div 2 \text{ 'undoes' } \times 2$$
$$y = 1$$

Check: if $y = 1$, L.H.S. $= 2 \times 1 + 5 = 7$
R.H.S. $= 7$
$\therefore$ L.H.S. $=$ R.H.S. and thus $y = 1$ is the correct solution.
Similarly, if $4x - 9 = 3$ we must first 'undo' the subtraction
$$\therefore 4x - 9 + 9 = 3 + 9 \qquad +9 \text{ 'undoes' } -9$$
$$4x = 12$$
Now we must 'undo' the multiplication
$$\frac{4x}{4} = \frac{12}{4} \qquad \div 4 \text{ 'undoes' } \times 4$$
$$\therefore x = 3$$
Check: if $x = 3$, L.H.S. $= 4 \times 3 - 9 = 12 - 9 = 3$
R.H.S. $= 3$
$\therefore$ L.H.S. $=$ R.H.S. and $x = 3$ is the correct solution.

Example 8.5
Solve the following equations:
(a) $4x + 3 = 19$
(b) $4 - 3y = 19$
(c) $3m + \frac{1}{5} = \frac{7}{5}$
(d) $2p - 0.6 = 4.2$

Solution
(a)
$$4x + 3 = 19$$
$$4x + 3 - 3 = 19 - 3$$
$$4x = 16$$
$$\frac{4x}{4} = \frac{16}{4}$$
$$x = 4$$
Check: if $x = 4$, L.H.S. $= 4 \times 4 + 3$
$$= 16 + 3$$
$$= 19$$
R.H.S. $= 19$
$\therefore$ L.H.S. $=$ R.H.S. and thus $x = 4$
is the correct solution.

(b)
$$4 - 3y = 19$$
$$4 - 4 - 3y = 19 - 4$$
$$-3y = 15$$
$$\frac{-3y}{-3} = \frac{15}{-3}$$
$$y = -5$$
Check: if $y = -5$,
L.H.S. $= 4 - 3 \times -5$
$$= 4 + 15$$
$$= 19$$
R.H.S. $= 19$
$\therefore$ L.H.S. $=$ R.H.S. and thus $y = -5$ is the correct solution.

(c)
$$3m + \frac{1}{5} = \frac{7}{5}$$
$$3m + \frac{1}{5} - \frac{1}{5} = \frac{7}{5} - \frac{1}{5}$$
$$3m = \frac{6}{5}$$
$$\frac{3m}{3} = \frac{\frac{6}{5}}{3}$$
$$m = \frac{6}{5} \div 3$$
$$m = \frac{6}{5} \times \frac{1}{3}$$
$$m = \frac{2}{5}$$
Check: if $m = \frac{2}{5}$
L.H.S. $= 3 \times \frac{2}{5} + \frac{1}{5}$
$$= \frac{6}{5} + \frac{1}{5} = \frac{7}{5}$$
R.H.S. $= \frac{7}{5}$
$\therefore$ L.H.S. $=$ R.H.S. and thus $m = \frac{2}{5}$ is the correct solution

(d)
$$2p - 0.6 = 4.2$$
$$2p - 0.6 + 0.6 = 4.2 + 0.6$$
$$2p = 4.8$$
$$\frac{2p}{2} = \frac{4.8}{2}$$
$$p = 2.4$$
Check: if $p = 2.4$
L.H.S. $= 2 \times 2.4 - 0.6$
$$= 4.8 - 0.6$$
$$= 4.2$$
R.H.S. $= 4.2$
$\therefore$ L.H.S. $=$ R.H.S. and thus $p = 2.4$ is the correct solution.

Exercise 8C

1. Solve each of the following equations
   (a) $3x + 5 = 8$
   (b) $7p + 4 = 25$
   (c) $2a + 9 = 15$
   (d) $11x + 34 = 1$
   (e) $5m - 3 = 12$
   (f) $7y - 5 = 9$
   (g) $6a - 7 = 13$
   (h) $8p - 4 = 15$
   (i) $4y + 5 = 1$
   (j) $3l + 9 = 6$
   (k) $2a + 8 = 16$
   (l) $7y + 4 = -3$
   (m) $3g + 4 = -5$
   (n) $5y + 6 = -4$
   (o) $2y - 3 = 5$
   (p) $4s - 8 = -16$
   (q) $8 + 2y = 6$
   (r) $8 + 4y = 12$
   (s) $4 + 8y = 6$
   (t) $7 + 3y = 9$
   (u) $3 - 2a = 5$
   (v) $4 - 5p = 9$
   (w) $7 - 3q = 9$
   (x) $8 - 2r = 11$
   (y) $6 - 5s = 15$
   (z) $9 - 3g = 15$

2. Solve each of the following
   (a) $2a + \frac{1}{4} = \frac{3}{4}$
   (b) $2a + \frac{1}{8} = \frac{3}{8}$
   (c) $5y + \frac{2}{3} = \frac{1}{3}$
   (d) $2p + \frac{1}{5} = \frac{3}{5}$
   (e) $2p - \frac{1}{7} = \frac{3}{7}$
   (f) $5y - \frac{1}{8} = \frac{5}{8}$
   (g) $7m - \frac{2}{5} = 1$
   (h) $8g - \frac{2}{3} = 1$
   (i) $5s + 0.2 = 0.8$
   (j) $3m + 0.5 = 1.7$
   (k) $4p + 0.3 = 1.5$
   (l) $8t + 0.4 = 2.8$
   (m) $4q - 0.5 = 1.1$
   (n) $3a - 1.4 = 0.1$
   (o) $2a - 1.3 = 1.1$
   (p) $2a - 2.6 = 2.4$

Consider the equation $4(2x + 5) = 4$.
It is simpler to expand the brackets on the left-hand side before attempting to solve for x.
$$4(2x + 5) = 4$$
gives $8x + 20 = 4$ when the brackets are expanded.
Now, solving the equation,
$$8x + 20 - 20 = 4 - 20 \qquad -20 \text{ 'undoes' } +20$$
$$8x = -16$$
$$\frac{8x}{8} = \frac{-16}{8} \qquad \div 8 \text{ 'undoes' } \times 8$$
$$x = -2$$
Check: if $x = -2$, L.H.S. $= 4(2 \times -2 + 5)$
$$= 4(-4 + 5)$$
$$= 4 \times 1$$
$$= 4$$
R.H.S. $= 4$
$\therefore$ L.H.S. $=$ R.H.S. and thus $x = -2$ is the correct solution.

Example 8.6

Solve each of the following equations:
(a) $2(3x - 4) = 5$
(b) $2(3 - a) = 7$
(c) $4 = 2(a + 1)$

Solution

(a) $2(3x - 4) = 5$
$$6x - 8 = 5 \qquad \text{using the distributive law}$$
$$6x - 8 + 8 = 5 + 8 \qquad +8 \text{ 'undoes' } -8$$
$$6x = 13$$
$$\frac{6x}{6} = \frac{13}{6} \qquad -6 \text{ 'undoes' } +6$$
$$x = \tfrac{13}{6}(= 2\tfrac{1}{6})$$

Check: if $x = \frac{13}{6}$, L.H.S. $= 2(3 \times \frac{13}{6} - 4)$
$$= 2(\frac{13}{2} - 4)$$
$$= 2(\frac{13}{2} - \frac{8}{2})$$
$$= 2 \times \frac{5}{2}$$
$$= 5$$
R.H.S. $= 5$

$\therefore$ L.H.S. $=$ R.H.S. and thus $x = \frac{13}{6}$ is the correct solution.

(b) $2(3 - a) = 7$

$6 - 2a = 7$ using the distributive law

$6 - 6 - 2a = 7 - 6$ -6 'undoes' $+6$

$-2a = 1$

$\dfrac{-2a}{-2} = \dfrac{1}{-2}$ $\div\,^-2$ 'undoes' $\times\,^-2$

$a = -\frac{1}{2}$

Check: if $a = -\frac{1}{2}$, L.H.S. $= 2(3 - -\frac{1}{2})$
$$= 2 \times 3\frac{1}{2}$$
$$= 7$$
R.H.S. $= 7$

$\therefore$ L.H.S. $=$ R.H.S. and thus $a = -\frac{1}{2}$ is the correct solution.

(c) $4 = 2(a + 1)$

$4 = 2a + 2$ using the distributive law

$4 - 2 = 2a + 2 - 2$ -2 'undoes' $+2$

$2 = 2a$

$\dfrac{2}{2} = \dfrac{2a}{2}$ $\div 2$ 'undoes' $\times 2$

$1 = a$ i.e. $a = 1$

Check: if $a = 1$, L.H.S. $= 4$
R.H.S. $= 2(1 + 1)$
$$= 2 \times 2 = 4$$

$\therefore$ L.H.S. $=$ R.H.S. and thus $a = 1$ is the correct solution.

Exercise 8D

1. Solve each of the following equations:

(a) $2(a + 4) = 6$ (b) $3(y + 5) = 9$

(c) $7(b - 3) = 14$ (d) $8(p - 4) = 8$

(e) $3(g + 5) = 4$ (f) $4(a + 7) = 6$

(g) $2(2a + 1) = 5$ (h) $6(p - 4) = 3$

(i) $3(3a - 4) = 7$ (j) $5(2m - 1) = 8$

(k) $7(2s - 1) = 12$ (l) $3(4t - 2) = 8$

(m) $3(2m - 5) = 9$ (n) $4(2h - 1) = 3$

(o) $2(4 - a) = 5$ (p) $5(3 - b) = 7$

(q) $6(2 - d) = 4$ (r) $7(5 - y) = 4$

2. Solve

(a) $6 = 2(a + 1)$ (b) $12 = 3(m + 4)$

(c) $7 = 5(g - 2)$ (d) $9 = 8(p - 4)$

(e) $8 = 6(x - 5)$ (f) $12 = 9(m - 4)$

(g) $4 = 2(z - 3)$ (h) $5 = 3(y - 7)$

(i) $9 = 4(2y + 3)$ (j) $11 = 5(2h - 5)$

(k) $7 = 6(3x + 5)$ (l) $9 = 2(3x - 6)$

(m) $6 = 2(1 - x)$ (n) $4 = 2(1 - 2x)$

(o) $7 = 3(2 - x)$ (p) $5 = 4(5 - x)$

Some equations have terms involving the pronumeral (unknown value) on both sides. Consider the equation

$$3a + 5 = 2a + 4$$

In order to solve the equation we must obtain a on its own on one side and the number terms on the other side. Since the *larger* pronumeral term, $3a$, is on the left-hand side, it is easier to collect the pronumeral terms on that side and thus we should subtract $2a$ from both sides of the equation. Thus, collecting the 'a' terms on the L.H.S.,

$$3a - 2a + 5 = 2a - 2a + 4$$
$$a + 5 = 4$$

Now, solving for a, 5 must be subtracted from both sides so that the number terms are on the R.H.S.

$$a + 5 - 5 = 4 - 5$$
$$a = -1$$

Check: if $a = -1$, L.H.S. $= 3 \times -1 + 5 = -3 + 5 = 2$
R.H.S. $= 2 \times -1 + 4 = -2 + 4 = 2$
$\therefore$ L.H.S. $=$ R.H.S. and thus $a = -1$ is the correct solution.

Example 8.7

Solve $5a + 3 = 2a + 8$

Solution

$$5a + 3 = 2a + 8$$
$$5a - 2a + 3 = 2a - 2a + 8 \qquad \text{collecting pronumeral terms on L.H.S.}$$
$$3a + 3 = 8$$
$$3a + 3 - 3 = 8 - 3 \qquad \text{collecting the number terms on R.H.S.}$$
$$3a = 5$$
$$\frac{3a}{3} = \frac{5}{3} \qquad \div 3 \text{ 'undoes' } \times 3$$
$$a = \tfrac{5}{3} (= 1\tfrac{2}{3})$$

Check: if $a = \tfrac{5}{3}$, L.H.S. $= 5 \times \tfrac{5}{3} + 3$ R.H.S. $= 2 \times \tfrac{5}{3} + 8$
$= \tfrac{25}{3} + 3$ $= \tfrac{10}{3} + 8$
$= \tfrac{25}{3} + \tfrac{9}{3}$ $= \tfrac{10}{3} + \tfrac{24}{3}$
$= \tfrac{34}{3}$ $= \tfrac{34}{3}$

$\therefore$ L.H.S. $=$ R.H.S. and thus $a = \tfrac{5}{3}$ is the correct solution.
Consider the equation $5a + 9 = 7a - 3$
It can be seen that the larger pronumeral term, $7a$, is on the right-hand side of the equation. It is easier, then, to collect the pronumeral terms on the right-hand side of the equation and the number terms on the left-hand side.

Thus, $5a + 9 = 7a - 3$
$$5a - 5a + 9 = 7a - 5a - 3 \qquad -5a \text{ 'undoes' } +5a$$
$$9 = 2a - 3$$
$$9 + 3 = 2a - 3 + 3 \qquad +3 \text{ 'undoes' } -3$$
$$12 = 2a$$
$$\frac{12}{2} = \frac{2a}{2} \qquad \div 2 \text{ 'undoes' } \times 2$$
$$6 = a \quad \text{i.e. } a = 6$$

Check: if $a = 6$, L.H.S. $= 5 \times 6 + 9 = 30 + 9 = 39$
R.H.S. $= 7 \times 6 - 3 = 42 - 3 = 39$
$\therefore$ L.H.S. $=$ R.H.S. and thus $a = 6$ is the correct solution.

Example 8.8

Solve $2a - 4 = 7a - 9$

Solution

$$2a - 4 = 7a - 9$$

Collect pronumeral terms on R.H.S. because $7a$ is larger than $2a$

$$2a - 2a - 4 = 7a - 2a - 9 \qquad\qquad -2a \text{ 'undoes' } +2a$$
$$-4 = 5a - 9$$
$$-4 + 9 = 5a - 9 + 9 \qquad\qquad +9 \text{ 'undoes' } -9$$
$$5 = 5a$$
$$\frac{5}{5} = \frac{5a}{5} \qquad\qquad \div 5 \text{ 'undoes' } \times 5$$
$$1 = a \quad \text{i.e. } a = 1$$

Check: if $a = 1$, L.H.S. $= 2 \times 1 - 4 = 2 - 4 = -2$
R.H.S. $= 7 \times 1 - 9 = 7 - 9 = -2$
∴ L.H.S. $=$ R.H.S. and thus $a = 1$ is the correct solution.

Exercise 8E

1. Solve each of the following equations
   (a) $2x = x + 5$
   (b) $3y = 2y + 7$
   (c) $5m = 3m + 4$
   (d) $8m = 4m + 12$
   (e) $2a = a - 5$
   (f) $3p = 2p - 7$
   (g) $6a = 4a - 8$
   (h) $8d = 5d - 9$
   (i) $2a = 3 - a$
   (j) $4g = 5 - g$
   (k) $3c = 7 - 4c$
   (l) $x = 4 - x$

2. Solve each of the following equations
   (a) $3a + 1 = 2a + 6$
   (b) $4g + 2 = 3g + 7$
   (c) $7y + 5 = 3y + 9$
   (d) $6p + 4 = 3p + 13$
   (e) $12s + 7 = 8s + 19$
   (f) $9m + 5 = 4m + 20$
   (g) $3p + 4 = 5p - 8$
   (h) $5y + 1 = 6y - 7$
   (i) $2d + 8 = 4d - 1$
   (j) $8x + 2 = 10x - 3$
   (k) $4s + 1 = 6s - 7$
   (l) $11h + 4 = 7h - 8$
   (m) $2a + 1 = 4 - a$
   (n) $5p + 6 = 12 - p$
   (o) $3d + 2 = 6 - d$
   (p) $7y + 3 = 27 - 5y$
   (q) $5 + 2d = 10 - 3d$
   (r) $4 + d = 8 - 3d$

3. Solve each of the following equations
   (a) $3a - 1 = 2a + 4$
   (b) $4b - 3 = 3b + 5$
   (c) $5d - 7 = 3d + 9$
   (d) $10y - 4 = 5y + 6$
   (e) $7p - 3 = 3p + 5$
   (f) $8g - 3 = 4g + 5$
   (g) $5y - 6 = 3y - 2$
   (h) $9h - 2 = 4h - 7$
   (i) $8l - 7 = 2l - 1$
   (j) $7m - 5 = 3m - 9$
   (k) $12d - 3 = 5d - 9$
   (l) $6p - 4 = 2p - 5$
   (m) $8s - 6 = 3s - 7$
   (n) $9g - 4 = 4g - 10$
   (o) $2a - 3 = 6 - a$
   (p) $5p - 4 = 2 - p$
   (q) $3m - 9 = 3 - m$
   (r) $2a - 3 = 7 - 3a$
   (s) $5l - 7 = 7 - l$
   (t) $7c - 6 = 5 - 2c$

4. Solve each of the following equations
   (a) $a + 4 = 2a + 3$
   (b) $5b + 6 = 6b - 5$
   (c) $3m + 5 = 5m + 7$
   (d) $3y + 8 = 6y - 1$
   (e) $7g + 3 = 9g - 5$
   (f) $2l - 4 = 5l - 1$
   (g) $8p + 7 = 10p - 5$
   (h) $9g - 3 = 16g + 11$
   (i) $2m - 3 = 4m - 5$
   (j) $8p - 13 = 11p + 5$
   (k) $6g - 3 = 9g + 12$
   (l) $7y - 3 = 11y + 9$

Consider now the equation $2(3a + 5) = 3(a + 7)$

We have seen in example 8.7 that it is easiest if the first step in solving an equation involving brackets is to expand the brackets using the Distributive Law.

$$\therefore\ 2(3a + 5) = 3(a + 7)$$
$$\text{gives}\quad 6a + 10 = 3a + 21$$

Now, as the larger pronumeral term is on the left-hand side, we will collect the pronumeral terms on that side.

$$6a - 3a + 10 = 3a - 3a + 21 \qquad -3a \text{ 'undoes' } +3a$$
$$3a + 10 = 21$$
$$3a + 10 - 10 = 21 - 10 \qquad -10 \text{ 'undoes' } + 10$$
$$3a = 11$$
$$\frac{3a}{3} = \frac{11}{3} \qquad \div 3 \text{ 'undoes' } \times 3$$
$$a = \tfrac{11}{3}(= 3\tfrac{2}{3})$$

Check: if $a = \tfrac{11}{3}$, L.H.S. $= 2(3 \times \tfrac{11}{3} + 5)$ \qquad R.H.S. $= 3(\tfrac{11}{3} + 7)$
$$= 2(11 + 5) \qquad\qquad = 3(\tfrac{11}{3} + \tfrac{21}{3})$$
$$= 2 \times 16 = 32 \qquad\quad = 3 \times \tfrac{32}{3} = 32$$

$\therefore$ L.H.S. $=$ R.H.S. and thus $a = \tfrac{11}{3}$ is the correct solution.

Example 8.9

Solve $3(a - 2) = 4(2a + 1)$

Solution

$$3(a - 2) = 4(2a + 1)$$
$$3a - 6 = 8a + 4 \qquad \text{using the Distributive Law}$$

Collecting pronumeral terms on R.H.S.,

$$3a - 3a - 6 = 8a - 3a + 4 \qquad -3a \text{ 'undoes' } +3a$$
$$-6 = 5a + 4$$
$$-6 - 4 = 5a + 4 - 4 \qquad -4 \text{ 'undoes' } +4$$
$$-10 = 5a$$
$$\frac{-10}{5} = \frac{5a}{5} \qquad \div 5 \text{ 'undoes' } \times 5$$
$$-2 = a \quad \text{i.e. } a = -2$$

Check: if $a = -2$, L.H.S. $= 3(-2 - 2) = 3 \times -4 = -12$
R.H.S. $- 4(2 \times -2 + 1) = 4 \times -3 = -12$
$\therefore$ L.H.S. $=$ R.H.S. and thus $a = -2$ is the correct solution.

Exercise 8F

Solve each of the following equations

1. $2(2a + 1) = 3(a + 4)$
2. $5(a + 3) = 4(a + 7)$
3. $3(3b + 4) = 4(2b + 5)$
4. $7(c - 3) = 2(3c - 1)$
5. $6(2x - 3) = 3(x - 2)$
6. $5(2 + 2y) = 6(3y - 2)$
7. $8(2 + p) = 5(3 + p)$
8. $7(2 + y) = 3(4 - 2y)$
9. $5(3a - 4) = 2(6a - 7)$
10. $3(3a - 5) = 4(2 - a)$
11. $3(a - 2) = 5(a + 4)$
12. $5(a - 4) = 3(2a + 1)$
13. $4(g - 3) = 6(g - 4)$
14. $3(p + 5) = 4(2p + 7)$

We have already seen that the inverse operation of division is multiplication.

Consider the expression $\frac{x}{3} + 1$. This expression has been formed by *first* dividing x by 3 and *then* adding 1. To 'undo' the expression we work in reverse; subtract 1 and then multiply by 3.

Hence, to solve, for example, the equation $\frac{x}{3} + 1 = 4$, the expression $\frac{x}{3} + 1$ must be 'undone' in the reverse order to that in which it was built up.

$$\frac{x}{3} + 1 - 1 = 4 - 1 \qquad\qquad -1 \text{ 'undoes' } +1$$

$$\frac{x}{3} = 3$$

$$\frac{x}{3} \times 3 = 3 \times 3 \qquad\qquad \times 3 \text{ 'undoes' } \div 3$$

$$x = 9$$

Check: if $x = 9$, L.H.S. $= \frac{9}{3} + 1 = 3 + 1 = 4$
R.H.S. $= 4$
$\therefore$ L.H.S. $=$ R.H.S. and thus $x = 9$ is the correct solution.

Example 8.10

Solve $\frac{x}{5} - 3 = 7$

Solution

$$\frac{x}{5} - 3 = 7$$

$$\frac{x}{5} - 3 + 3 = 7 + 3 \qquad\qquad +3 \text{ 'undoes' } -3$$

$$\frac{x}{5} = 10$$

$$\frac{x}{5} \times 5 = 10 \times 5 \qquad\qquad \times 5 \text{ 'undoes' } \div 5$$

$$x = 50$$

Check: if $x = 50$, L.H.S. $= \frac{50}{5} - 3 = 10 - 3 = 7$
R.H.S. $= 7$
$\therefore$ L.H.S. $=$ R.H.S. and thus $x = 50$ is the correct solution.

Exercise 8G

Solve each of the following equations:

1. $\frac{x}{2} + 3 = 4$

2. $\frac{a}{5} + 2 = 7$

3. $\frac{m}{3} + 1 = 4$

4. $\frac{p}{4} + 1 = 3$

5. $\frac{s}{4} + 2 = -5$

6. $\frac{y}{9} + 4 = -2$

7. $\frac{a}{7} - 1 = 3$

8. $\frac{b}{4} - 5 = 3$

9. $\frac{g}{4} - 3 = -1$

10. $\frac{s}{2} - 4 = -3$

11. $2 + \frac{a}{3} = 3$

12. $3 + \frac{m}{4} = 5$

13. $5 - \frac{y}{7} = 2$

14. $5 - \frac{h}{4} = 3$

15. $2 = \dfrac{y}{3} + 1$ 16. $4 = \dfrac{a}{3} + 2$

17. $5 = \dfrac{s}{4} - 1$ 18. $9 = \dfrac{t}{2} - 2$

Some equations involve both multiplication and division.

Consider now the equation $\dfrac{2x}{3} = 4$.

In solving the equation, the left-hand side must be 'undone' in the reverse order to that in which it was built up.

Thus, solving $\dfrac{2x}{3} = 4$

$$\dfrac{2x}{3} \times 3 = 4 \times 3 \qquad\qquad\qquad \times 3 \text{ 'undoes' } \div 3$$

$$2x = 12$$

$$\dfrac{2x}{2} = \dfrac{12}{2} \qquad\qquad\qquad\qquad \div 2 \text{ 'undoes' } \times 2$$

$$x = 6$$

Check: if $x = 6$, L.H.S. $= \dfrac{2 \times 6}{3} = \tfrac{12}{3} = 4$

R.H.S. $= 4$

$\therefore$ L.H.S. $=$ R.H.S. and $x = 6$ is the correct solution.

Example 8.11

Solve: (a) $\dfrac{4x}{5} = -2$ (b) $\dfrac{2x}{3} = \dfrac{1}{4}$

Solution

(a)

$$\dfrac{4x}{5} = -2$$

$$\dfrac{4x}{5} \times 5 = -2 \times 5 \qquad\qquad\qquad \times 5 \text{ 'undoes' } \div 5$$

$$4x = -10$$

$$\dfrac{4x}{4} = \dfrac{-10}{4} \qquad\qquad\qquad\qquad \div 4 \text{ 'undoes' } \times 4$$

$$x = -\tfrac{10}{4}$$

$$\therefore x = -\tfrac{5}{2}(= 2\tfrac{1}{2})$$

Check: if $x = -\tfrac{5}{2}$, then L.H.S. $= \dfrac{4 \times -\tfrac{5}{2}}{3}$

$$= -\tfrac{10}{5}$$

$$= -2$$

R.H.S. $= -2$

$\therefore$ L.H.S. $=$ R.H.S. and $x = -\tfrac{5}{2}$ is the correct solution.

(b) $\dfrac{2x}{3} = \tfrac{1}{4}$

$$\dfrac{2x}{3} \times 3 = \tfrac{1}{4} \times 3 \qquad\qquad\qquad \times 3 \text{ 'undoes' } \div 3$$

$$2x = \tfrac{3}{4}$$

$$\frac{2x}{2} = \frac{\tfrac{3}{4}}{2} \qquad\qquad\qquad \div 2 \text{ 'undoes' } \times 2$$

$$x = \tfrac{3}{4} \div 2$$

$$x = \tfrac{3}{4} \times \tfrac{1}{2}$$

$$\therefore\; x = \tfrac{3}{8}$$

Check: if $x = \tfrac{3}{8}$, then L.H.S. $= \dfrac{2 \times \tfrac{3}{8}}{3} = \dfrac{\tfrac{3}{4}}{3} = \tfrac{1}{4}$

$$\text{R.H.S.} = \tfrac{1}{4}$$

$\therefore$ L.H.S. $=$ R.H.S. and $x = \tfrac{3}{8}$ is the correct solution.

Exercise 8H

Solve each of the following equations

1. $\dfrac{2a}{3} = 4$

2. $\dfrac{4x}{5} = 8$

3. $\dfrac{6h}{7} = 12$

4. $\dfrac{3a}{8} = 9$

5. $\dfrac{4m}{3} = 6$

6. $\dfrac{8y}{5} = 10$

7. $\dfrac{3p}{4} = -2$

8. $\dfrac{5z}{6} = -1$

9. $\dfrac{7c}{2} = -4$

10. $\dfrac{6d}{5} = -8$

11. $\dfrac{4p}{3} = \dfrac{1}{3}$

12. $\dfrac{3s}{4} = \dfrac{1}{4}$

13. $\dfrac{2g}{3} = \dfrac{1}{4}$

14. $\dfrac{5y}{3} = \dfrac{1}{2}$

15. $\dfrac{7m}{2} = -\dfrac{1}{3}$

16. $\dfrac{3p}{4} = -\dfrac{1}{2}$

17. $\dfrac{2s}{3} = -\dfrac{1}{2}$

18. $\dfrac{4m}{5} = -\dfrac{1}{4}$

Consider the equation $\dfrac{x + 1}{3} = 2$.

The left-hand side of the equation has been built up by adding 1 to x and then dividing the result by 3. It must be 'undone' in the reverse order.

Thus
$$\frac{x + 1}{3} = 2$$

$$\frac{(x + 1)}{3} \times 3 = 2 \times 3 \qquad\qquad \times 3 \text{ 'undoes' } \div 3$$

Note: Brackets are placed around $(x + 1)$ to indicate that the whole of the L.H.S. is multiplied by 3.

$$x + 1 = 6$$

$$x + 1 - 1 = 6 - 1 \qquad\qquad -1 \text{ 'undoes' } +1$$

$$x = 5$$

Check: if $x = 5$, then L.H.S. $= \dfrac{5 + 1}{3} = \tfrac{6}{3} = 2$

$$\text{R.H.S.} = 2$$

$\therefore$ L.H.S. $=$ R.H.S. and thus $x = 5$ is the correct solution.

Example 8.12

Solve: $\dfrac{2x + 1}{3} = 5$

Solution

$$\dfrac{2x + 1}{3} = 5$$

$$\dfrac{(2x + 1)}{3} \times 3 = 5 \times 3 \qquad\qquad \times 3 \text{ 'undoes' } \div 3$$

$$2x + 1 = 15$$

$$2x + 1 - 1 = 15 - 1 \qquad\qquad -1 \text{ 'undoes' } +1$$

$$2x = 14$$

$$\dfrac{2x}{2} = \dfrac{14}{2} \qquad\qquad \div 2 \text{ 'undoes' } \times 2$$

$$x = 7$$

Check: if $x = 7$, then L.H.S. $= \dfrac{2 \times 7 + 1}{3} = \frac{15}{3} = 5$

R.H.S. $= 5$

∴ L.H.S. $=$ R.H.S. and thus $x = 5$ is the correct solution.

Exercise 8I

1. Solve each of the following equations:

   (a) $\dfrac{a + 1}{2} = 3$

   (b) $\dfrac{g + 5}{2} = 4$

   (c) $\dfrac{y + 2}{3} = 1$

   (d) $\dfrac{p + 4}{3} = 3$

   (e) $\dfrac{g + 5}{7} = 2$

   (f) $\dfrac{s + 2}{5} = 7$

   (g) $\dfrac{s + 3}{4} = 5$

   (h) $\dfrac{y + 3}{4} = 7$

   (i) $\dfrac{m + 2}{5} = -1$

   (j) $\dfrac{a + 2}{2} = -1$

   (k) $\dfrac{a + 3}{3} = -2$

   (l) $\dfrac{m + 4}{3} = -2$

   (m) $\dfrac{s + 5}{4} = -4$

   (n) $\dfrac{y + 5}{6} = -3$

   (o) $\dfrac{y + 6}{3} = -2$

   (p) $\dfrac{a + 3}{2} = -4$

2. Solve each of the following equations:

   (a) $\dfrac{a - 1}{2} = 3$

   (b) $\dfrac{b - 1}{3} = 1$

   (c) $\dfrac{s - 2}{3} = 2$

   (d) $\dfrac{g - 2}{4} = 2$

   (e) $\dfrac{y - 1}{3} = 4$

   (f) $\dfrac{s - 3}{3} = 3$

   (g) $\dfrac{p - 1}{2} = 5$

   (h) $\dfrac{g - 4}{2} = 5$

   (i) $\dfrac{a - 3}{2} = -1$

   (j) $\dfrac{t - 3}{5} = -2$

   (k) $\dfrac{m - 4}{3} = -2$

   (l) $\dfrac{a - 4}{6} = -2$

   (m) $\dfrac{p - 3}{6} = -2$

   (n) $\dfrac{m - 3}{5} = -2$

(o) $\dfrac{s - 5}{4} = -4$ (p) $\dfrac{y - 4}{4} = -2$

(q) $\dfrac{3 - a}{2} = 6$ (r) $\dfrac{3 - t}{2} = 5$

(s) $\dfrac{4 - g}{3} = 6$ (t) $\dfrac{5 - y}{4} = 1$

3. Solve each of the following equations:

(a) $\dfrac{2a + 1}{3} = 3$ (b) $\dfrac{2b + 3}{5} = 3$

(c) $\dfrac{3a + 2}{4} = 2$ (d) $\dfrac{5b + 1}{3} = 2$

(e) $\dfrac{5m + 1}{3} = 7$ (f) $\dfrac{2m + 1}{5} = 3$

(g) $\dfrac{2 + 3a}{5} = 4$ (h) $\dfrac{6 + 4a}{6} = 3$

(i) $\dfrac{3 + 5a}{9} = 2$ (j) $\dfrac{4 + 3a}{4} = 4$

(k) $\dfrac{2a - 4}{3} = 6$ (l) $\dfrac{3g - 2}{2} = 5$

(m) $\dfrac{3a - 4}{8} = 2$ (n) $\dfrac{5p - 3}{2} = 9$

(o) $\dfrac{4 - 2a}{3} = 6$ (p) $\dfrac{7 - 3g}{2} = 5$

(q) $\dfrac{4 - 3a}{2} = 1$ (r) $\dfrac{3 - 2q}{4} = 2$

Consider now the equation $\dfrac{2x}{3} + 1 = 3$.

The order in which operations have been performed in order to build up the left-hand side is: x has been multiplied by 2, then divided by 3, and then 1 has been added. In solving the equation, these operations must be undone in the reverse order: 1 must be subtracted, then the result multiplied by 3 and then divided by 2.

Thus, solving the equation

$$\dfrac{2x}{3} + 1 - 1 = 3 - 1 \qquad\qquad -1 \text{ 'undoes' } +1$$

$$\dfrac{2x}{3} = 2$$

$$\dfrac{2x}{3} \times 3 = 2 \times 3 \qquad\qquad \times 3 \text{ 'undoes' } \div 3$$

$$2x = 6$$

$$\dfrac{2x}{2} = \dfrac{6}{2} \qquad\qquad \div 2 \text{ 'undoes' } \times 2$$

$$x = 3$$

Check: if $x = 3$, then L.H.S. $= \dfrac{2 \times 3}{3} + 1 = 2 + 1 = 3$

R.H.S. $= 3$

$\therefore$ L.H.S. $=$ R.H.S. and thus $x = 3$ is the correct solution.

Example 8.13

Solve   (a) $\dfrac{4x}{5} - 2 = 5$ (b) $2 - \dfrac{3x}{2} = 4$

Solution

(a) $\dfrac{4x}{5} - 2 = 5$

$\dfrac{4x}{5} - 2 + 2 = 5 + 2$ $+2$ 'undoes' -2

$\dfrac{4x}{5} = 7$

$\dfrac{4x}{5} \times 5 = 7 \times 5$ $\times 5$ 'undoes' $\div 5$

$4x = 35$

$\dfrac{4x}{4} = \dfrac{35}{4}$ $\div 4$ 'undoes' $\times 4$

$x = \tfrac{35}{4}(= 8\tfrac{3}{4})$

Check: if $x = \tfrac{35}{4}$, then L.H.S. $= \dfrac{4 \times \frac{35}{4}}{5} - 2$

$= \tfrac{35}{5} - 2$

$= 7 - 2$

$= 5$

R.H.S. $= 5$

$\therefore$ L.H.S. $=$ R.H.S. and thus $x = \tfrac{35}{4}$ is the correct solution.

(b) $2 - \dfrac{3x}{2} = 4$

$2 - 2 - \dfrac{3x}{4} = 4 - 2$ -2 'undoes' $+2$

$-\dfrac{3x}{2} = 2$

$-\dfrac{3x}{2} \times 2 = 2 \times 2$ $\times 2$ 'undoes' $\div 2$

$-3x = 4$

$\dfrac{-3x}{-3} = \dfrac{4}{-3}$ $\div -3$ 'undoes' $\times -3$

$x = -\tfrac{4}{3}(= -1\tfrac{1}{3})$

Check: if $x = -\tfrac{4}{3}$, then L.H.S. $= 2 - \dfrac{3 \times \frac{-4}{3}}{2}$

$= 2 - \tfrac{-4}{2}$

$= 2 - {}^{-}2$

$= 4$

R.H.S. $= 4$

$\therefore$ L.H.S. $=$ R.H.S. and thus $x = -\tfrac{4}{3}$ is the correct solution.

Exercise 8J

Solve each of the following equations:

1. $\dfrac{2x}{3} + 1 = 5$

2. $\dfrac{4x}{5} + 2 = 6$

3. $\dfrac{3y}{4} + 4 = 7$

4. $\dfrac{5a}{3} + 3 = 13$

5. $\dfrac{2x}{3} - 1 = 5$

6. $\dfrac{4z}{9} - 1 = 9$

7. $\dfrac{7g}{3} - 4 = -2$

8. $\dfrac{3y}{7} - 1 = -4$

9. $5 + \dfrac{2g}{3} = 3$

10. $3 + \dfrac{4s}{5} = 5$

11. $6 - \dfrac{3h}{2} = 4$

12. $4 - \dfrac{5y}{2} = 8$

13. $5 - \dfrac{4p}{3} = 5$

14. $7 - \dfrac{7y}{3} = 7$

Exercise 8K

Solve each of the following equations:

1. $a + 5 = 9$

2. $3 + b = 12$

3. $m - 7 = 4$

4. $g - 3 = 11$

5. $3m = 9$

6. $7m = 14$

7. $\dfrac{g}{3} = 5$

8. $\dfrac{y}{5} = -4$

9. $\dfrac{2s}{5} = -6$

10. $\dfrac{3p}{4} = 2$

11. $2s + 3 = 9$

12. $4a + 3 = -5$

13. $3g - 7 = -4$

14. $5y - 2 = -3$

15. $3(a + 2) = 5$

16. $7(2 + b) = 9$

17. $4(2 - b) = 3$

18. $6(1 - c) = 12$

19. $3a + 5 = a + 7$

20. $5a - 4 = 3a + 7$

21. $4a + 6 = 8(a + 5)$

22. $2b - 3 = 5(4 + b)$

23. $\dfrac{a + 1}{3} = 4$

24. $\dfrac{b + 5}{7} = 1$

25. $\dfrac{c}{3} + 1 = 4$

26. $\dfrac{b}{7} + 5 = 1$

27. $\dfrac{2a + 3}{5} = 1$

28. $\dfrac{4a - 3}{2} = 5$

29. $\dfrac{5d}{6} + 3 = 7$

30. $\dfrac{3h}{4} - 5 = 2$

8.1.3   Applications of Equations

Equations may be used to help us solve certain problems. Consider the problem 'What do I add to 5 to give 11?' To solve this problem, we should first represent the unknown number by a pronumeral, say y, and then form an equation which may be solved to give us the value of the pronumeral.

What do I add to 5 to give 11?

$y \quad + \quad 5 \quad = \quad 11$

Solving the equation

$$y + 5 - 5 = 11 - 5 \qquad\qquad -5 \text{ 'undoes' } +5$$
$$y = 6$$

Thus the number I must add to 5 to give 11 is 6.

Similarly, the problem 'From what number must I subtract 3 to give 5?' may be solved using an equation. Again using y for the unknown number,

From what number must I subtract 3 to give 5?

$y \qquad\qquad - \quad 3 \quad = \quad 5$

Solving the equation

$$y - 3 + 3 = 5 + 3 \qquad \qquad +3 \text{ 'undoes' } -3$$
$$y = 8$$

Thus the number is 8.

Consider now the problem 'Three more than y is 6. What is the value of y?'

Three more than y means that 3 is added to y; i.e. $y + 3$.

Three more than y is 6

$$y + 3 \qquad = 6$$

Solving the equation,

$$y + 3 - 3 = 6 - 3 \qquad \qquad -3 \text{ 'undoes' } +3$$
$$y = 3$$

Thus the value of y is 3.

Consider now the problem 'Five less than p is 9. What is the value of p?' Five less than p means that five is subtracted from p; i.e. $p - 5$.

Five less than p is nine

$$p - 5 = 9$$

Solving the equation

$$p - 5 + 5 = 9 + 5 \qquad \qquad +5 \text{ 'undoes' } -5$$
$$p = 14$$

Thus the value of p is 14

Example 8.14

If I double a certain number and subtract 7, the result is 9. What is the number?

Solution

Let the number be p.

I double the number and subtract 7. The result is 9

$$2 \times p \qquad \quad - \quad 7 \quad = \quad 9$$

Thus the equation is $2p - 7 = 9$.

Solving the equation

$$2p - 7 + 7 = 9 + 7 \qquad \qquad +7 \text{ 'undoes' } -7$$
$$2p = 16$$
$$\frac{2p}{2} = \frac{16}{2} \qquad \qquad \div 2 \text{ 'undoes' } \times 2$$
$$p = 8$$

Thus the number is 8.

Exercise 8L

1. Write each of the following as an equation using the pronumeral m for the unknown number, and then find the value of the number.

   (a) What do I add to 4 to give 7?   (b) What do I add to 3 to give 10?

   (c) What do I add to 6 to give 15?   (d) What do I add to 9 to give 17?

   (e) What do I add to 8 to give 19?   (f) What do I add to 6 to give 21?

   (g) What do I add to 9 to give 15?   (h) What do I add to 11 to give 22?

   (i) What do I add to 16 to give 25?   (j) What do I add to 14 to give 23?

   (k) From what number do I subtract 8 to give 4?

   (l) From what number do I subtract 3 to give 5?

   (m) From what number do I subtract 9 to give 11?

   (n) From what number do I subtract 6 to give 5?

   (o) From what number do I subtract 7 to give 8?

   (p) From what number do I subtract 12 to give 9?

   (q) From what number do I subtract 15 to give 7?

   (r) From what number do I subtract 23 to give 12?

   (s) From what number do I subtract 17 to give 2?

   (t) From what number do I subtract 11 to give 9?

2. Write each of the following as an equation using the pronumeral, *y*, for the unknown number and then solve the equation.
   (a) Twice what number is 6?
   (b) Twice what number is 28?
   (c) Three times what number is 27?
   (d) Three times what number is 15?
   (e) What number must be multiplied by 4 to give 16?
   (f) What number must be multiplied by 8 to give 32?
   (g) What number must be multiplied by 5 to give 35?
   (h) What number must be multiplied by 9 to give 63?
   (i) What number must be multiplied by 11 to give -132?
   (j) What number must be multiplied by 12 to give -108?
   (k) What number must be divided by 3 to give 4?
   (l) What number must be divided by 2 to give 6?
   (m) What number must be divided by 5 to give 5?
   (n) What number must be divided by 6 to give 7?
   (o) What number must be divided by 7 to give -9?
   (p) What number must be divided by 9 to give -7?
   (q) A half of what number is 7?
   (r) A quarter of what number is 7?
   (s) A fifth of what number is 3?
   (t) A tenth of what number is 2?

3. For each of the following, write an equation and then solve it.
   (a) Three more than *a* is five.
   (b) Six more than *m* is twelve.
   (c) Eleven more than *p* is five.
   (d) Five more than *s* is seven.
   (e) Eight more than *y* is four.
   (f) Ten more than *g* is three.
   (g) Five more than *a* is one.
   (h) Six more than *t* is two.
   (i) Two less than *a* is five.
   (j) Seven less than *g* is two.
   (k) Four less than *b* is eight.
   (l) Three less than *c* is five.
   (m) Seven less than *y* is nine.
   (n) Five less than *p* is eight.
   (o) Twelve less than *m* is four.
   (p) Six less than *g* is eleven.
   (q) Four less than *y* is nine.
   (r) Eight less than *t* is five.
   (s) Ten less than *z* is four.
   (t) Twelve less than *x* is six.

4. I think of a number and add 5. If the result is 12, what is the number?

5. I think of a number and add 7. If the result is 18, what is the number?

6. I think of a number and add 6. If the result is 12, what is the number?

7. I think of a number and add 3. If the result is 8, what is the number?

8. I think of a number and subtract 6. If the result is 5, what is the number?

9. I think of a number and subtract 2. If the result is 7, what is the number?

10. I think of a number and subtract 5. If the result is -2, what is the number?

11. I think of a number and subtract 7. If the result is -4, what is the number?

12. I think of a number and double it. If the result is 22, what is the number?

13. I think of a number and double it. If the result is 12, what is the number?

14. If I multiply a certain number by 8, the result is 32. What is the number?

15. If I multiply a certain number by 5, the result is 25. What is the number?

16. If I multiply a certain number by 3, the result is -12. What is the number?

17. If I double a number and add 5, the result is 11. What is the number?

18. If I double a number and add 4, the result is 8. What is the number?

19. If I multiply a number by 3 and add 5, the result is 17. What is the number?

20. If I multiply a number by 8 and add 6, the result is 30. What is the number?

21. If I add 5 to three times a number, the result is 14. What is the number?

22. If I add 7 to twice a certain number, the result is 19. What is the number?

23. If I subtract 6 from five times a certain number, the result is 19. What is the number?

24. If I subtract 8 from three times a certain number, the result is 7. What is the number?

25. Two less than three times a certain number is seven. What is the number?

26. Five less than four times a certain number is seven. What is the number?

27. Three less than four times a certain number is nine. What is the number?

28. Six less than eight times a certain number is two. What is the number?

Consider now the following problem. 'A rectangle is 10 cm long and its perimeter is 30 cm. How wide is the rectangle?'

Let us call the width of the rectangle, w.

Drawing a diagram, we have

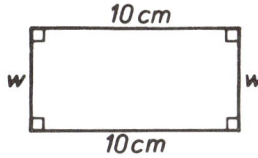

Fig 8.2

Perimeter = distance around the rectangle
$$= 10 + w + 10 + w$$
$$= 20 + 2w$$

However, we have been told that the perimeter is 30 cm and thus $20 + 2w = 30$.

Solving this equation we have

$$20 - 20 + 2w = 30 - 20 \qquad -20 \text{ 'undoes' } +20$$
$$2w = 10$$
$$\frac{2w}{2} = \frac{10}{2} \qquad \div 2 \text{ 'undoes' } \times 2$$
$$w = 5$$

Thus the width of the rectangle is 5 cm.

Example 8.15

Fig 8.3

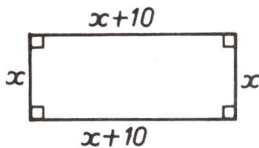

A rectangle is 10 cm longer than it is wide. If the perimeter (the distance around the rectangle) is 48 cm, find the width and the length of the rectangle.

Solution

Let the width be x cm. Since the length is 10 cm longer than the width, the length will be $(x + 10)$ cm.

Perimeter $= (x + 10) + x + (x + 10) + x$
$$= 4x + 20$$

But perimeter $= 48$ cm.

Thus
$$4x + 20 = 48$$
$$\therefore 4x + 20 - 20 = 48 - 20 \qquad -20 \text{ 'undoes' } +20$$
$$4x = 28$$
$$\frac{4x}{4} = \frac{28}{4} \qquad \div 4 \text{ 'undoes' } \times 4$$
$$x = 7$$

$\therefore$ width $= 7$ cm
length $= x + 10 = 17$ cm.

Exercise 8M

1.

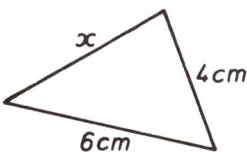

Two sides of a triangle are 4 cm and 6 cm respectively. If the perimeter of the triangle is 15 cm, what is the length of the third side?

2.

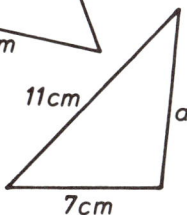

Two sides of a triangle are 7 cm and 11 cm respectively. If the perimeter of the triangle is 26 cm, what is the length of the third side?

3.

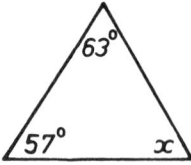

The three angles of a triangle add up to 180°. If two of the angles are 57° and 63° respectively find the third angle.

4.

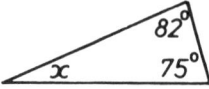

The three angles of a triangle add up to 180°. If two of the angles are 75° and 82° respectively, find the third angle.

5.

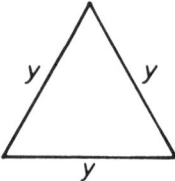

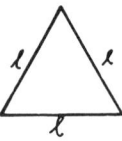

An equilateral triangle has three equal sides. If the perimeter of an equilateral triangle is 30 cm, what is the length of each side?

6.

An equilateral triangle has three equal sides. If the perimeter of an equilateral triangle is 21 cm, what is the length of each side?

7.

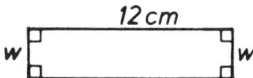

The length of a rectangle is 12 cm and its perimeter is 30 cm. What is its width?

8.

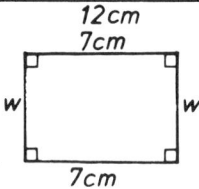

The length of a rectangle is 7 cm and its perimeter is 24 cm. What is its width?

9.

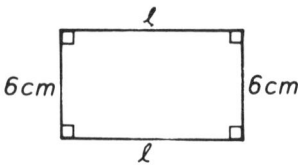

A rectangle is 6 cm wide and its perimeter is 32 cm. What is its length?

10.

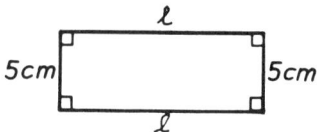

The perimeter of a rectangle is 36 cm. If its width is 5 cm, what is its length?

11.

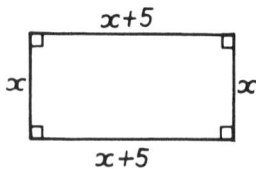

A rectangle is 5 cm longer than it is wide. If the perimeter is 30 cm, find the width and the length of the rectangle.

12.

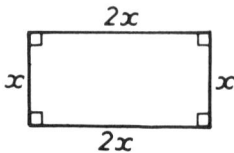

A rectangle is twice as long as it is wide. If the perimeter is 24 cm, find the width and the length of the rectangle.

13.

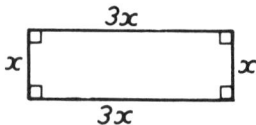

A rectangle is three times as long as it is wide. If the perimeter is 16 cm, find the width and length of the rectangle.

8.2   Inequalities

Any number can be represented on a number line. All numbers to the *right* of a particular number are *greater* than that number.

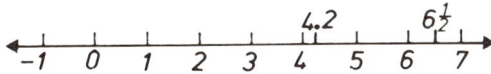

Fig 8.4

For example, all numbers to the right of 2 are greater than 2.

4 is greater than 2	i.e. $4 > 2$
7 is greater than 2	i.e. $7 > 2$
$6\frac{1}{2}$ is greater than 2	i.e. $6\frac{1}{2} > 2$
4.2 is greater than 2	i.e. $4.2 > 2$

The set of numbers greater than 2 is *an infinite set* containing whole numbers, mixed numbers and decimals.

In set notation, the set of numbers greater than 2 may be written as

{	a	:	a	>	2	}
the	any	such	this	is greater	the number 2	
set	number	that	number	than		
of						

On a number line the set $\{a:a > 2\}$ is shown as

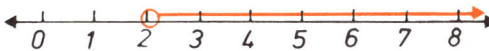

Fig 8.5

Notice that an open circle is placed around the 2 to indicate that the number 2 is not included, but every number to the right of it is.

Similarly, any number to the *left* of a particular number is *smaller* than that number.

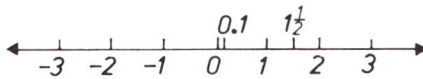

Fig 8.6

All numbers to the left of 3 are smaller than (less than) 3.

1 is less than 3	i.e. $1 < 3$
-2 is less than 3	i.e. $-2 < 3$
$1\frac{1}{2}$ is less than 3	i.e. $1\frac{1}{2} < 3$
0.1 is less than 3	i.e. $0.1 < 3$

In set notation, the set of numbers less than 3 may be written as:

{	a	:	a	<	3	}
the	any	such	this	is less	the number 3	
set	number	that	number	than		
of						

On a number line the set $\{a:a < 3\}$ is shown as

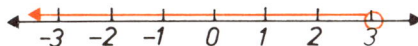

Fig 8.7

Notice again that an open circle is placed around the number 3 to indicate that the number itself, 3, is not included. An expression such as $\{a:a < 3\}$ is called an *inequality*.

Example 8.16

For each of the following sets list four elements and then show the set on a number line:

(a) $\{a: a > 1\}$ (b) $\{a: a < 5\}$

Solution

(a) $\{a: a > 1\}$ This is the set of numbers greater than 1
$\{1.01, 3, 10\frac{1}{2}, 998\}$

> means numbers to the right of 1 on the number line.

Fig 8.8

(b) $\{a: a < 5\}$ This is the set of numbers less than 5
$\{-105, -16.3, 0, 4.99\}$

< means numbers to the left of 5 on the number line

Fig 8.9

Exercise 8N

1. Copy and fill in > or < for each of the following:
   (a) 6 □ 8   (b) 4 □ 2
   (c) 5 □ −3   (d) −2 □ 4
   (e) −3 □ −4   (f) −5 □ −7
   (g) −8 □ −5   (h) 6 □ −3

2. State whether each of the following is true (T) or false (F)
   (a) $6 > 3$ (b) $5 < 7$
   (c) $9 < 8$ (d) $4 > 5$
   (e) $-2 > -1$ (f) $-3 > -2$
   (g) $-3 < -1$ (h) $-4 < -1$
   (i) $-5 > 5$ (j) $6 > -6$

3. For each of the following list four elements of the set:
   (a) $\{a: a > 2\}$ (b) $\{a: a > 7\}$
   (c) $\{a: a > 4\}$ (d) $\{a: a > 3\}$
   (e) $\{a: a > -5\}$ (f) $\{a: a > -1\}$
   (g) $\{a: a > -2\}$ (h) $\{a: a > 0\}$

4. Show each of the sets in question 3 on a separate number line.

5. For each of the following list four elements of the set:
   (a) $\{a: a < 4\}$ (b) $\{a: a < 5\}$
   (c) $\{a: a < 6\}$ (d) $\{a: a < 3\}$
   (e) $\{a: a < -2\}$ (f) $\{a: a < -1\}$
   (g) $\{a: a < -4\}$ (h) $\{a: a < -5\}$

6. Show each of the sets in question 5 on a separate number line.

7. For each of the following, state the inequality:

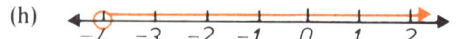

8. Given $4, -3, -1, 0, 6, 9, 1$, state which of these numbers belong to:
   (a) $\{a: a > 1\}$ (b) $\{a: a < 3\}$
   (c) $\{a: a < 0\}$ (d) $\{a: a > -1\}$

Consider the set of numbers consisting of the number 2 *and* all the numbers to the right of 2 on a number line. This set may be described as the set of numbers *greater than or equal to 2*. Using the mathematical symbol $\geq$ for greater than or equal to, we may say $\{a: a \geq 2\}$. On a number line, a closed circle is used to indicate that 2 is included along with all the numbers to the right of 2.

Fig 8.10

Similarly, the set of numbers less than or equal to 3 contains the number 3 and all numbers to the left of 3 on the number line.

Fig 8.11

Example 8.17

Show on a number line
(a) $\{a: a \geq -3\}$ (b) $\{a: a \leq 0\}$

Solution

(a) $\{a: a \geq -3\}$ i.e. the set of numbers consisting of -3 and all numbers to the right of -3 on the number line.

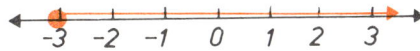

Fig 8.12

(b) $\{a: a \leq 0\}$ i.e. the set of numbers consisting of 0 and all numbers to the left of 0 on the number line.

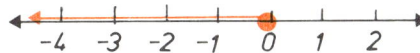

Fig 8.13

It is important to remember the inequality symbols.

$>$	'is greater than'
$4 > 2$	4 is greater than 2
$<$	'is less than'
$-3 < -1$	-3 is less than -1
$\geq$	'is greater than or equal to'
$a \geq 2$	a is greater than or equal to 2
$\leq$	'is less than or equal to'
$a \leq -1$	a is less than or equal to -1

Exercise 8O

1. Given $-3, -2, 0, 1, 3, 5$, which of these numbers belong to:
   (a) $\{a: a \geq 1\}$ (b) $\{a: a \geq -2\}$
   (c) $\{a: a \leq -2\}$ (d) $\{a: a \leq 1\}$

2. On a number line, show
   (a) $\{a: a \geq 1\}$ (b) $\{a: a \geq 3\}$
   (c) $\{a: a \geq 4\}$ (d) $\{a: a \geq 5\}$
   (e) $\{a: a \geq -2\}$ (f) $\{a: a \geq -1\}$
   (g) $\{a: a \geq 0\}$ (h) $\{a: a \geq -4\}$

3. On a number line, show
   (a) $\{a: a \leq 3\}$ (b) $\{a: a \leq 1\}$
   (c) $\{a: a \leq 5\}$ (b) $\{a: a \leq 2\}$
   (e) $\{a: a \leq -3\}$ (f) $\{a: a \leq -1\}$
   (g) $\{a: a \leq 0\}$ (h) $\{a: a \leq -4\}$

Consider the set $\{a: a \geq 1\}$
Representing this set on a number line, we have

Fig 8.14

This is the set of all numbers (integers, decimals and fractions) greater than or equal to 1.
The set $\{a: 1 \leq a\}$ is the set containing 1 and all numbers which are greater than 1. Representing this on a number line, we have

Fig 8.15

Notice that the set of numbers obtained is the same as the set $\{a: a \geq 1\}$.
I.e. $\{a: 1 \leq a\}$ is equivalent to $\{a: a \geq 1\}$.
Now consider the set $\{a: a \leq 4\}$.
Representing this set on a number line, we have

Fig 8.16

If we represent both the sets $\{a: 1 \leq a\}$ and $\{a: a \leq 4\}$ on the same number line, we have

Fig 8.17

——— $\{a: 1 \leq a\}$ - - - $\{a: a \leq 4\}$

It can be seen that the numbers 1 and 4 and all the numbers in between are contained in both the sets of numbers; that is, in the intersection of the sets.
Thus $\{a: 1 \leq a\} \cap \{a: a \leq 4\}$ is shown on a number line as

Fig 8.18

In set notation, this set of numbers can be written as

{	a	:	$1 \leq a \leq 4$	}
the	all	such	the number lies between	
set	numbers	that	1 and 4 inclusive	
of				

Example 8.18
Show on a number line $\{x: -2 \leq x \leq 2\}$.

Solution
$\{x: -2 \leq x \leq 2\}$ This is the set of numbers between, and including, -2 and 2.

Fig 8.19

Exercise 8P
Show each of the following on a number line:

1. $\{a: -1 \leq a \leq 1\}$
2. $\{a: -3 \leq a \leq 4\}$
3. $\{a: -3 < a < 2\}$
4. $\{a: -2 < a < 5\}$
5. $\{a: -4 \leq a \leq 3\}$
6. $\{a: -4 \leq a \leq 0\}$
7. $\{a: -2 \leq a \leq 0\}$
8. $\{a: -3 \leq a \leq -1\}$
9. $\{a: -3 \leq a \leq 3\}$
10. $\{a: -5 \leq a \leq 2\}$
11. $\{a: 0 < a \leq 5\}$
12. $\{a: 0 \leq a < 10\}$
13. $\{a: 0 \leq a \leq 8\}$
14. $\{a: 0 \leq a \leq 3\}$

Revision Exercises for Chapters 6–8

1. Evaluate each of the following given $a = 7$, $b = 5$ and $c = -3$.
   (a) $a + b$ (b) $a + c$
   (c) $b + a$ (d) $c + a$
   (e) $a - b$ (f) $b - c$
   (g) $b - a$ (h) $c - b$
   (i) $a^2 - c$ (j) $b^2 - a^2$
   (k) $ab + c$ (l) $ac - b$
   (m) $a(b + c)$ (n) $b(a + c)$
   (o) $c(a - b)$ (p) $a(b - c)$

2. Simplify
   (a) $3 \times 5a$ (b) $8 \times 9g$
   (c) $6s \times {}^-5s$ (d) ${}^-3t \times 2t$
   (e) $2a^2 \times 4a$ (f) $3p \times 5p$
   (g) $2p^2 \div 2$ (h) $18m^2 \div 6$
   (i) $-2b \div b$ (j) $-6s^2 \div s$

3. Simplify
   (a) $6m^2 + 4m^2$ (b) $3gh + 4gh$
   (c) $7pq - 5pq + 7q$ (d) $a^2 - 4ab + 9a^2$
   (e) $2a^2b - 3ab^2 - 4ab^2 - a^2b$ (f) $2pq^2 + 5pq - 4pq - 3pq^2$

4. Expand each of the following
   (a) $7(p + 4)$ (b) $9(a + 7)$
   (c) $12(2 - g)$ (d) $3(5 - s)$
   (e) $a(b - c)$ (f) $t(t - s)$
   (g) $3(2a - 4b - c)$ (h) $5(4p - 3q + 5r)$
   (i) $-2(a + 4)$ (j) $-3(s + 5)$
   (k) $-2b(b - 1)$ (l) $-3s(s - 4)$

5. Simplify
   (a) $2(a + 4) + 3(a + 2)$ (b) $7(s + 3) + 2(s + 5)$
   (c) $5(h - 4) + 2(3 + h)$ (d) $7(5 + p) + 3(2 + p)$
   (e) $2(4s - 3) - 3(s - 1)$ (f) $5(2g - 3) - 3(3g + 4)$

6. Find the factors of
   (a) 12 (b) 15
   (c) ab (d) $6pq$

7. Find the highest common factors of
   (a) $3a$ and 6 (b) $9y$ and 18
   (c) $4p$ and $10p$ (d) $6g$ and $8g$
   (e) $2y$ and $2y^2$ (f) $3p^2$ and $3p$

8. Factorise by taking out the highest common factor
   (a) $4m + 4$ (b) $9h + 9$
   (c) $6s - 3$ (d) $8t - 4$
   (e) $a^2 - 4a$ (f) $m^2 - 4m$
   (g) $7ad + 14ag$ (h) $8pq + 16qr$
   (i) $-5x + 15$ (j) $-7y + 21$
   (k) $-7 + 14d$ (l) $-3 + 12p$
   (m) $-2m - 4n$ (n) $-5t - 10s$
   (o) $-2p^2 - 2p$ (p) $-8m^2 - m$

9. Say whether the following are vector or scalar quantities:
   (a) The number of bones in your body.
   (b) Moving a rook in a game of chess.
   (c) The boat was 10 km from the lighthouse.
   (d) The boat was 10 km due East of the lighthouse.
   (e) The boat was travelling SE at 8 km per hour.

10.

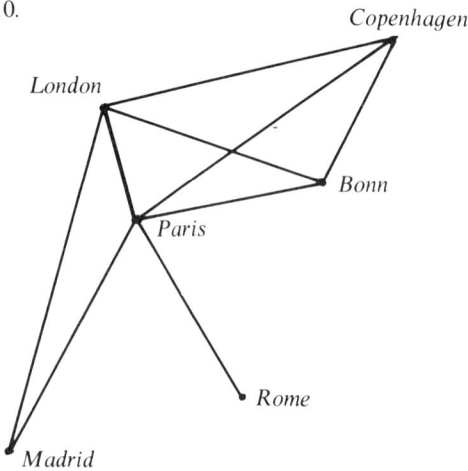

Using the initial letters of the cities on this route map, write down the vectors representing journeys from:
(a) Paris to Rome
(b) Paris to Bonn
(c) Bonn to Paris
(d) Madrid to Paris
(e) Copenhagen to Paris

(f) Christopher the pilot flies from Paris to London and then to Copenhagen. Describe this journey using vectors. What single vector is this equivalent to?

11.

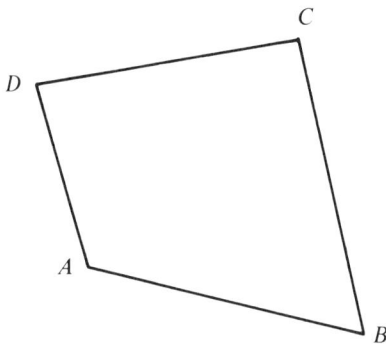

Using this diagram write down the single vectors equal to:
(a) $\vec{AB} + \vec{BC}$
(b) $\vec{BC} + \vec{CD}$
(c) $\vec{AB} + \vec{BC} + \vec{CD}$
(d) $\vec{DC} + \vec{CB} + \vec{BA} + \vec{AD}$
(e) $\vec{AB} + \vec{BA}$

12. Four posts, A, B, C and D, are in a straight line as shown below:

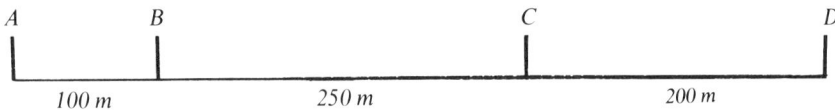

(a) Express the following in terms of $\vec{AB}$:
(i) $\vec{BC}$
(iii) $\vec{AC}$
(v) $\vec{AD}$
(ii) $\vec{CD}$
(iv) $\vec{BD}$
(vi) $4\vec{BC}$

(b) Express the following in terms of $\vec{CD}$:
(i) $\vec{AB}$
(iii) $\vec{BA}$
(v) $\vec{AD}$
(ii) $\vec{BC}$
(iv) $\vec{CB}$
(vi) $\vec{DA}$

13. Solve each of the following equations:
(a) $x + 9 = 12$
(c) $p - 6 = 4$
(b) $a + 5 = 7$
(d) $s - 3 = 11$

(e) $2a = 16$

(f) $3t = 15$

(g) $\dfrac{m}{5} = -4$

(h) $\dfrac{y}{3} = 11$

(i) $g - 3 = {}^-5$

(j) $t - 7 = {}^-4$

(k) $m + 2 = {}^-1$

(l) $s + 1 = {}^-4$

14. Solve each of the following equations:

(a) $3a + 1 = 13$

(b) $5b - 3 = 7$

(c) $2 - 5g = {}^-8$

(d) $6 - 7t = {}^-8$

(e) $2(m + 5) = 12$

(f) $3(s - 2) = 8$

(g) $2a + 4 = a + 9$

(h) $4m - 3 = 3m + 2$

(i) $4(a - 3) = 3(a - 6)$

(j) $7(p - 1) = 6(p + 1)$

(k) $\dfrac{a}{3} + 1 = 4$

(l) $1 - \dfrac{m}{2} = 5$

(m) $3 = 2 - \dfrac{a}{5}$

(n) $8 = \dfrac{m}{2} - 2$

(o) $\dfrac{2a}{3} = {}^-5$

(p) $\dfrac{a + 2}{3} = 4$

(q) $\dfrac{y + 3}{4} = 5$

(r) $\dfrac{2a}{3} + 1 = {}^-3$

15. I think of a number and add 6. If the result is 9, what is the number?

16. I think of a number and double it. If the result is ${}^-10$, what is the number?

17. If I subtract 5 from twice a certain number, the result is 15. What is the number?

18.

l cm

6cm 6cm

l cm

The perimeter of a rectangle is 34 cm. If the width is 6 cm, what is the length of the rectangle?

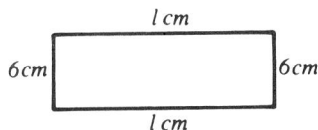

19. State whether each of the following is true (T) or false (F):

(a) $6 > -2$

(b) $-3 < -1$

(c) $-4 > -1$

(d) $6 < -6$

20. On a number line, show:

(a) $\{a : a > 4\}$

(b) $\{a : a \leq 3\}$

(c) $\{a : a \geq 9\}$

(d) $\{a : a < -2\}$

(e) $\{a : -2 \leq a \leq 2\}$

(f) $\{a : -2 < a < 2\}$

9.1 Revision of Definitions

A *line* is an infinite set of points extending indefinitely in opposite directions.

A *ray* is part of a line which extends indefinitely in one direction.

An *interval* is part of a line which is finite. Another name given to it is a *line segment*. It is named by its 2 end points; e.g. $\overline{CD}$.

9.2 Angles

An *angle* is formed between 2 rays intersecting at a *vertex*.

VERTEX

Fig 9.1

It is usually named by 3 points.
The angle may be named in any order as long as the vertex point is written between the points describing the rays.
E.g.

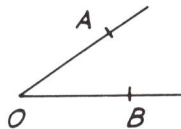

Fig 9.2 $\angle AOB$ or $\angle BOA$

The amount of turning between the 2 rays is measured in degrees on a protractor.
An angle can measure between $0°$ and $360°$.

9.2.1 Types of Angles

Angles may be classified according to their size or according to their relationship with other angles.

9.2.1.1 Acute, Obtuse and Reflex

We have already seen that a *right angle* is an angle of measure $90°$, a *straight angle* is an angle of measure $180°$ and a full turn is an angle of measure $360°$.
An *acute angle* is an angle whose measure is between $0°$ and $90°$.

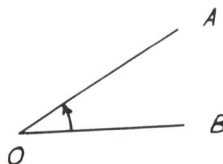

Fig 9.3

An *obtuse angle* is an angle whose measure is between $90°$ and $180°$.

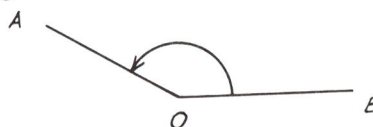

Fig 9.4

A *reflex angle* is an angle whose measure is between 180° and 360°.

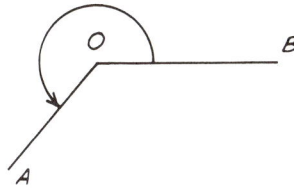

Fig 9.5

9.2.1.2. Complementary and Supplementary

Two angles whose measures add up to 90° are said to be *complementary*.

E.g.

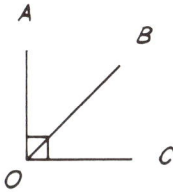

Fig 9.6

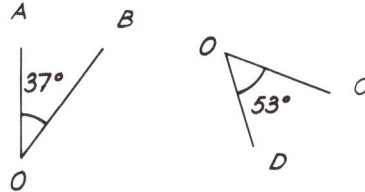

Since $\angle AOC$ is 90°,
$\angle AOB$ and $\angle BOC$ are complementary
angles.

Since $\angle AOB + \angle COD = 37° + 53° = 90°$,
$\angle AOB$ and $\angle COD$ are complementary
angles.

If two angles are complementary, we can say that one angle is the *complement* of the other.
Thus if $\angle XOY = 42°$, its complement has measure 48°.

Two angles whose measures total 180° are said to be *supplementary*.

E.g.

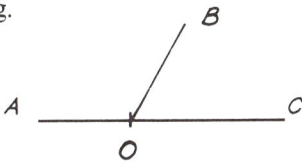

Fig 9.7

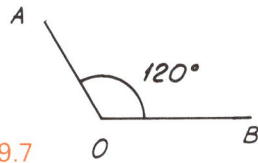

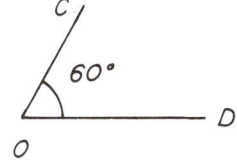

Since $\angle AOC$ is a straight angle (180°),
$\angle AOB$ and $\angle BOC$ are supplementary
angles.

Since $\angle AOB + \angle COD = 120° + 60° = 180°$,
$\angle AOB$ and $\angle COD$ are supplementary
angles.

If two angles are supplementary, we can say that one angle is the *supplement* of the other.
Thus if $\angle XOY = 40°$, its supplement has measure 140°.

Example 9.1

Find the value of the pronumeral.

(a)

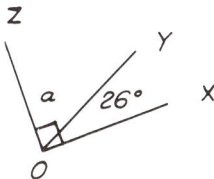

(b)

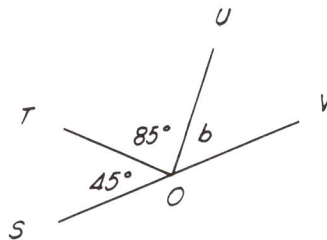

Solution

(a) $\angle XOY$ and $\angle YOZ$ form a right angle and are therefore complementary angles which add up to 90°.

Thus $a + 26° = 90°$

We can find a by solving this equation using algebra.

$$a + 26° - 26° = 90° - 26°$$
$$a = 64°$$

(b) $\angle SOU$ and $\angle UOV$ form a straight line and are therefore supplementary angles which add to $180°$.

$$\text{But} \quad \angle SOU = 45° + 85° = 130°$$
$$\therefore b + 130° = 180°$$
$$b + 130° - 130° = 180° - 130°$$
$$b = 50°$$

Example 9.2

(a) Find the complement, x, of $62°$.
(b) Find the supplement, x, of $86°$.

Solution

(a) Since complementary angles add to $90°$
$$x + 62° = 90°$$
$$x + 62° - 62° = 90° - 62°$$
$$x = 28°$$
The complement of $62°$ is $28°$.

(b) Since supplementary angles add to $180°$
$$x + 86° = 180°$$
$$x + 86° - 86° = 180° - 86°$$
$$x = 94°$$
The supplement of $86°$ is $94°$.

9.2.1.3   Vertically Opposite and Adjacent

If two lines AB and CD intersect at a point, O, four angles are formed.

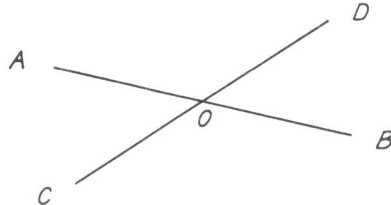

Fig 9.8

The angles which are side by side and have a common ray are called *adjacent angles*.
E.g. $\angle AOD$ and $\angle DOB$ are adjacent angles since they lie side by side with common ray OD.
Similarly $\angle COB$ and $\angle BOD$ are adjacent angles since they lie side by side with common ray OB.

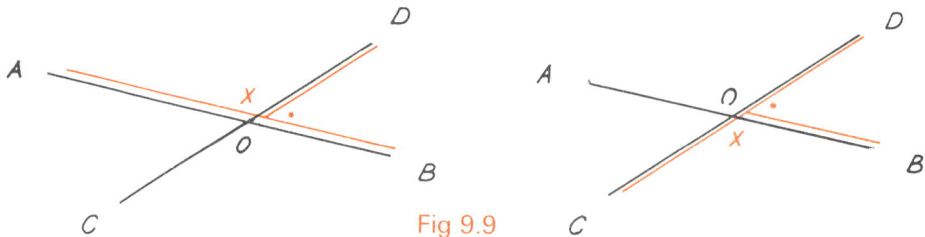

Fig 9.9

Example 9.3

Name the angles adjacent to $\angle ROQ$ and hence find the value of x.

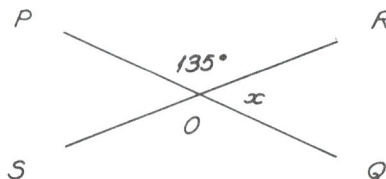

Solution

$\angle POR$ is adjacent to $\angle ROQ$ with common ray OR.

$\angle QOS$ is adjacent to $\angle ROQ$ with common ray OQ.

Since $\angle POR$ and $\angle ROQ$ are adjacent angles lying on line PQ

$$135° + x = 180°$$

$-135°$ both sides

$$135° + x - 135° = 180° - 135°$$
$$x = 45°$$

When two lines AB and CD intersect at O, the angles which are on opposite sides of the vertex O and have no common ray are called *vertically opposite angles*.

E.g. $\angle AOC$ and $\angle DOB$ are vertically opposite since they lie on opposite sides of vertex O and have no common ray. Similarly $\angle AOD$ and $\angle COB$ are also vertically opposite.

Fig. 9.10

Example 9.4

Find the value of y and x in each of the following:

(a)

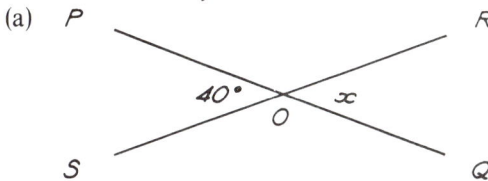

(b)

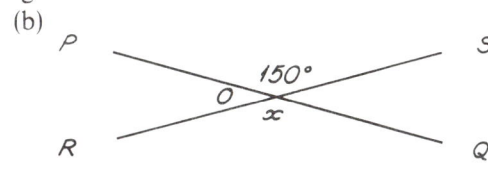

Solution

(a) To find angle x we need to consider $\angle POR$ which is adjacent to both 40° and x.

$\angle POR$ is the supplement of the angle marked 40° as they are on a straight line $\overline{SR}$.

$\angle POR$ is also the supplement of the angle marked x as they are on a straight line $\overline{PQ}$.

$\therefore$ the angle marked 40° and the angle marked x must have the same magnitude as they are both supplementary to $\angle POR$. Thus $x = 40°$.

(b) To find x we need to consider $\angle SOQ$ which is adjacent to both 150° and the angle marked x.

$\angle SOQ$ is the supplement of the angle marked 150° as they are on a straight line $\overline{PQ}$.

$\angle SOQ$ is also the supplement of the angle marked x as they are on a straight line $\overline{RS}$.

$\therefore$ the angle marked 150° and the angle marked x must have the same magnitude as they are both supplementary to $\angle SOQ$. Thus $x = 150°$.

Notice that in both the above examples, the angle of measure x is vertically opposite to $\angle POS$ and in both cases the value of x was equal to the measure of $\angle POS$.

> In general, when lines intersect at a point, the vertically opposite angles are equal.

Example 9.5

(a)

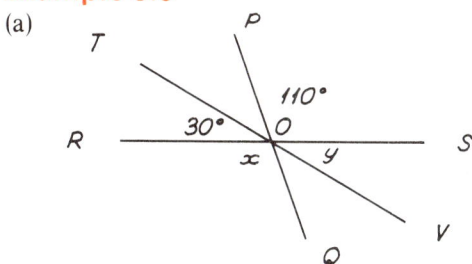

(b)

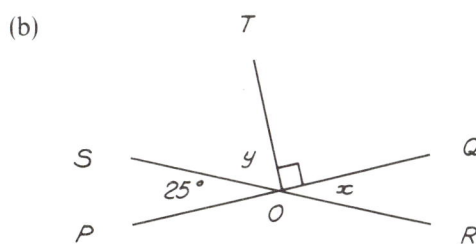

Solution

(a) Since the angle of measure x is vertically opposite the angle of measure $110°$, $x = 110°$. Similarly, the angle of measure y is vertically opposite the angle of measure $30°$. Hence $y = 30°$.

(b) Since the angle of measure x is vertically opposite the angle of measure $25°$, $x = 25°$.

Since $\angle POQ$ is a straight angle

$$25° + y + 90° = 180°$$
$$y + 115° = 180°$$

$-115°$ both sides

$$y + 115° - 115° = 180° - 115°$$
$$y = 65°$$

Summary of Types of Angles

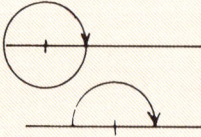

	A full turn has a measure of $360°$.
	Straight angles have a measure of $180°$.
	Right angles have a measure of $90°$.
	Acute angles have measure of $0°$ to $90°$.
	Obtuse angles have measure of $90°$ to $180°$.
	Reflex angles have measure of $180°$ to $360°$.
	Complementary angles are angles whose sum is $90°$.
	Supplementary angles are angles whose sum is $180°$.
	Adjacent angles are side-by-side and have a common ray.
	Vertically opposite angles are equal.

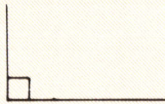

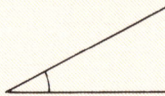

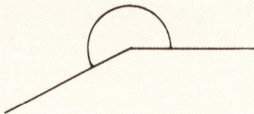

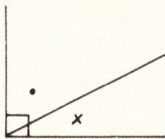

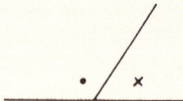

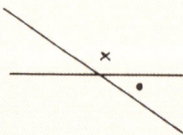

Exercise 9A

1. Name each of the following angles using the labels given; and state what type of angle each one is.

(a)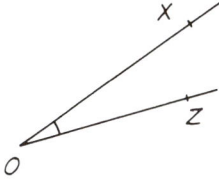

(b)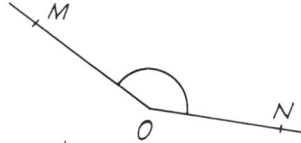

(c)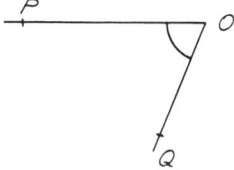

(d)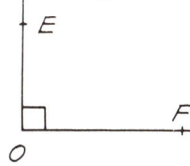

(e)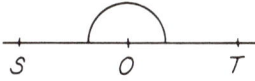

(f)

(g)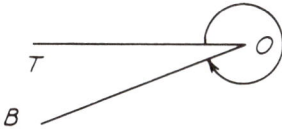

(h)

(i)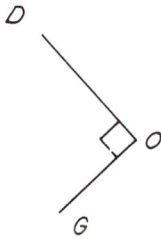

(j)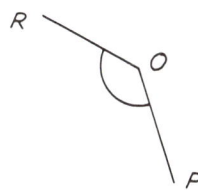

2. State whether the following angles are supplementary (S) or complementary (C).

(a) 32°, 58°
(b) 76°, 14°
(c) 87°, 93°
(d) 107°, 73°
(e) 43°, 12°, 35°
(f) 59°, 68°, 53°
(g) 21°, 76°, 83°
(h) 31°, 43°, 16°
(i) 19°, 84°, 7°, 70°
(j) 8°, 56°, 12°, 14°
(k) 32°, 19°, 23°, 16°
(l) 19°, 53°, 63°, 45°

3. Find the value of the pronumeral and give a reason for your answer.

(a)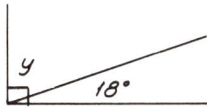

(b)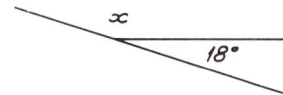

(c)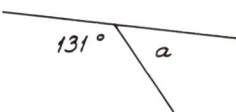

(d)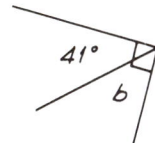

(e)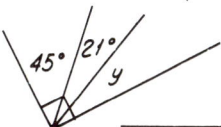

(f)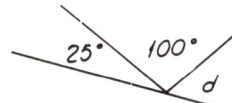

(g)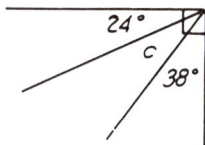

(h)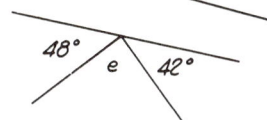

4. Find the value of the pronumerals in the following diagrams:

(a)
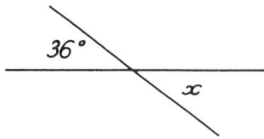

(b)
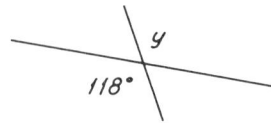

(c)
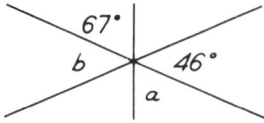

(d)
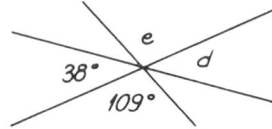

(e)
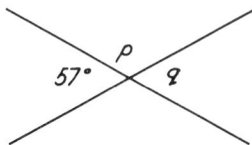

(f)
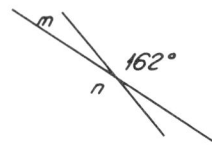

(g)
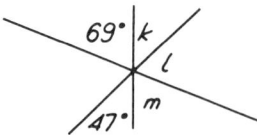

(h)
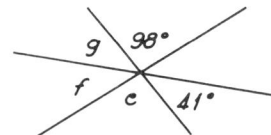

(i)
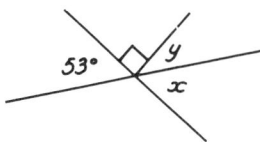

(j)
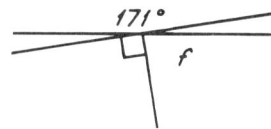

(k)
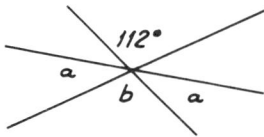

(l)
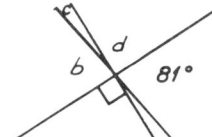

5. Find the value of the pronumeral and state a reason for your answer.

(a)
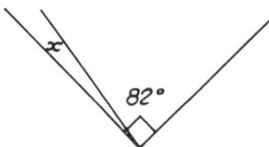

(b)
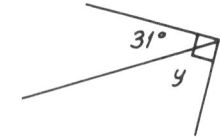

(c)
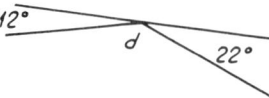

(d)
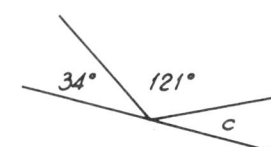

(e)
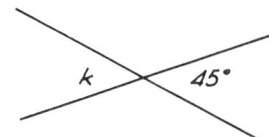

(f)
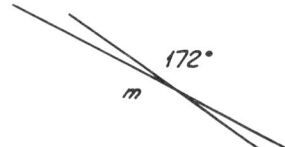

9.3 Parallel Lines

Lines which run in exactly the same direction are 'parallel' lines.

Fig 9.11  a pair of parallel lines.

Looking around, you can see many examples of parallel lines.

Fig 9.12

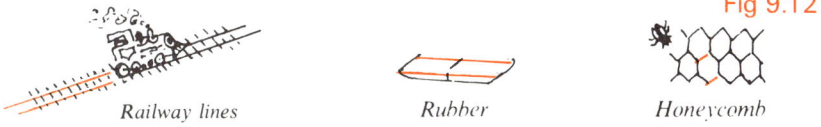

Railway lines   *Rubber*   *Honeycomb*

We indicate parallel lines on a diagram by using arrows:

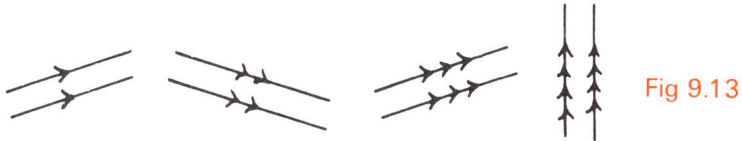 Fig 9.13

Exercise 9 B

1. List as many pairs of lines as you can, which you think might be parallel, from the following diagram:

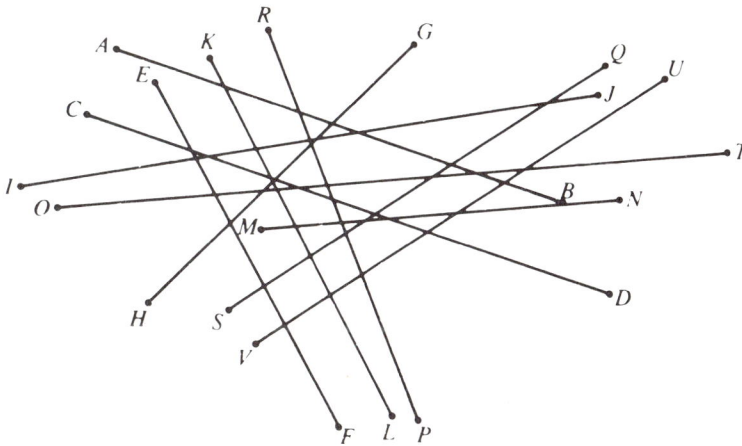

2. Use arrows to show the pairs of parallel lines in each of the following plane figures.
   E.g.

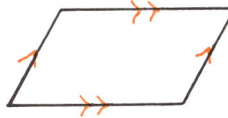

(a)

(b)

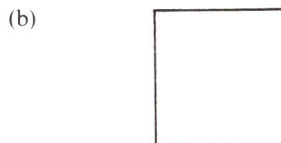

(c)

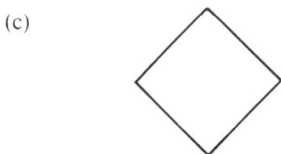

(d)

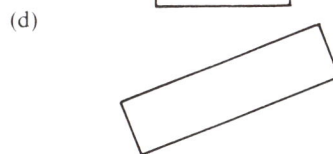

(e)

(f)

9.3.1   Construction of Parallel Lines

(a) Place a ruler on a piece of paper in any position.

(b) Place a set square against the ruler and draw a line $\overline{AB}$ across the top of the set square.

(c) Slide the set square down the ruler, being careful not to move the position of the ruler, and rule another line $\overline{CD}$ on the top of the set square.

(d) $\overline{AB}$ and $\overline{CD}$ are parallel lines.

This process can be repeated as many times as required to form as many parallel lines as needed.

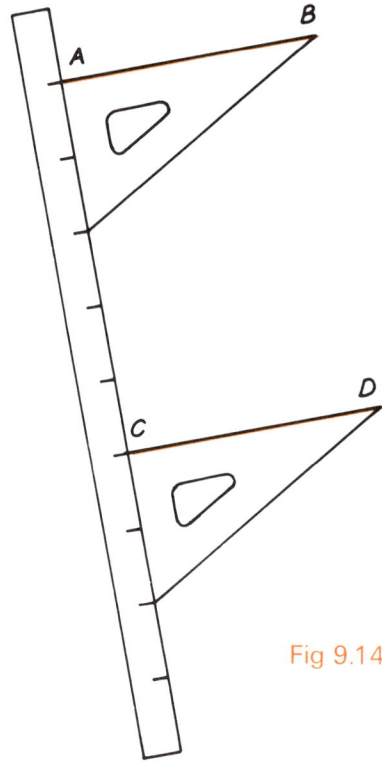

Fig 9.14

9.3.2   Angles Associated with a Transversal

A *transversal* is a line which intersects with a set of parallel lines.

9.3.2.1   Corresponding Angles

Consider the construction of parallel lines in fig 9.14. The ruler can be seen to represent a transversal.

Therefore the angle formed between the line at the top of the set square and the ruler is equivalent to the corresponding angle of the set square.

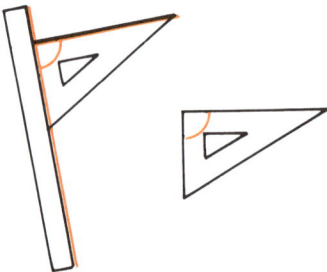

Fig 9.15

Furthermore, the angle formed between the ruler and set square when constructing the second parallel line is equivalent to the corresponding angle of the set square and so must be identical to the first angle formed in the same position.

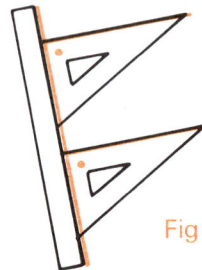

Fig 9.16

These angles are called *corresponding angles* as they are in the same corresponding position with respect to the parallel line and the transversal.

> When parallel lines are cut by a transversal, *corresponding angles* are equal.

Example 9.6

Find the value of the pronumeral in each of the following:

(a)

(b)

Solution

(a) Since the angles labelled b and $60°$ are in corresponding positions with respect to the parallel lines and transversal, $b = 60°$ (corresponding angles).

(b) Similarly $a = 120°$ (corresponding angles).

Exercise 9C

1. Copy the following diagrams and mark the corresponding angle to x:

(a)

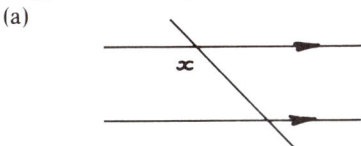

(b)

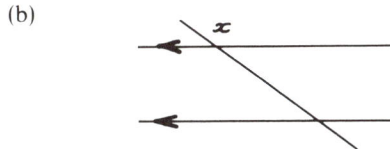

(c)

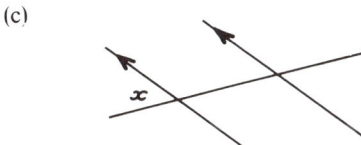

(d)

(e)

(f)

(g)

(h)

2. Find the value of the pronumeral in each of the following:

(a)

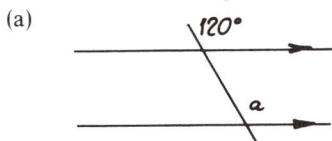

(b)

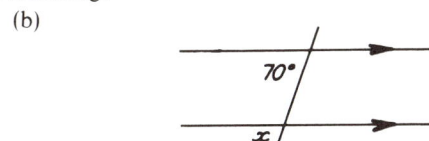

(c)

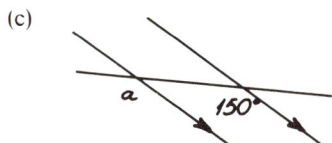

(d)

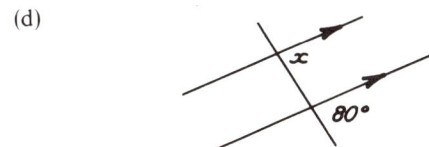

3. Name all the pairs of corresponding angles in the following diagrams:

(a)

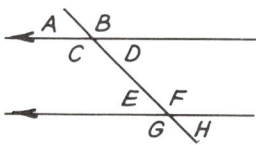

(b)

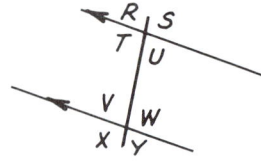

9.3.2.2   Alternate Angles

Example 9.7

Find the value of *a* in the following diagram

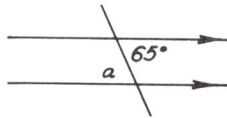

Solution

The angle marked ➤ is vertically opposite 65°
and therefore has the same magnitude.
The angle marked *a* is corresponding to the
marked angle and therefore has a magnitude of
65°.

$$a = 65°$$

OR

The angle marked ◄, is corresponding to 65°
and therefore has the same magnitude.
The angle marked *a* is vertically opposite the
marked angle and thus is equal to 65°.

$$a = 65°$$

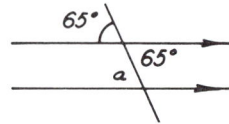

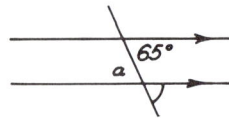

Angles which are between the parallel lines but on alternate sides of the transversal are called
alternate angles.

> When parallel lines are cut by a transversal *alternate angles* are equal.

Exercise 9D

1. Copy the following diagrams, marking the angle which is alternate to angle *a*.

(a)

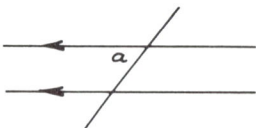

(b)

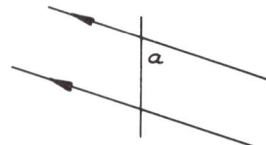

(c)

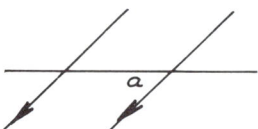

(d)

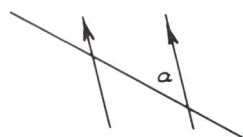

(e)

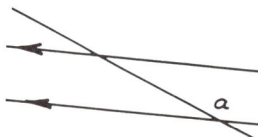

(f)

2. Find the value of the pronumeral:

(a)

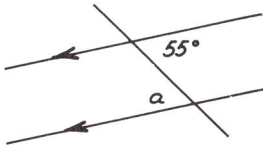

(b)

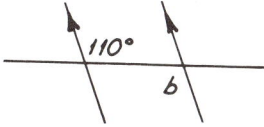

(c)

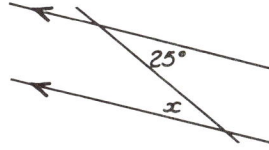

(d)

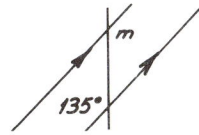

9.4 Polygons

We saw in Book 1 that polygons are plane, closed figures with straight sides (triangles, quadrilaterals, etc).

9.4.1 Angle Sums of Polygons

If we were to draw a triangle and label each angle with a letter and then tear off each corner, we could place them side by side with their vertices together. We would then find that they would form a straight angle.

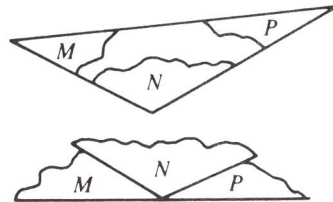

Fig 9.17

Since we know that a straight angle is exactly 180°, we can say that the angles in a triangle add up to 180°.

> The angle sum of a triangle is 180°.

Remember that angles which fit together so that they are adjacent angles and form a straight angle are called supplementary angles. Thus, the angles of a triangle are supplementary.
We can find the angle sum of a quadrilateral in the same way as we did for the triangle.

Fig 9.18

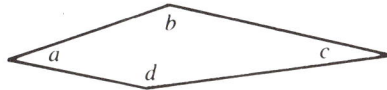

When we place the four angles together, they form the same angle as a full turn which is 360°.
Thus the angles in a quadrilateral add up to 360°.

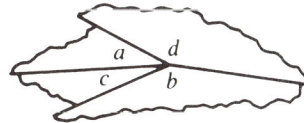

Fig 9.19

> The angle sum of a quadrilateral is 360°.

It is also interesting to note that by drawing just one line from vertex to vertex in the quadrilateral, it is split into two triangles.

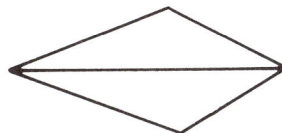

Fig 9.20

Thus, if we say that a quadrilateral is made up of two triangles and since we know that the angle sum of a triangle is 180°, we again find that the angle sum of a quadrilateral is 2 × 180° or 360°.

Example 9.8

(a) Draw an example of the polygon shown here, tear off all the corners and place them together to form full turns and straight angles and thus find the angle sum of each polygon.

(b) For the same polygon, draw straight lines from vertex to vertex, splitting the polygon into triangles and thus find the angle sum of the polygon.

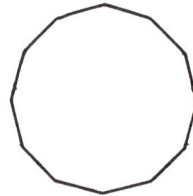

Solution

(a) The polygon drawn is called a *dodecagon* because it has *twelve* sides. If we tear off each corner and place them together to find the total size of the angles, we will find the angle sum of the polygon.

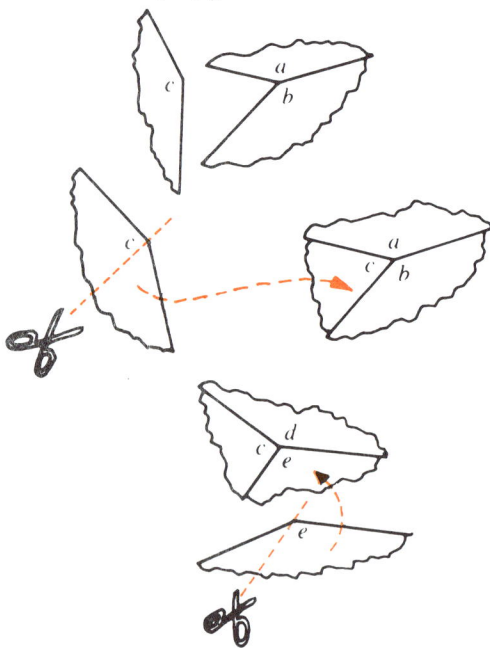

When we put the first two angles together, we find that they almost form a full turn and the third angle will not fit in the space. So we will cut the third angle in two from the vertex to the torn edge so that it fits in the space.

When we place the remainder of the third angle with the fourth angle, there is not enough space to fit the whole fifth angle so it is also cut.

We continue putting the angles together, cutting some of them, until we have used up all the angles.

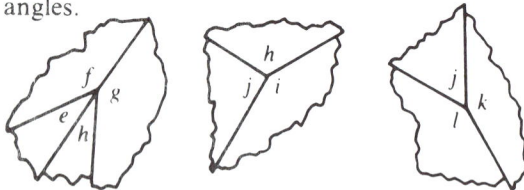

When we count all the groups of angles we find that there are five full turns formed.

Since the number of degrees in a full turn is $360°$, then we can say that the angle sum of a dodecagon is five full turns or $5 \times 360° = 1800°$

(b) The easiest way to split the polygon into triangles is to start from one vertex and, missing the two vertices on either side of this one, join it to each of the other vertices using straight lines. Counting the triangles thus formed, we find there are ten. Since the angle sum of a triangle is $180°$, then the angle sum of a dodecagon is $10 \times 180° = 1800°$. This is the same answer as was found by the above method.

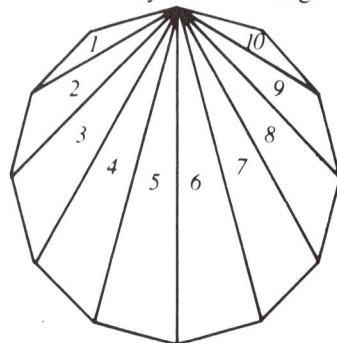

Exercise 9 E

1. Draw an example of each of the following figures or trace those given. Then, for each shape, tear off each corner and place them together to form straight angles. Thus find the angle sum of each polygon.

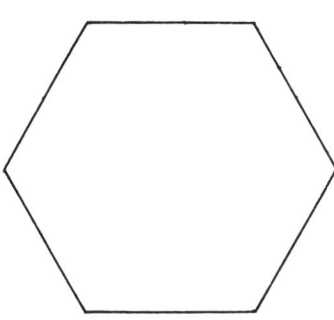 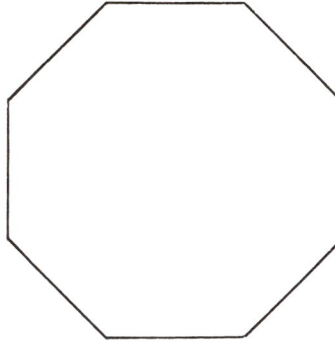 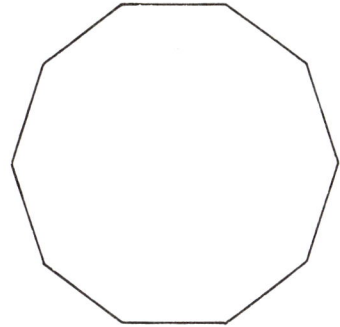

(a) hexagon   (b) octagon   (c) decagon

2. Draw or trace each of the following polygons. Then, by drawing as few straight lines as possible from vertex to vertex, split the polygon into triangles. Thus find the angle sum of each polygon.

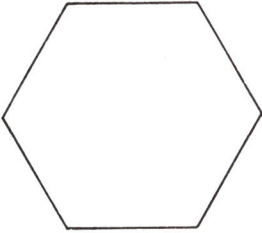 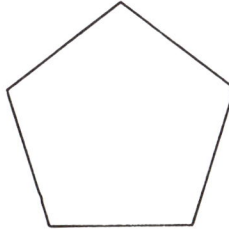 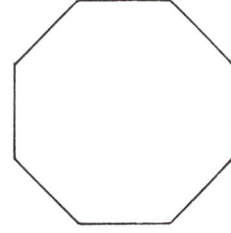

(a) hexagon   (b) pentagon   (c) octagon

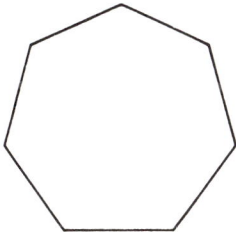 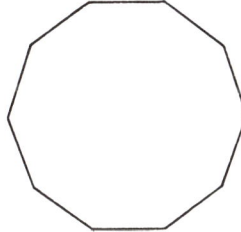 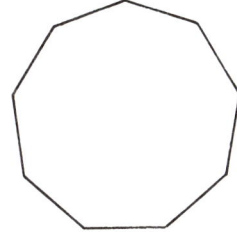

(d) heptagon   (e) decagon   (f) nonagon

9.4.2   Properties of Triangles

Triangles are named by either the properties of their sides OR the properties of their angles, OR both.

9.4.2.1   Side-Named Triangles

(a)

Fig 9.21

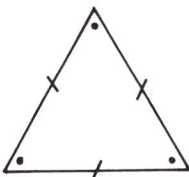

A triangle with *all* sides of *equal* length is an *equilateral triangle*. Thus all angles are of equal measure.

(b)

Fig 9.22

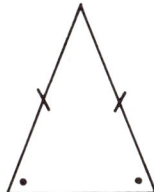

A triangle with *2* sides of *equal* length is an *isosceles triangle*. Furthermore the 2 angles opposite the equal sides are also equal.

(c)

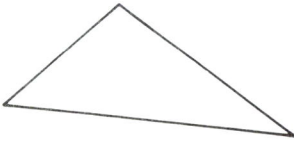

Fig 9.23

A triangle with *all* sides of *different* length is a *scalene triangle*. Thus no angles have equal magnitude.

9.4.2.2   Angle-Named Triangles

(a)

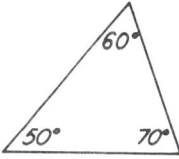

Fig 9.24

A triangle with all angles of measure less than 90° is an *acute-angled triangle*.

(b)

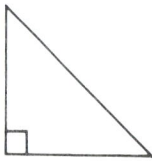

Fig 9.25

A triangle with one angle measuring exactly 90° is a *right-angled triangle*.

(c)

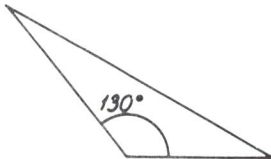

Fig 9.26

A triangle with one angle measuring greater than 90° is an *obtuse-angled triangle*.

Example 9.9

Find the value of the pronumeral and name the triangle:

(a)

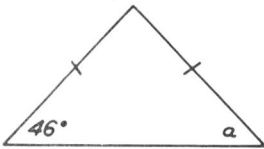

(b)

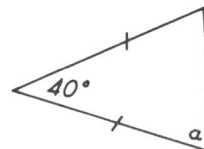

Solution

(a)

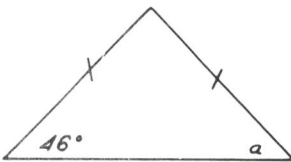

An isosceles triangle has 2 sides of equal length, thus the triangle is an *isosceles triangle*. An isosceles triangle also has 2 equal angles opposite the equal sides, $\therefore a = 46°$.

(b)

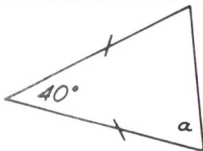

The given triangle is *isosceles* as 2 sides are equal.
Also the 2 angles opposite the equal sides are equal, thus the missing angle must equal a also.

Using the angle sum of a triangle,
$$40° + a + a = 180°$$
$$40° - 40 + 2a = 180° \quad - 40$$
$$2a = 140°$$
$$\frac{2a}{2} = \frac{140°}{2}$$
$$a = 70°$$

Exercise 9 F

1. Give the
   (i) side name and
   (ii) the angle name
   of the following triangles.

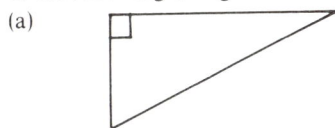

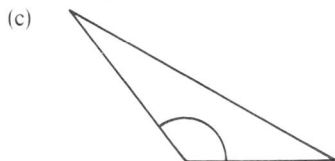

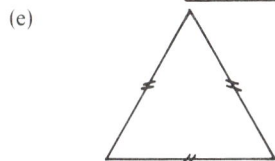

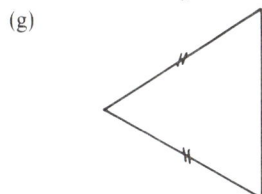

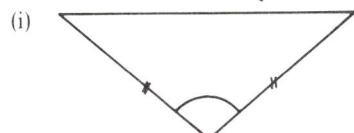

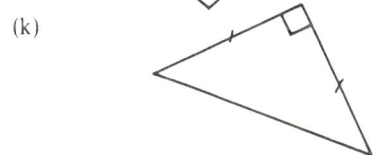

(a)

(b)

(c)

(d)

(e)

(f)

(g)

(h)

(i)

(j)

(k)

(l)

2. Find the value of the pronumeral:

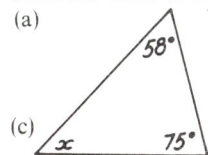

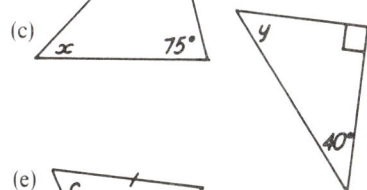

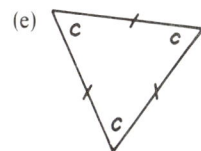

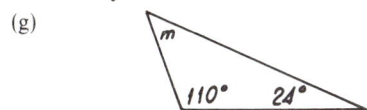

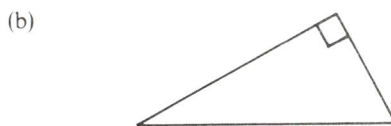

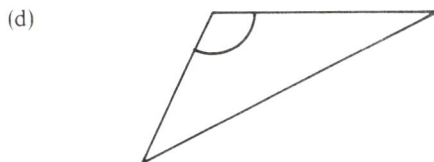

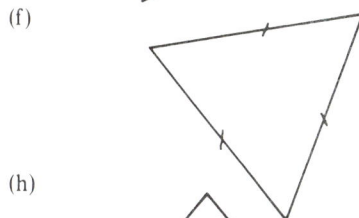

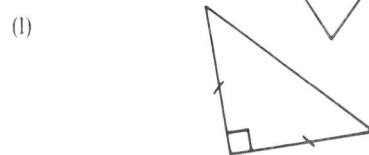

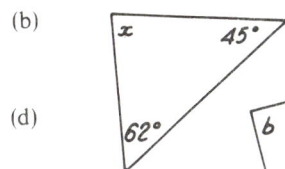

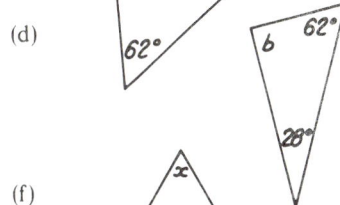

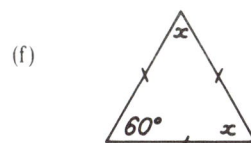

(a) 58° 75° x

(b) x 45° 62°

(c) y 40°

(d) b 62° 28°

(e) c c c

(f) x 60° x

(g) m 110° 24°

(h) 41° n 34°

9.4.2.3　Exterior Angle of a Triangle

An exterior angle of a triangle is formed when
one of the sides is extended (produced).
It is the angle between the extended side and
the adjacent side of the triangle.

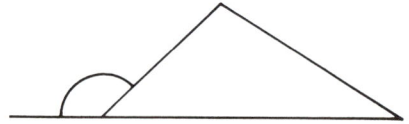

Fig 9.27

Example 9.10

Find the value of the pronumeral in each of the following:

(a)

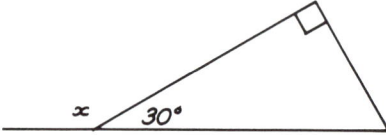

(b)

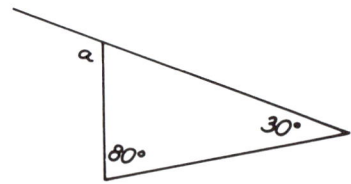

Solution

(a)

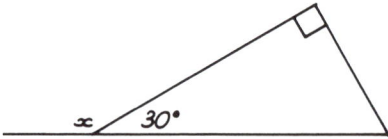

x and $30°$ are on a straight line and therefore
supplementary (add to $180°$).
$$x + 30° = 180°$$
$$x + 30° - 30° = 180° - 30°$$
$$x = 150°$$

(b)

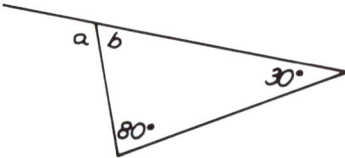

The angle sum of a triangle is $180°$.
Thus, let the unmarked angle be b.
$$b + 30° + 80° = 180°$$
$$b + 110° - 110° = 180° - 110°$$
$$b = 70°$$

Furthermore the angles marked a and b are supplementary angles and thus add to $180°$.
$$a + b = 180°$$
$$a + 70° - 70° = 180° - 70°$$
$$a = 110°$$

The measure of the exterior angle of a triangle is equal to the sum of the two interior
opposite angles.

Exercise 9G

Find the value of the pronumeral in each of the following:

1.

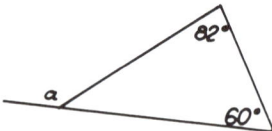

2.

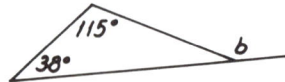

3.

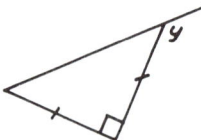

4.

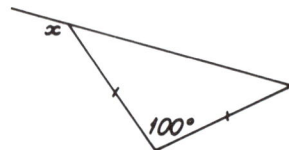

5.

6.

7.

8.

9.

10.

11.

12.

9.5   Quadrilaterals

Remember that quadrilaterals are four-cornered, or four-sided figures.

Quadrilateral   Fig 9.28

There are six main types of quadrilaterals, which have special properties.

Rectangle

Square

Parallelogram

Rhombus   Fig 9.29   *Kite*

Trapezium

The easiest way to describe the six types of quadrilaterals shown is:

1 Rectangle   —has all angles equal, each measures 90°.

2 Square   —is a rectangle, with all sides equal.

3 Parallelogram—has two pairs of parallel sides.

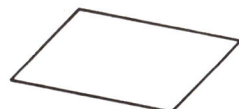

4 Rhombus   —is a parallelogram with all sides equal.

5 Kite   —has two pairs of adjacent sides equal.

6 Trapezium   —has one pair of parallel sides.

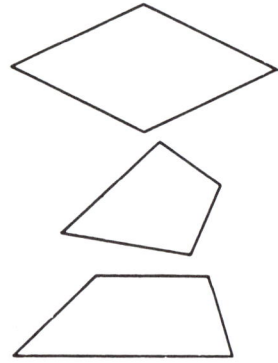

Fig 9.30

Exercise 9H

Name each of the following quadrilaterals:

1.

2.

3.

4.

5.

6.

7.

8.

9.

10.

11.

12.

9.5.1 Angle Sums of Quadrilaterals

Remember that the sum of angles of a quadrilateral is 360°.

Consider a quadrilateral, for which we know the size of three of the angles, and wish to find the size of the fourth angle. Let the unknown angle be x.

Fig 9.31

We can see that the known angles are 30° + 160° + 50°.

$$\therefore\ x = 360° - (30° + 160° + 50°)$$
$$x = 360° - 240°$$
$$x = 120°$$

Example 9.11

Find the angle labelled 'a' in the following quadrilateral:

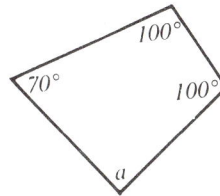

Solution

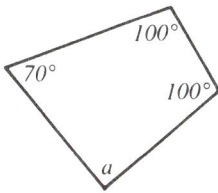

The known angles are 70° + 100° + 100°. Since the four angles in a quadrilateral add to 360°, then we subtract the sum of the known angles from 360°, to find the unknown angle:

$$a = 360° - (70° + 100° + 100°)$$
$$a = 360° - (270°)$$
$$a = 90°$$

Exercise 9I

Find the angle labelled a in each of the following:

1.

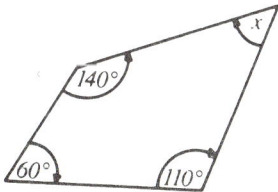

2.

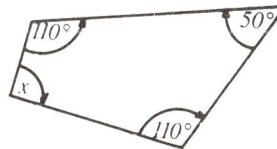

3.

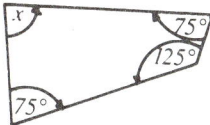

4.

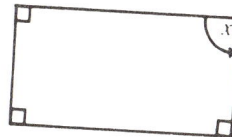

5.

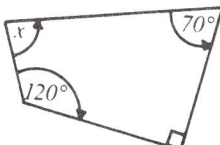

6.

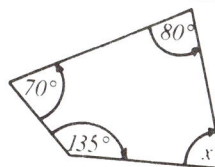

7.

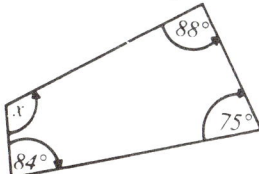

8.

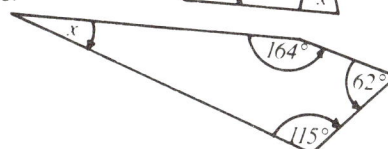

9.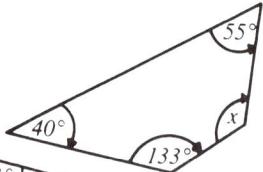

10.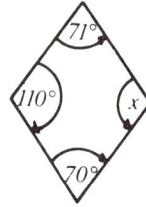

11.

12.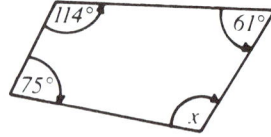

13.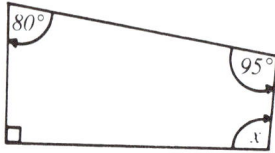

14.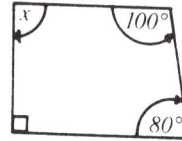

Exercise 9J

1.
   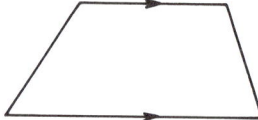   

Parallelogram   Trapezium   Rhombus

Kite   Square   Rectangle

Using the information given in the diagrams above, copy and complete the diagram below by listing the quadrilaterals which have the given property.

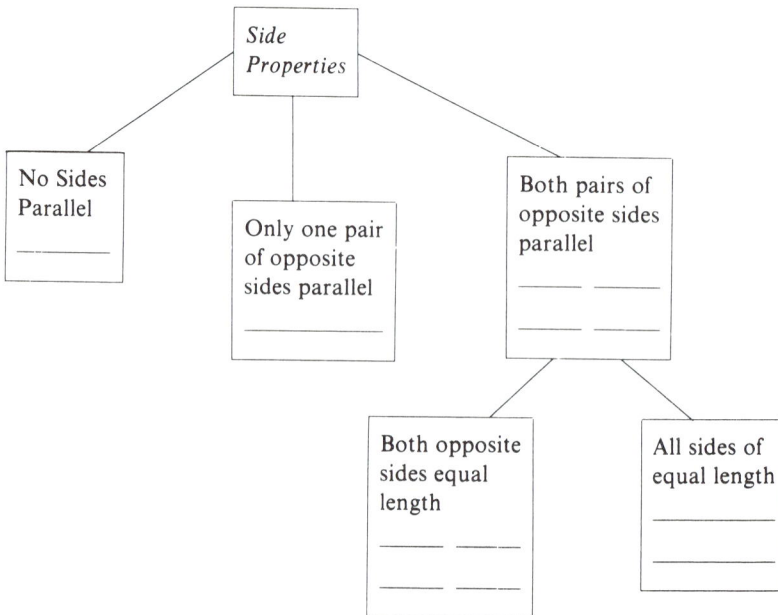

2.

Parallelogram

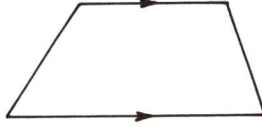
Trapezium

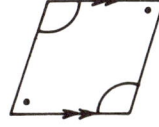
Rhombus

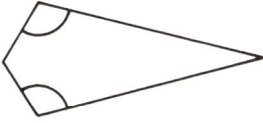
Kite

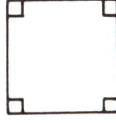
Square

Rectangle

Using the information given in the diagrams above, copy and complete the diagram below by listing the quadrilaterals which have the given properties.

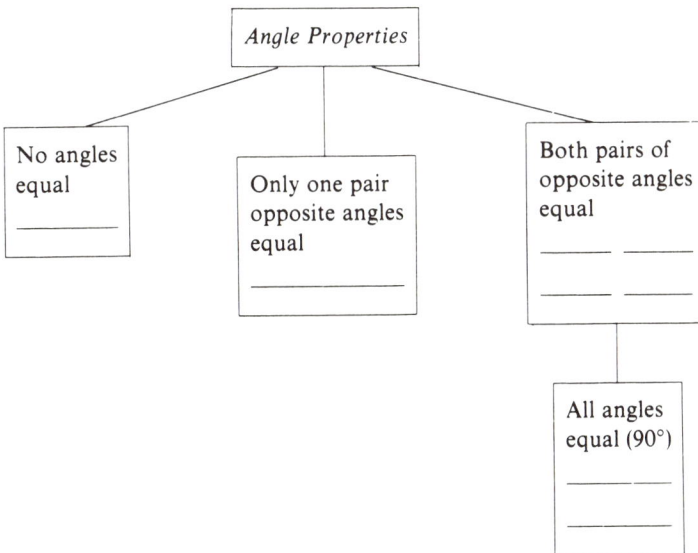

Quadrilaterals, like triangles, have side properties and angle properties, however as they have 4 sides and 4 angles we can draw lines between opposite vertices forming *diagonals*.

Exercise 9 K

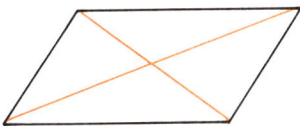
Parallelogram

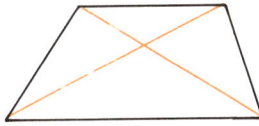
Trapezium

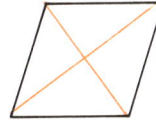
Rhombus

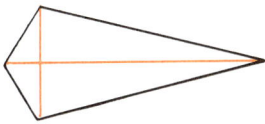
Kite

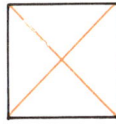
Square

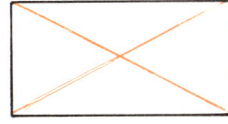
Rectangle

Using the information in the diagrams above and measuring where necessary, fill in the table below.

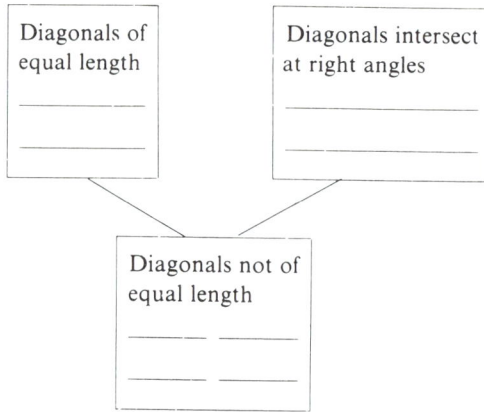

Diagonals of equal length	Diagonals intersect at right angles

Diagonals not of equal length

Example 9.12

Find the value of the pronumerals in each of the following:

(a)

(b)

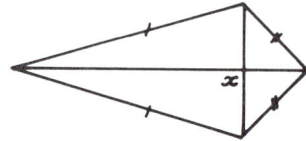

(c)

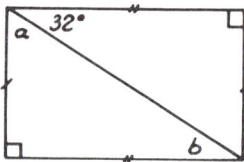

(d)

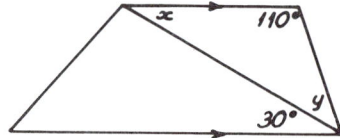

Solution

(a)

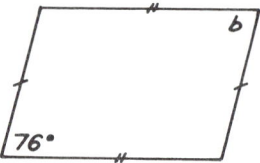

The figure is a parallelogram therefore both pairs of opposite angles are equal.
$b = 76°$

(b)

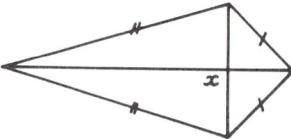

The figure is a kite, therefore the diagonals intersect at right angles.
$x = 90°$

(c)

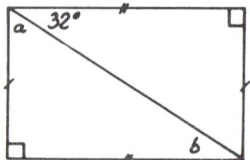

The figure is a rectangle, therefore each angle measures 90°.
Thus $a + 32° = 90°$
$a + 32° - 32° = 90° - 32°$
$a = 58°$

The diagonal acts as a transversal cutting the top and bottom sides of the rectangle, which are parallel. Thus, b is on the alternate side of the transversal to the angle of 32°. Alternate angles are equal so $b = 32°$.

(d)

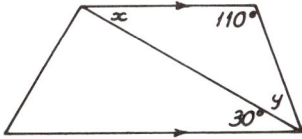

The figure is a trapezium since one pair of opposite sides is parallel. The diagonal acts as a transversal cutting the parallel sides.

Thus x is alternate to the angle measuring 30°.

$$\therefore x = 30°.$$

To find y we will need to consider the triangle with angles x, y and 110°.

We know the angle sum of a triangle is 180°.

We also know that $x = 30°$

$$\therefore y + 30° + 110° = 180°$$
$$y + 140° = 180°$$
$$y + 140° - 140° = 180° - 140°$$
$$y = 40°$$

Exercise 9L

1. Using the information given in each of the following diagrams,
   (i) state the properties shown in the diagram and hence,
   (ii) name the shape.

   (a)

   (b)

   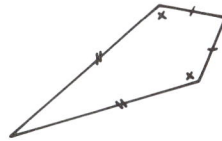

   (c)

   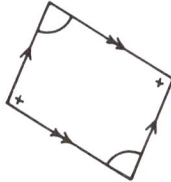

   (d)

   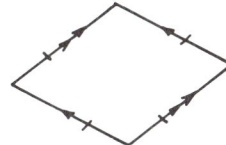

   (e)

   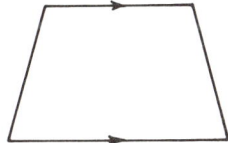

   (f)

   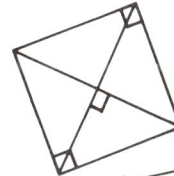

   (g)

   (h)

   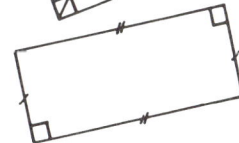

   (i)

   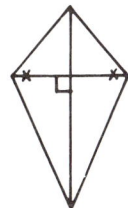

   (j)

   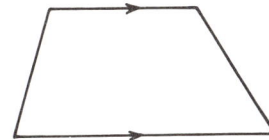

   (k)

   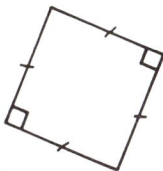

   (l)

   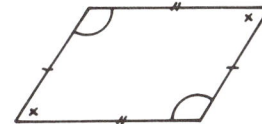

   (m)

   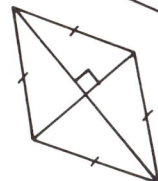

   (n)

   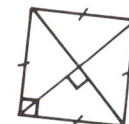

2. Name the following shapes and find the value of the pronumerals:

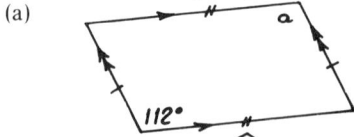

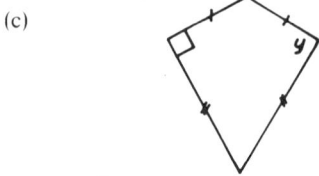

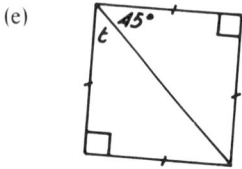

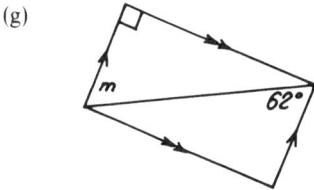

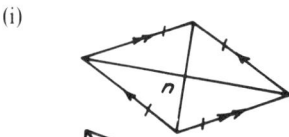

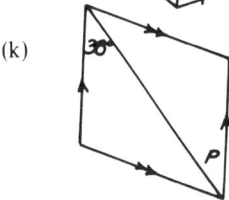

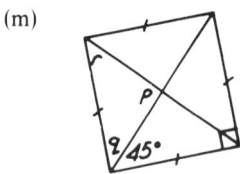

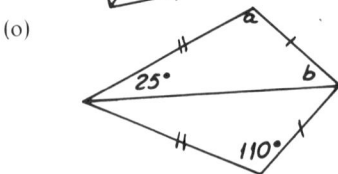

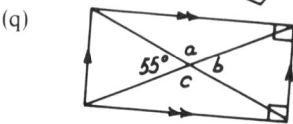

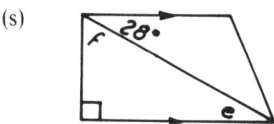

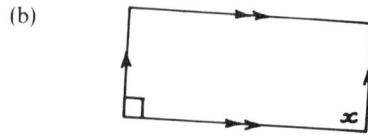

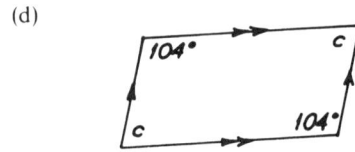

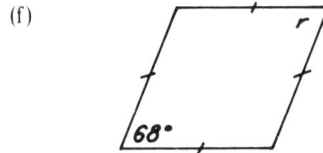

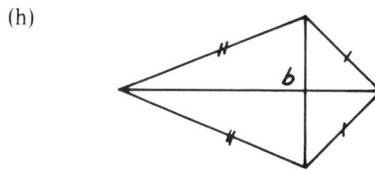

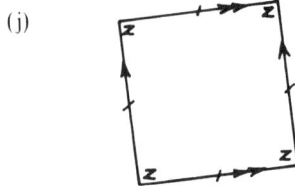

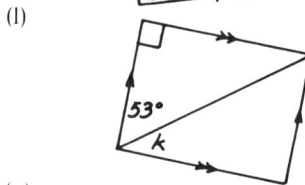

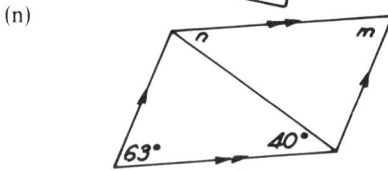

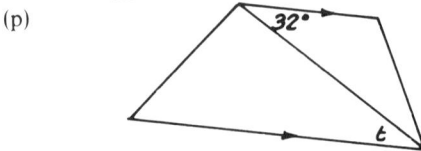

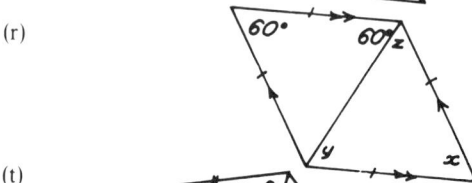

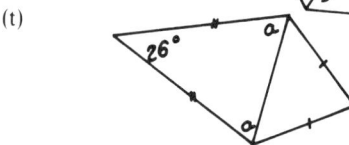

(a)

112°

(b)

x

(c)

y

(d)

104°   *c*

c   104°

(e)

45°
t

(f)

r

68°

(g)

m

62°

(h)

b

(i)

n

(j)

z   *z*

z   *z*

(k)

30°

P

(l)

53°
k

(m)

r

P

q 45°

(n)

n   *m*

63°   40°

(o)

a

25°   *b*

110°

(p)

32°

t

(q)

a
55°   *b*
c

(r)

60°   60° *z*

y

x

(s)

28°
f

e

(t)

26°   *a*

a

9.6 Circles

A *circle* is a set of points which are a fixed distance from a known point called the 'centre'.

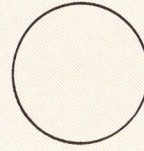

The *radius* of a circle is the fixed distance from the centre to any point on the circle.

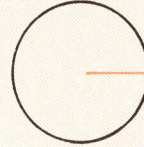

The *circumference* of a circle is its 'perimeter'; that is, the distance around the border.

A *chord* is a line joining any two points on the circle. This forms two *segments*, the smaller being called the minor and the larger, the major.

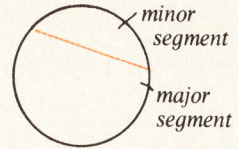

minor segment

major segment

A chord which passes through the centre point is the longest chord and is called the *diameter*.

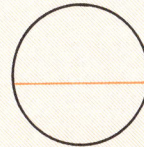

The diameter divides the circle into two equal segments called *semi-circles*.

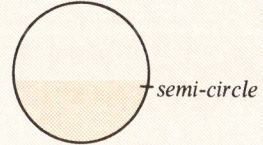

semi-circle

If two radii are drawn, they form two *sectors*, the smaller being called the minor and the larger, the major.

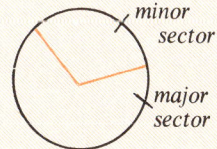

minor sector

major sector

If two diameters are drawn at right angles, the sectors formed are called *quadrants*.

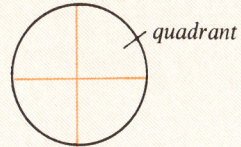

quadrant

Exercise 9M

1. Draw circles of radii 4 cm, 6 cm and 28 mm. In each one draw and measure a diameter.
2. Draw a circle of radius 8 cm. Using a protractor, draw sectors of 90°, 70° and 45° on your circle. Measure the size of the angle of the remaining sector.

9.7   Topology

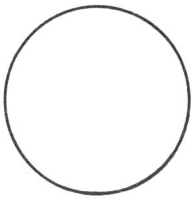

Fig 9.32 (a)

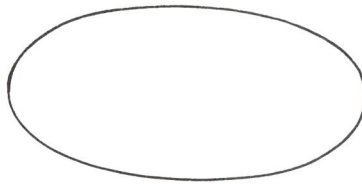

Fig 9.32 (b)

Fig 9.32 (c)

The three drawings in fig 9.32 have some common features.
They are all curves and they all have an inside and an outside; that is they are all *closed* curves.

We can turn fig 9.32(a) into fig 9.32(b) by stretching it:

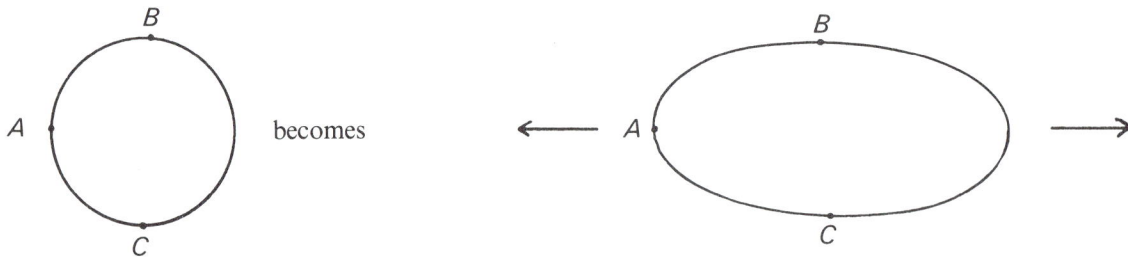

becomes

We have transformed fig 9.32(a) into fig 9.32(b).

We can transform fig 9.32(a) into fig 9.32(c) by stretching it in 4 directions at once:

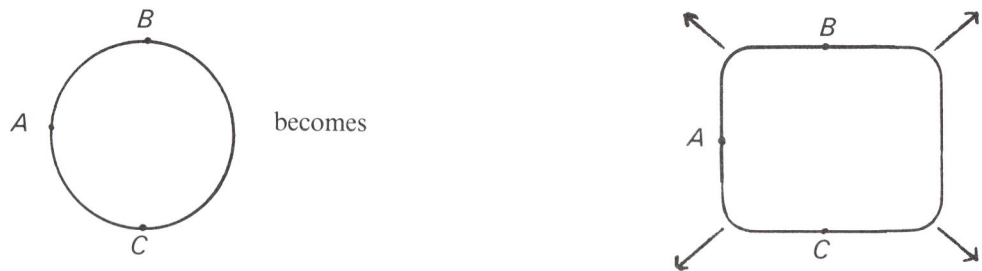

becomes

In topology we study such changes. It is often called 'rubber sheet geometry'.

9.7.1 Topological Transformations

Here are some more examples of topological transformations of a circle drawn on a rubber sheet:

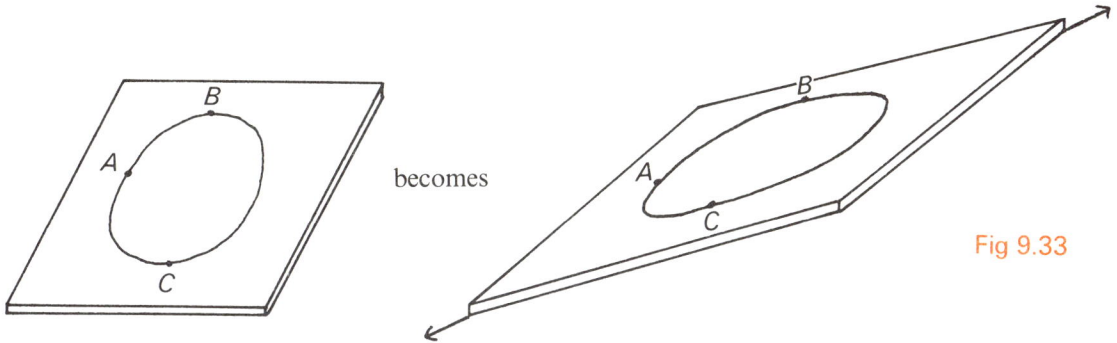

becomes

Fig 9.33

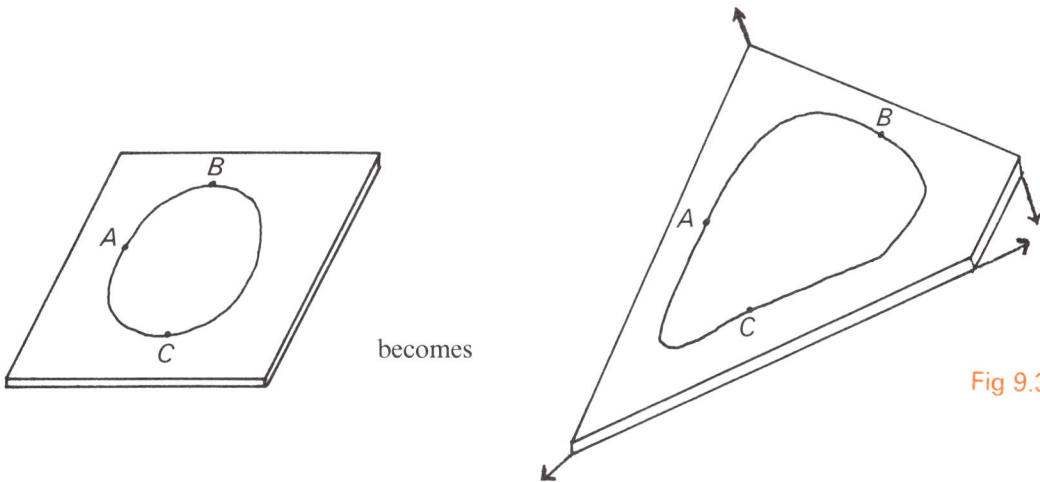

becomes

Fig 9.34

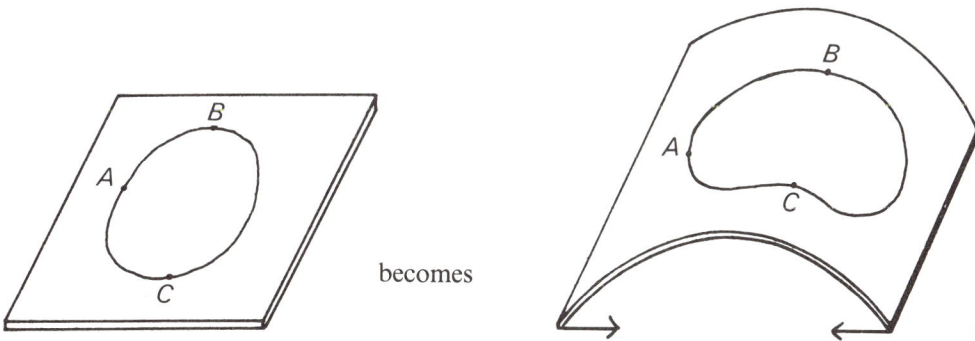

becomes

Fig 9.35

Any new shapes we get by bending or stretching the rubber sheet will be topologically equivalent to the circle.

The distance between points changes and the shape of the curve changes but the order of the points does not change. *A* is still between *B* and *C*.

A curve is closed if it begins and ends at the same point.
A curve is simple if it does not cross itself.

Any simple closed curved is topologically equivalent to the circle.

Exercise 9 N

1. Which of the following are closed curves?

(a)    (b)

(c) 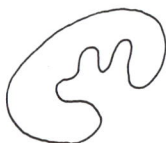   (d)

(e)    (f)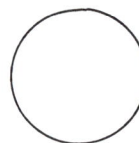

2. Which of the following are simple curves?

(a)    (b)

(c)    (d)

(e) 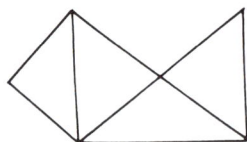   (f)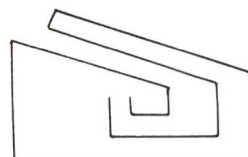

3. Which of the following are simple closed curves?

(a)    (b)

(c)

(d)

(e)

(f)

(g)

(h)

(i)

(j)

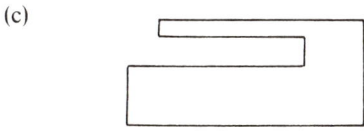

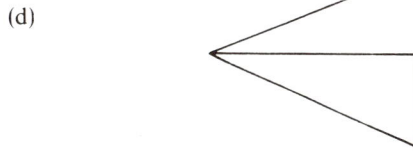

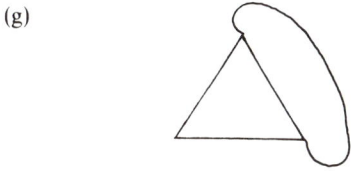

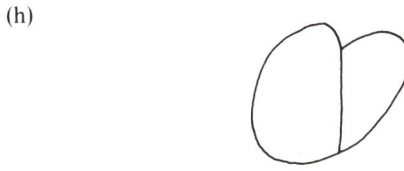

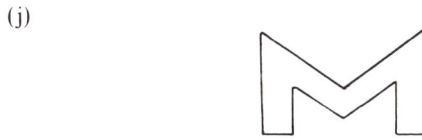

9.7.2 Regions, Nodes, Arcs and Networks

A simple closed curve divides the plane into two parts, inside the curve and outside the curve.

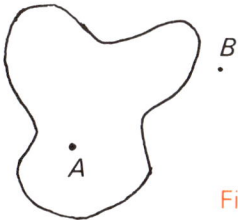

B

The point *A* is inside the curve

The point *B* is outside the curve

A

Fig 9.36

Exercise 9O

In the following drawings of simple closed curves, find whether the points *A* and *B* are in the same region of the plane (i.e. inside or outside the curve):

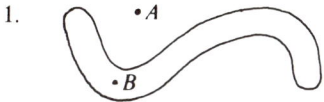

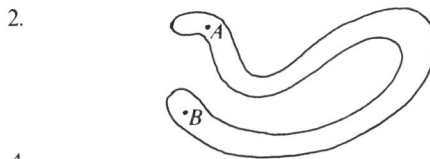

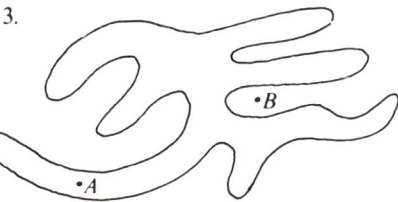

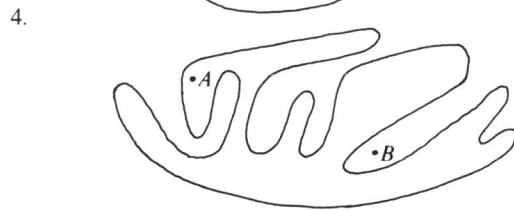

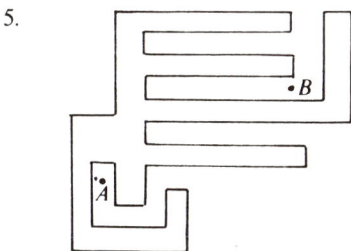

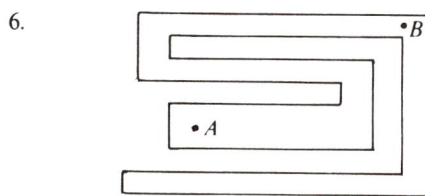

1. •*A*

   •*B*

2. •*A*

   •*B*

3. •*B*

   •*A*

4. •*A*

   •*B*

5. •*B*

   •*A*

6. •*B*

   •*A*

Any area which is enclosed by one or more curves is called a *region*.

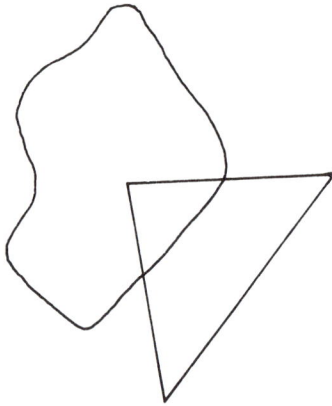

Three regions

1.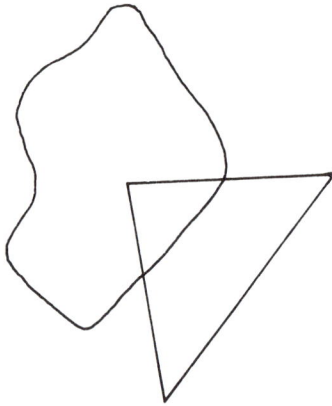
2.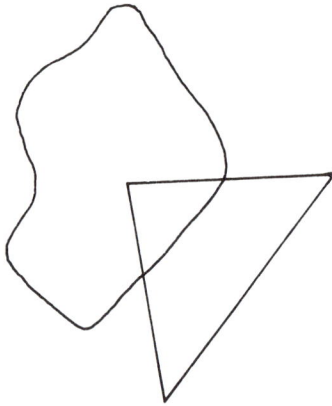
3.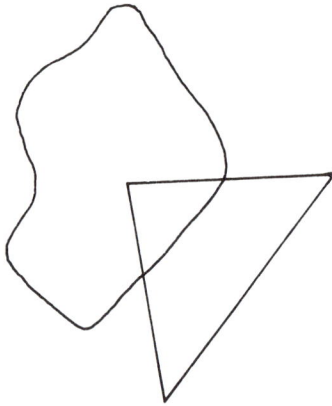

Fig 9.37

Where two curves cross is called a *node*. We also get a node at the end of a curve.

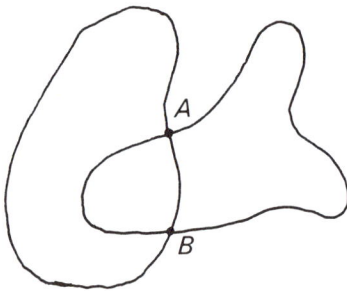

Three regions

Two nodes — *A* and *B*

Fig 9.38

Part of a curve which joins together two nodes is called an *arc*.

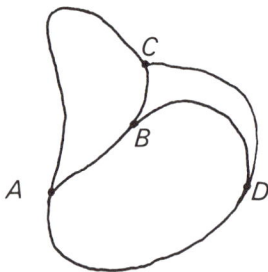

Three regions

Four nodes

Six arcs — *AC*, *AB*, *BC*, *BD*, *AD* and *CD*

Fig 9.39

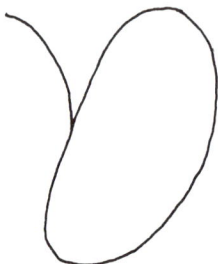

One region

Two nodes

Two arcs

Fig 9.40

Exercise 9P

For each of the following diagrams say:
    (i) how many regions
    (ii) how many nodes
    (iii) how many arcs

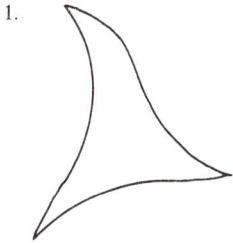

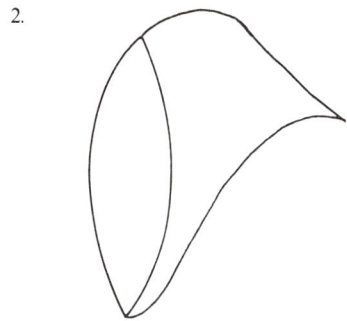

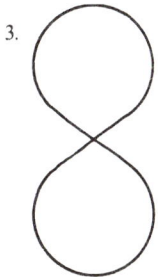

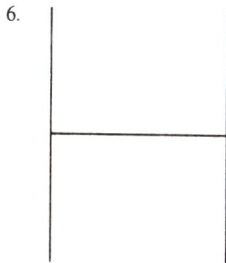

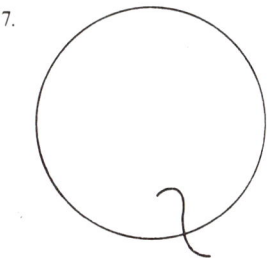

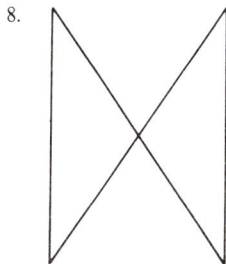

1.

2.

3.

4.

5.

6.

7.

8.

Each node has at least one arc going to it. The *order* of a node is the number of arcs which are going to it.

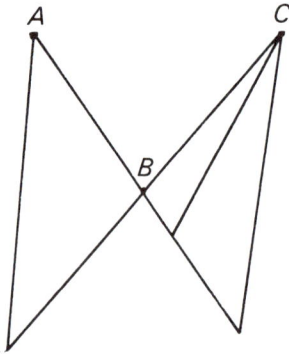

A is a node of order 2

B is a node of order 4

C is a node of order 3

Fig 9.41

If a node has even order, as with *A* and *B*, it is called an even node.
Other nodes are called odd nodes.

Exercise 9Q

For each of these networks, count the number of regions, nodes and arcs. Put your answers in a table like this:

EVEN NODES	ODD NODES	TOTAL NODES	REGIONS	ARCS

1.

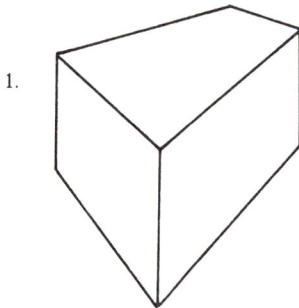

2.

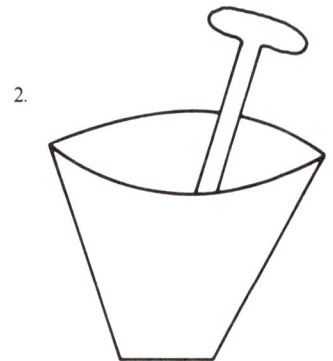

3.

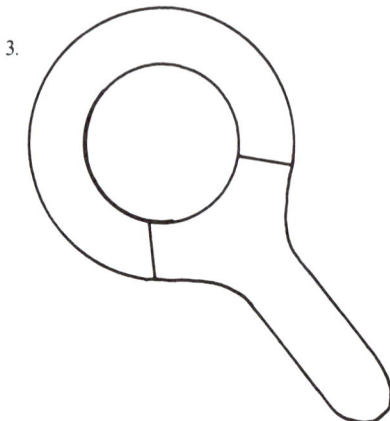

4.

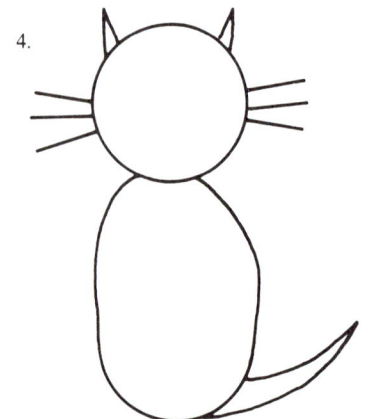

5.

6.

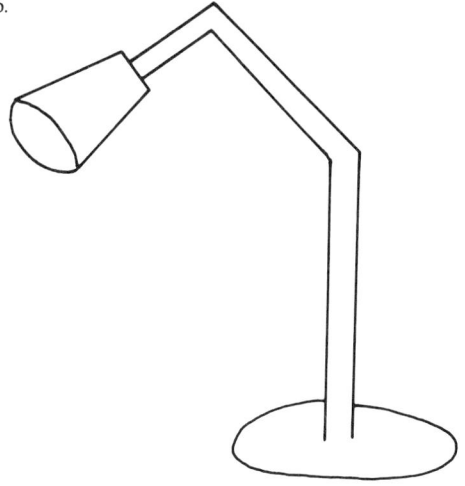

10 The Cartesian Plane

10.1 Using Directed Numbers for Position

The mathematician always tries to express things in the simplest shorthand manner. Thus, to describe the position of a point a mathematician might use the system of directed numbers (see Chapter 3), because a directed number is a number which denotes *both distance* and *direction* from a certain point. Remember that, on a number line (see fig. 10.1), a *positive* (+) number is one that lies to the right (East) of zero, and a *negative* (−) number is one that lies to the left (West) of zero.

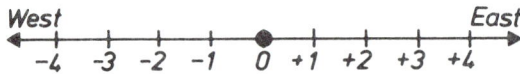

West ← −4 −3 −2 −1 0 +1 +2 +3 +4 → East   Fig 10.1

This idea of East being positive (+) and West being negative (−) would fit in well with the use of compass bearings. However, what about the directions North and South? Just as East and West are opposites of each other, so are North and South. We can thus use the idea of positive and negative for North and South. Taking North being positive and South being negative we obtain two number lines at right angles to each other intersecting at the zero points (*Origin*).

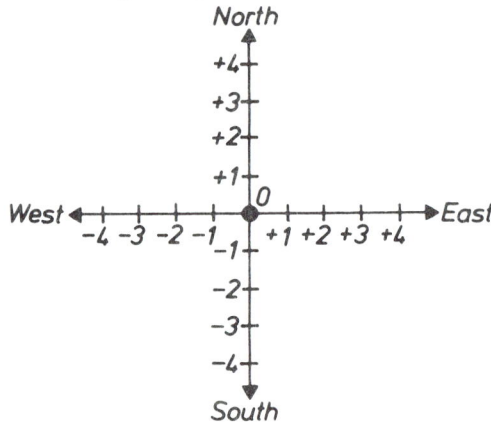

North
+4
+3
+2
+1
0
West ← −4 −3 −2 −1 +1 +2 +3 +4 → East   Fig 10.2
−1
−2
−3
−4
South

The position of a point on a plane surface can be described giving two measurements, one taken from each of two direction axes. Generally these two axes are at right angles to one another and the point at which the axes intersect is the *reference point* or *origin* of the system.

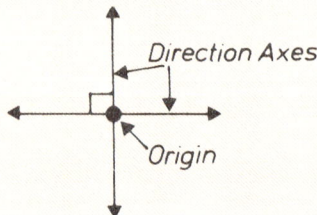

Direction Axes

Fig 10.3

Origin

Using this directed number system to describe the position of the point given in fig. 10.4,

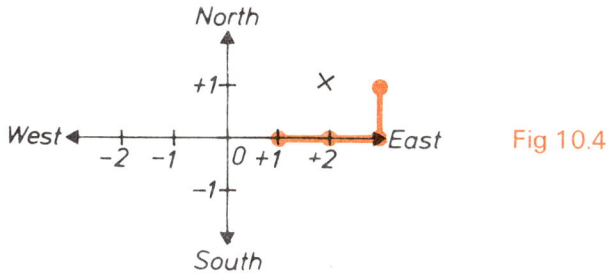

Fig 10.4

X is 2 steps East $(+2)$ and 1 step North $(+1)$ of the origin. Thus, we say that the position of the point is $(+2, +1)$. The position is given as an *ordered pair* of directed numbers, 'ordered' meaning that the East/West instruction is always written *first* and the North/South instruction written *second*.

Now, consider the point given in fig. 10.5.

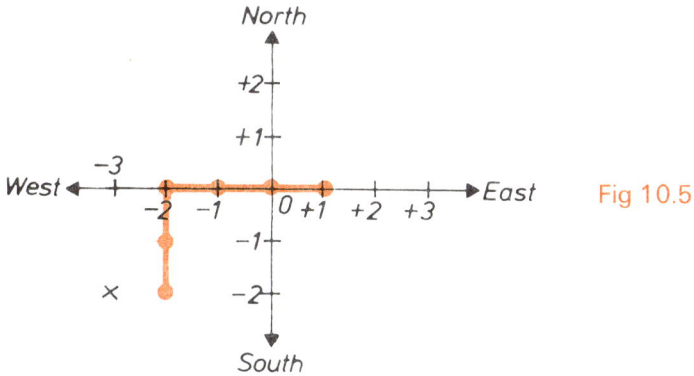

Fig 10.5

Here, the position of the point is $(-3, -2)$; that is, 3 steps *West* $(-)$ and 2 steps *South* $(-)$ of the origin.

For the point in fig. 10.6,

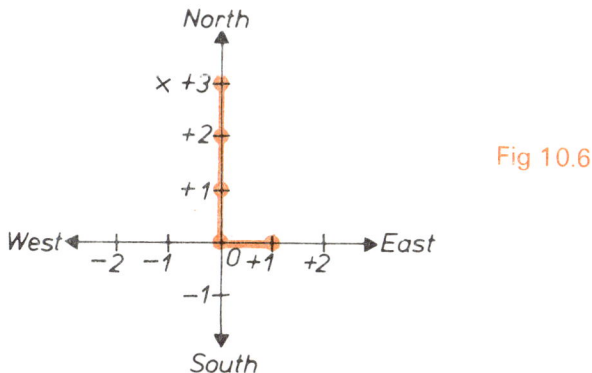

Fig 10.6

The position of the point is $(-1, +3)$; that is, 1 step *West* $(-)$ and 3 steps *North* $(+)$ of the origin. This system of ordered pairs of directed numbers is the simplest way to explain the position of a point on a plane.

Example 10.1

Taking the directions East and North as being positive and West and South being negative, describe the position of Town B from Town A as an ordered pair of directed numbers:

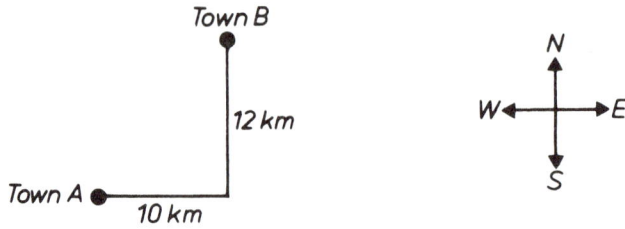

Fig 10.7

Solution

Since Town *A* is the reference point, it should represent the origin of the direction axes. The unit of measurement is kilometres.

Position of Town *B* is (+ 10, + 12).

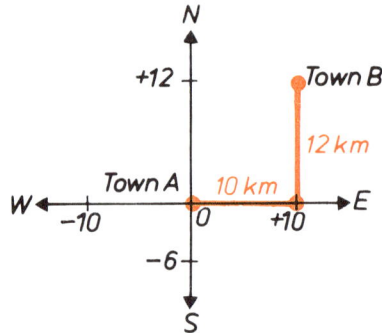

Fig 10.8

Exercise 10A

Taking the directions East and North as being positive and West and South being negative, describe the position of Town *B* from Town *A* as an ordered pair of directed numbers.

In each ordered pair give the East/West position *first*.

All distances are measured in km. Place *N, S, E, W*, axes on each diagram below, so that Town *A* is the reference point.

10.2 The Cartesian Plane

One method of describing the position of a point on a plane surface is, first, to draw two directed number lines at right angles to each other intersecting at the zero points. This intersection point is called the *origin* (denoted *0*) of the system and is the reference point for describing the position of any other point on the plane. Then, to state the position of a point on the plane, we give its horizontal and vertical distance and direction from the origin (see 10.1); that is, how far to the right or left of the origin and how far up or down from the origin does the point lie?

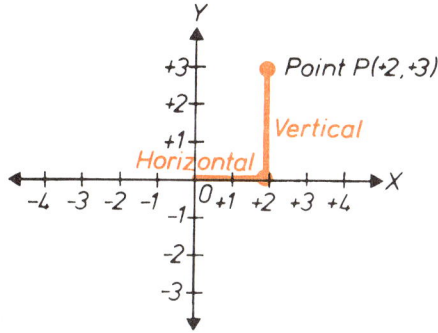

Fig 10.9

This method of describing the position of a point on a plane was that used by a man called Déscartes, after whom the system was named. In his honour, the plane on which the axes (number lines) are drawn is called the *Cartesian Plane*. The horizontal axis is called the X-axis, with the vertical axis called the Y-axis (see fig 10.9). This naming of the axes allows the system to refer to any plane and thus not be restricted in its use (unlike the N, S, E, W, axes in 10.1).

10.2.1 Definitions

Referring to the Cartesian Plane given in fig 10.9,

1. The X-axis is a horizontal number line while the Y-axis is a vertical number line.
2. The X and Y-axes intersect at a point called the *origin* (denoted by the letter *0*). The origin is the reference point for describing the position of any point on the plane.
3. On the X-axis any point to the *right* of the origin has a *positive* director sign ($+$), while any point to the *left* of the origin has a *negative* director sign ($-$).
4. On the Y-axis any point *above* the origin has a *positive* director sign ($+$), while any point *below* the origin has a *negative* director sign ($-$).
5. Both the X and Y-axes extend indefinitely either side of the origin.
6. Since the position of a point on the plane is given in terms of its horizontal and vertical distance and direction from the origin, any point can be seen as being part of a rectangular grid network on the plane (see fig 10.10 below).

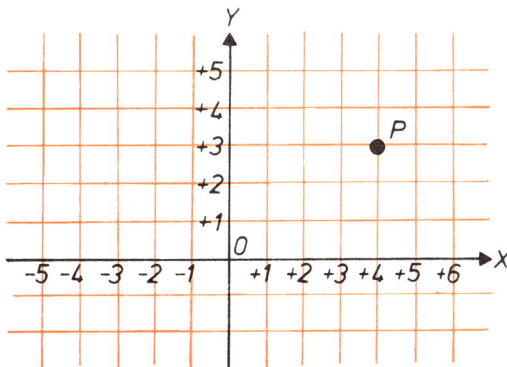

Fig 10.10

10.2.2   The Position of a Point on the Plane

As already stated, the position of a point on the Cartesian Plane is given in terms of its horizontal and vertical distance and direction from the origin. Thus, to name the position of a point we need two directed measurements. These two measurements are given in the form of an *ordered pair* (see 10.1). The *first* measurement given is called the *X-coordinate* or *abcissa* and measures the *horizontal* distance and direction of the point from the origin (see fig 10.11 (a)), that is, how far to the left or right of the origin is the point? The *second* measurement is called the *Y-coordinate* or *ordinate* and measures the *vertical* distance and direction of the point from the origin (see fig 10.11 (b)); that is, how far above or below the origin is the point?

Fig 10.11 (a) and (b) below show the two steps required to arrive at the point $P(+2, +3)$.

First move 2 units to the *right* (+)   of the origin,

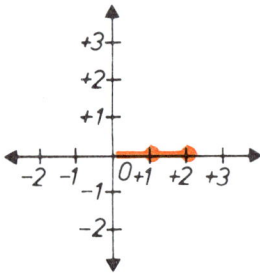

Second, move 3 units *upwards* (+).

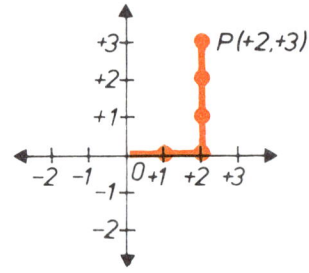

THEN

Fig 10.11 (a)

Fig 10.11 (b)

From fig 10.11 (b) it can be seen that the point P is 2 units to the right of the origin *and* 3 units up from the origin.

Rather than say 'the point P has position $(+2, +3)$', we simply say 'the *coordinates* of P are $(+2, +3)$'.

The *coordinates* of any point on the Cartesian plane are the *ordered pair* of directed numbers which describe the position of the point on the plane.

The coordinates of the origin, *0*, are (0, 0).

Example 10.2

Give the (i)   *X*-coordinate
         (ii)   *Y*-coordinate
         (iii) coordinates
of the points marked on the Cartesian Plane below.

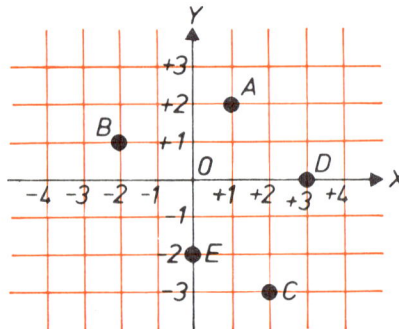

Fig 10.12

Solution

For the point marked A:

(i) the X-coordinate is $+1$; i.e., one unit to the right of the origin.

(ii) the Y-coordinate is $+2$; i.e., 2 units above the origin.

(iii) the coordinates are $(+1, +2)$.

For the point marked B:

(i) X-coordinate is -2.

(ii) Y-coordinate is $+1$.

(iii) coordinates are $(-2, +1)$.

For the point marked C:

(i) X-coordinate is $+2$.

(ii) Y-coordinate is -3.

(iii) coordinates are $(+2, -3)$.

For the point marked D:

(i) X-coordinate is $+3$.

(ii) Y-coordinate is 0; i.e., the point is neither above nor below the origin.

(iii) coordinates are $(+3, 0)$.

For the point marked E:

(i) X-coordinate is 0; i.e., the point is neither to the left nor to the right of the origin.

(ii) Y-coordinate is -2.

(iii) coordinates are $(0, -2)$.

Usually, the positive director signs are left off so that any number without a director sign is intended to be positive. For example, the point $(3, 1)$ should be read as $(+3, +1)$ and the point $(-2, 5)$ read as $(-2, +5)$.

A pictorial summary of the Cartesian Plane and coordinate system:

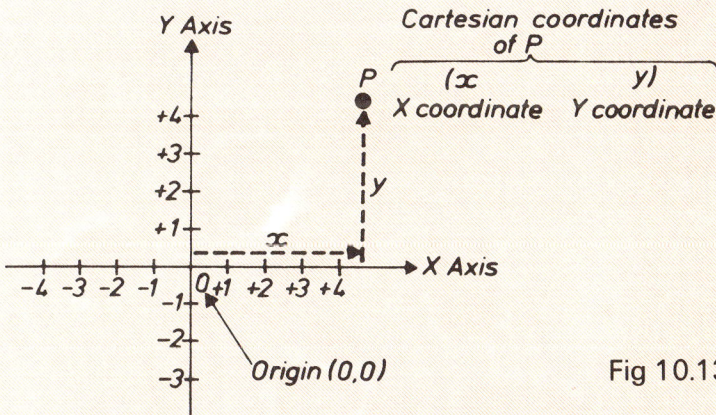

Fig 10.13

Exercise 10 B

1. Give the (i) *X*-coordinate (ii) *Y*-coordinate (iii) set of Cartesian coordinates for each of the points *A* to *L* inclusive marked on the Cartesian Plane below.

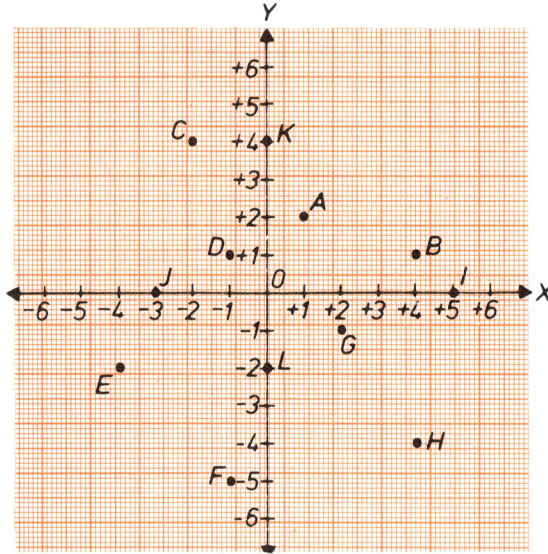

Fig 10.14

2. Give the Cartesian coordinates of the points *A* to *L* inclusive in fig 10.15 below.

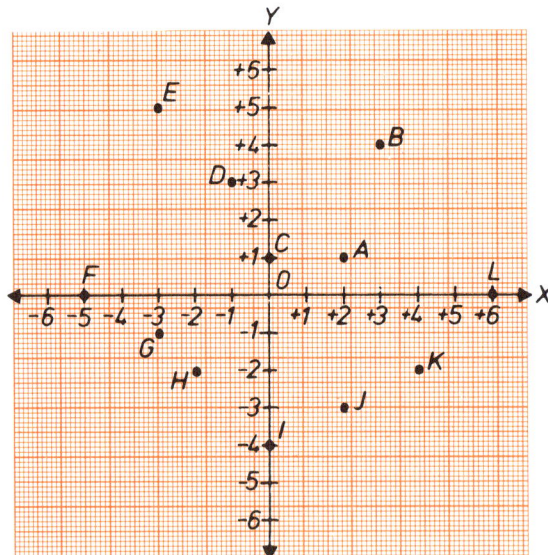

Fig 10.15

3. On a piece of graph paper draw up a set of Cartesian Axes so that the *X* and *Y*-axes can each be marked off from -5 to $+5$. On this graph paper mark the following points:
   (a) *A* $(+3, +1)$
   (b) *B* $(+4, +5)$
   (c) *C* $(+2, 0)$
   (d) *D* $(0, 0)$
   (e) *E* $(-4, +2)$
   (f) *F* $(-3, +1)$
   (g) *G* $(-2, 0)$
   (h) *H* $(-4, 0)$
   (i) *I* $(+3, -2)$
   (j) *J* $(+2, -5)$
   (k) *K* $(0, -4)$
   (l) *L* $(0, +1)$
   (m) *M* $(-4, -1)$
   (n) *N* $(-3, -2)$

4. Draw a set of *X*–*Y* axes on a piece of graph paper with both the *X* and the *Y*-axes extending from -5 to $+5$. Mark in the following points:
   (a) *M* $(5, 3)$
   (b) *N* $(4, -2)$

(c) $P(-1,3)$
(e) $R(0,2)$
(g) $T(1,-4)$
(i) $V(-3,-1)$

(d) $Q(-3,-5)$
(f) $S(5,0)$
(h) $U(-2,1)$
(j) $W(-1,0)$

5. By drawing a set of $X-Y$ axes, a plane is effectively broken up into 4 quarters (quadrants) as shown in fig 10.16 below.

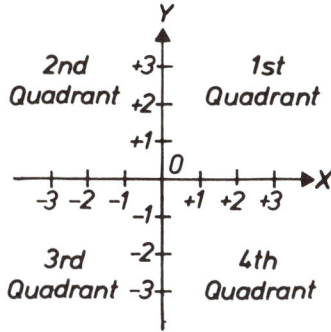

Fig 10.16

Using fig 10.16, copy and complete the following table:

Point	X-coord positive or negative	Y-coord positive or negative	Quadrant
$A(+3,+2)$	positive	positive	1st
$B(-1,+4)$	negative	positive	2nd
$C(+2,-3)$			
$D(-4,-1)$			
$E(2,-5)$			
$F(-3,1)$			
$G(5,-2)$			
$H(-6,-3)$			
$I(-2,4)$			
$J(1,5)$			
$K(4,-1)$			

6. (a) Draw up a set of $X-Y$ axes on a piece of graph paper and mark in the following points:
(i) $A(0,4)$ (ii) $B(2,1)$ (iii) $C(5,1)$
(iv) $D(3,-2)$ (v) $E(4,-5)$ (vi) $F(0,-3)$
(vii) $G(-4,-5)$ (viii) $H(-3,-2)$ (ix) $I(-5,1)$
(x) $J(-2,1)$
(b) Join the points given in (a) by a series of straight lines; that is, join A to B, B to C, C to D, etc, finishing with J back to A.
(c) Describe the figure obtained by joining the dots given.

7. (a) Draw up a set of $X-Y$ axes on a piece of graph paper and mark in the following points:
(i) $A(5,2)$ (ii) $B(8,4)$ (iii) $C(5,1)$
(iv) $D(5,-4)$ (v) $E(3,-4)$ (vi) $F(3,-1)$
(vii) $G(-2,-1)$ (viii) $H(-2,-4)$ (ix) $I(-4,-4)$

(x) $J(-4, 1)$ (xi) $K(-5, 3)$ (xii) $L(-6, 2)$
(xiii) $M(-8, 2)$ (xiv) $N(-8, 3)$ (xv) $P(-6, 5)$
(xvi) $Q(-5, 5)$ (xvii) $R(-5, 6)$ (xviii) $S(-4, 5)$
(xix) $T(-2, 2)$ (xx) $U(5, 2)$

(b) Join the points given in (a) by a series of straight lines; that is, join A to B, B to C, C to D, etc, finishing with T to U.

(c) Describe the figure obtained by joining the dots given.

8. (a) On a set of $X-Y$ axes, mark in the following points:
   (i) $A(-3, -6)$ (ii) $B(-2, -4)$ (iii) $C(-1, -2)$
   (iv) $D(0, 0)$ (v) $E(1, 2)$ (vi) $F(2, 4)$
   (vii) $G(3, 6)$ (viii) $H(4, 8)$

   (b) Join the points given in (a); that is, join A to B to C to D, etc, finishing with G to H.

   (c) Describe the pattern obtained by joining the given points.

9. (a) On a set of $X-Y$ axes, mark in the following points:
   (i) $A(-4, -2)$ (ii) $B(-3, -1)$ (iii) $C(-2, 0)$
   (iv) $D(-1, 1)$ (v) $E(0, 2)$ (vi) $F(1, 3)$
   (vii) $G(2, 4)$ (viii) $H(3, 5)$

   (b) Join the points given in (a); that is, join A to B to C to D, etc, finishing with G to H.

   (c) Describe the pattern obtained by joining the given points.

10.3   Linear Patterns and Relationships

The last four questions in the preceding exercise (10 B) involved plotting points on the Cartesian Plane and then observing the pattern formed by these points. In particular, the last two questions gave rise to *linear* patterns; that is, the points plotted all lay in the same straight line. In this section we will be looking further into linear patterns.

Example 10.3

The set of points $\{(-3, -3)\ (-2, -2)\ (-1, -1)\ (0, 0)\ (1, 1)\}$ represents a linear pattern on the Cartesian Plane. By plotting these points and using a ruler, find the coordinates of two more points in the pattern.

Solution

$\{(-3, -3)(-2, -2)(-1, -1)(0, 0)(1, 1)\}$

Fig 10.17

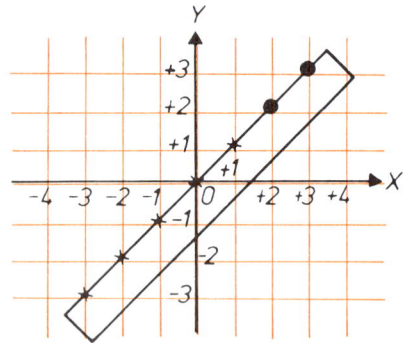

Using the straight edge of a ruler and the given points, it can be seen that two points in the pattern could be $(2, 2)$ and $(3, 3)$.

In example 10.3 above, need we have used a ruler to find the coordinates of the additional points or is there another way? In this case, the pattern is obvious merely by looking at the coordinates of the given points. In each pair of coordinates the X-coordinate and Y-coordinate are equal:

$$(-3, -3)\quad (-2, -2)\quad (-1, -1)\quad (0, 0)\quad (1, 1)$$
$$\;x = y\qquad\;\;x = y\qquad\;\;x = y\qquad x = y\quad\;x = y$$

If the pattern is obvious simply by looking at the *relationship* (connection) between each pair of x and y given, then it is unnecessary to plot the points on graph paper in order to find any additional points.

Exercise 10 C

1. State whether or not the following Cartesian patterns are linear patterns.

(a)

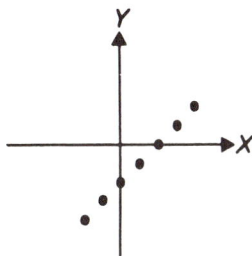

(b)

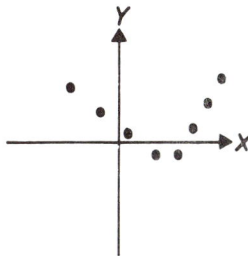

(c)

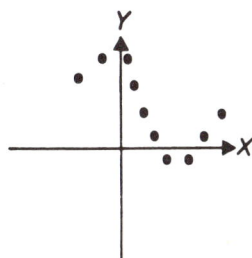

(d)

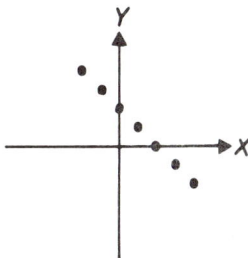

(e)

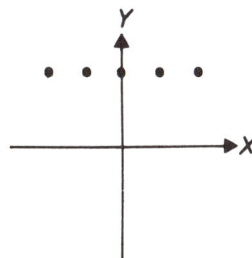

(f)

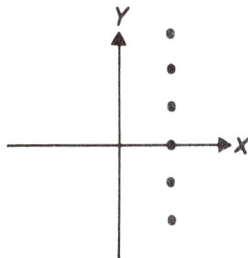

(g)

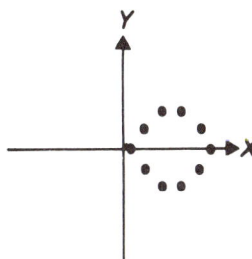

(h)

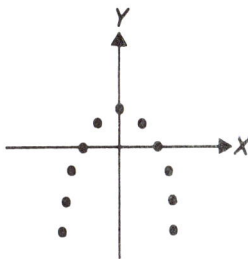

2. Plot each of the following sets of points on separate pieces of graph paper and then state whether or not the patterns formed are linear.

   (a) $\{(-2, -4), (-1, -2), (0, 0), (1, 2), (2, 4)\}$
   (b) $\{(-2, 4), (-1, 1), (0, 0), (1, 1), (2, 4)\}$
   (c) $\{(-2, -8), (-1, -1), (0, 0), (1, 1), (2, 8)\}$
   (d) $\{(-2, -1), (-1, 0), (0, 1), (1, 2), (2, 3)\}$
   (e) $\{(-2, 3), (-1, 3), (0, 3), (1, 3), (2, 3)\}$
   (f) $\{(-1, -2), (-1, -1), (-1, 0), (-1, 1), (-1, 2)\}$
   (g) $\{(-2, -5), (-1, -3), (0, -1), (1, 1), (2, 3)\}$
   (h) $\{(-2, 4), (-1, 3), (0, 2), (1, 1), (2, 0)\}$
   (i) $\{(-2, -8), (-1, -2), (0, 0), (1, -2), (2, -8)\}$
   (j) $\{(-2, 0), (-1, 1), (-1, -1), (0, 2), (0, -2)\}$

3. Each of the following sets of points represents a linear pattern on the Cartesian Plane. By plotting each set of points and using a ruler, find the coordinates of two more points in the pattern.

   (a) $\{(-3, -6), (-2, -4), (-1, -2), (0, 0), (1, 2)\}$
   (b) $\{(-3, 3), (-2, 2), (-1, 1), (0, 0), (1, -1)\}$
   (c) $\{(-3, 9), (-2, 6), (-1, 3), (0, 0), (1, -3)\}$
   (d) $\{(-3, -15), (-2, -10), (-1, -5), (0, 0), (1, 5)\}$
   (e) $\{(-3, -2), (-2, -1), (-1, 0), (0, 1), (1, 2)\}$
   (f) $\{(-3, -5), (-2-4), (-1, -3), (0, -2), (1, -1)\}$
   (g) $\{(-3, -6), (-2, -5), (-1, -4), (0, -3), (1, -2)\}$
   (h) $\{(-3, -1), (-2, 0), (-1, 1), (0, 2), (1, 3)\}$

(i) $\{(-3, 4), (-2, 3), (-1, 2), (0, 1), (1, 0)$
(k) $\{(-3, -7), (-2, -5), (-1, -3), (0, -1), (1, 1)\}$
(m) $\{(-3, 9), (-2, 7), (-1, 5), (0, 3), (1, 1)\}$

(j) $\{(-3, 0), (-2, -1), (-1, -2), (0, -3), (1, -$
(l) $\{(-3, -7), (-2, -4), (-1, -1), (0, 2), (1,5)\}$
(n) $\{(-3, 8), (-2, 5), (-1, 2), (0, -1), (1, -4)\}$

4. (a) *Without* plotting the linear patterns given below, try to find, by inspection, the connection relationship) between each x and y-coordinate in each of the patterns.

Hence, or otherwise, find the coordinates of two more points for each pattern.

(i) $\{(-3, -6), (-2, -4), (-1, -2), (0, 0), (1, 2), (2, 4), (3, 6)\}$
(ii) $\{(-3, -9), (-2, -6), (-1, -3), (0, 0), (1, 3), (2, 6), (3, 9)\}$
(iii) $\{(-3, 3), (-2, 2), (-1, 1), (0, 0), (1, -1), (2, -2), (3, -3)\}$
(iv) $\{(-3, 6), (-2, 4), (-1, 2), (0, 0), (1, -2), (2, -4), (3, -6)\}$
(v) $\{(-3, -2), (-2, -1), (-1, 0), (0, 1), (1, 2), (2, 3), (3, 4)\}$
(vi) $\{(-3, -1), (-2, 0), (-1, 1), (0, 2), (1, 3), (2, 4), (3, 5)\}$
(vii) $\{(-3, 2), (-2, 3), (-1, 4), (0, 5), (1, 6), (2, 7), (3, 8)\}$
(viii) $\{(-3, -5), (-2, -4), (-1, -3), (0, -2), (1, -1), (2, 0), (3, 1)\}$
(ix) $\{(-3, -6), (-2, -5), (-1, -4), (0, -3), (1, -2), (2, -1), (3, 0)\}$
(x) $\{(-3, 4), (-2, 3), (-1, 2), (0, 1), (1, 0), (2, -1), (3, -2)\}$

(b) Plot the complete linear patterns obtained in (a).

10.3.1   Finding Rules for Linear Relationships

> A *linear relationship* between x and y in an ordered number pair gives a set of points which, when plotted, form a linear pattern.

Instead of *listing* a sequence of points which form a linear relationship, it is simpler to state a *rule* which describes all points in the relationship.

Consider the linear relationship given in example 10.3. Listing the points in the relationship we have $(-3, -3)(-2, -2)(-1, -1)(0, 0)(1, 1)(2, 2)$ and $(3, 3)$. However, it would be far simpler to say that each point conforms to the rule $y = x$.

Example 10.4

Find the rule which describes the linear relationship given by the set of points $(-3, -2)(-2, -1)$ $(-1, 0)(0, 1)(1, 2)(2, 3)$.

Solution

It may be easier to find the rule for this linear relationship by concentrating first on the points whose coordinates are positive.

Consider the points $(0, 1)(1, 2)(2, 3)$ first.

In each pair of coordinates the y value is one more than the x value.

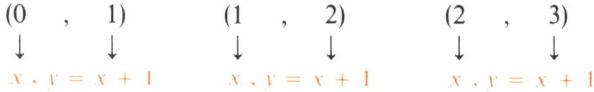

$$(0 \quad , \quad 1) \qquad (1 \quad , \quad 2) \qquad (2 \quad , \quad 3)$$
$$\downarrow \qquad \downarrow \qquad\quad \downarrow \qquad \downarrow \qquad\quad \downarrow \qquad \downarrow$$
$$x, y = x + 1 \qquad x, y = x + 1 \qquad x, y = x + 1$$

Now check that the points with negative coordinates also conform to this rule:

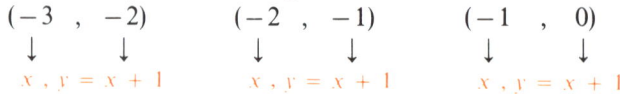

$$(-3 \quad , \quad -2) \qquad (-2 \quad , \quad -1) \qquad (-1 \quad , \quad 0)$$
$$\downarrow \qquad \downarrow \qquad\quad \downarrow \qquad \downarrow \qquad\quad \downarrow \qquad \downarrow$$
$$x, y = x + 1 \qquad x, y = x + 1 \qquad x, y = x + 1$$

These points do agree with the rule, and so the general rule for this particular linear relationship is

$$y = x + 1.$$

It may also be a help to actually plot the given points so that you have a pictorial view of the relationship. Plotting the given points,

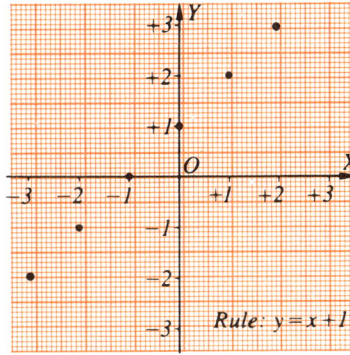

Fig 10.18

If the linear relationship shown in fig 10.18 above were redrawn but with a straight line passing through the given points (see fig 10.19), what effect on the relationship would this have? In fact, the rule, $y = x + 1$, would be the same. However, the full line means that we want not only points with whole number coordinates but also every other point in between. For example, the full line includes points such as $(1.5, 2.5)$ and $(0.7, 1.7)$. The only restriction is that the y coordinate must be one more than the x coordinate:

$$(0.7 \quad , \quad 1.7)$$
$$\downarrow \qquad \downarrow$$
$$x, y = x + 1$$

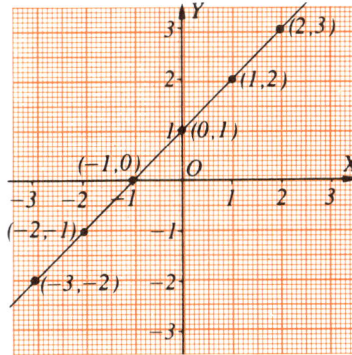

Fig 10.19

The complete rule for this relationship can be set out in the form:

$$\{ \qquad (x, y) \qquad : \qquad y = x + 1 \}$$
$$\uparrow \qquad \uparrow \qquad \uparrow \qquad \uparrow$$

the set of   *all* points   such that   the rule is $y = x + 1$

Example 10.5
Find the rule for the linear relationship given in fig 10.20 below.

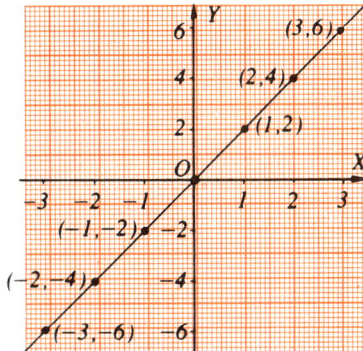

Fig 10.20

Solution

Looking at the points $(1, 2)$ $(2, 4)$ and $(3, 6)$ we see that, in each case, the y value is double the x value:

$$(1 \ , \ 2) \qquad (2 \ , \ 4) \qquad (3 \ , \ 6)$$
$$\downarrow \quad \downarrow \qquad \downarrow \quad \downarrow \qquad \downarrow \quad \downarrow$$
$$x, \ y = 2x \qquad x, \ y = 2x \qquad x, \ y = 2x$$

This rule also applies to the other points $(-3, -6)$, $(-2, -4)$, $(-1, -2)$ and $(0, 0)$. Since we want *all* points (full line) that conform to the rule $y = 2x$, the relationship can be described by:

$$\{(x, y): y = 2x\}$$

Exercise 10D

1. Find the rule which describes the linear relationship given by each of the following sets of points:
   (a) $\{(-3, -6), (-2, -4), (-1, -2), (0, 0), (1, 2), (2, 4), (3, 6)\}$
   (b) $\{(-3, -9), (-2, -6), (-1, -3), (0, 0), (1, 3), (2, 6), (3, 9)\}$
   (c) $\{(-3, 3), (-2, 2), (-1, 1), (0, 0), (1, -1), (2, -2), (3, -3)\}$
   (d) $\{(-3, 6), (-2, 4), (-1, 2), (0, 0), (1, -2), (2, -4), (3, -6)\}$
   (e) $\{(-3, -2), (-2, -1), (-1, 0), (0, 1), (1, 2), (2, 3), (3, 4)\}$
   (f) $\{(-3, -1), (-2, 0), (-1, 1), (0, 2), (1, 3), (2, 4), (3, 5)\}$
   (g) $\{(-3, 2), (-2, 3), (-1, 4), (0, 5), (1, 6), (2, 7), (3, 8)\}$
   (h) $\{(-3, -5), (-2, -4), (-1, -3), (0, -2), (1, -1), (2, 0), (3, 1)\}$
   (i) $\{(-3, -6), (-2, -5), (-1, -4), (0, -3), (1, -2), (2, -1), (3, 0)\}$
   (j) $\{(-3, 4), (-2, 3), (-1, 2), (0, 1), (1, 0), (2, -1), (3, -2)\}$
   (k) $\{(-3, -7), (-2, -6), (-1, -5), (0, -4), (1, -3), (2, -2), (3, -1)\}$
   (l) $\{(-3, 4), (-2, 5), (-1, 6), (0, 7), (1, 8), (2, 9), (3, 10)\}$
   (m) $\{(-3, 5), (-2, 4), (-1, 3), (0, 2), (1, 1), (2, 0), (3, -1)\}$
   (n) $\{(-3, 0), (-2, -1), (-1, -2), (0, -3), (1, -4), (2, -5), (3, -6)\}$
   (o) $\{(-3, -5), (-2, -3), (-1, -1), (0, 1), (1, 3), (2, 5), (3, 7)\}$
   (p) $\{(-3, -11), (-2, -8), (-1, -5), (0, -2), (1, 1), (2, 4), (3, 7)\}$

2. Find the rule which describes each of the following linear relationships. Give answer in full set notation form.

(a)

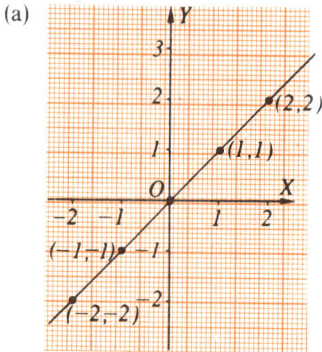

(b)

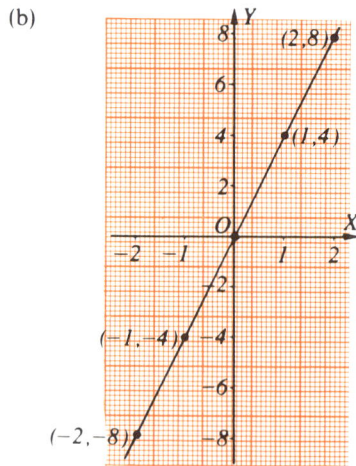

(c)

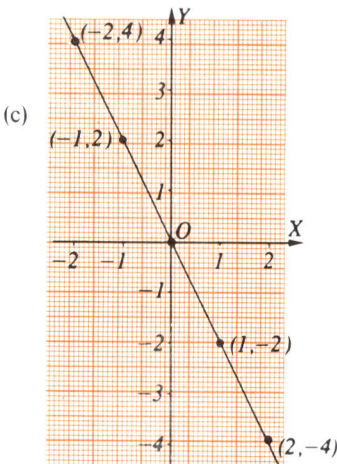

(d)

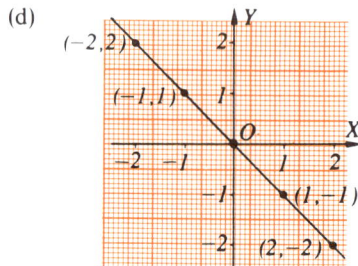

(e)

(2,5)
(1,4)
(0,3)
(−1,2)
(−2,1)

(f)

(2,6)
(1,5)
(0,4)
(−1,3)
(−2,2)

(g)

(2,1)
(1,0)
(0,−1)
(−1,−2)
(−2,−3)

(h)

(2,−3)
(1,−4)
(0,−5)
(−1,−6)
(−2,−7)

(i)

(−2,5)
(−1,4)
(0,3)
(1,2)
(2,1)

(j)

(−2,1)
(−1,0)
(0,−1)
(1,−2)
(2,−3)

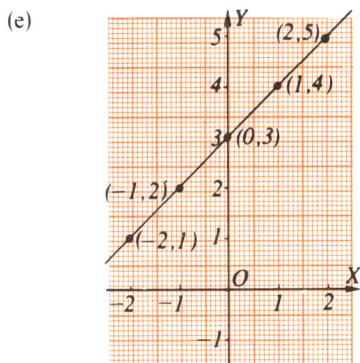

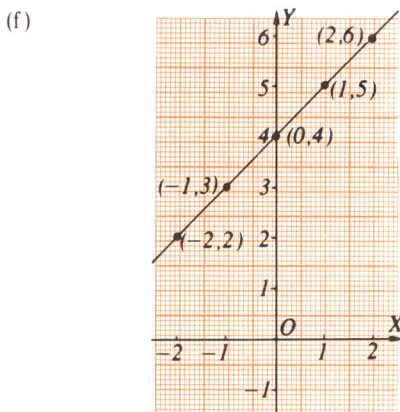

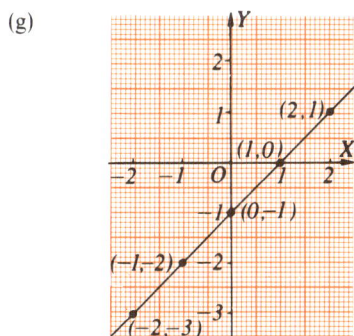

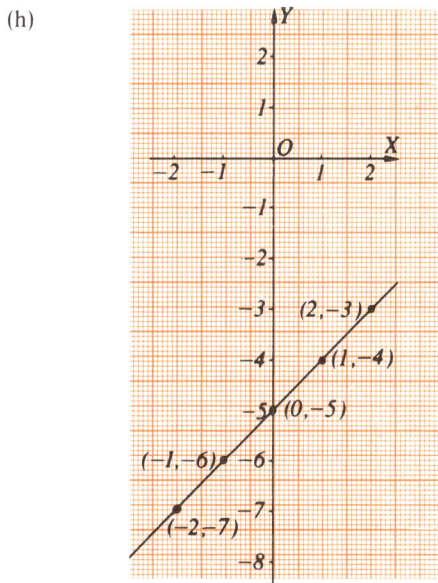

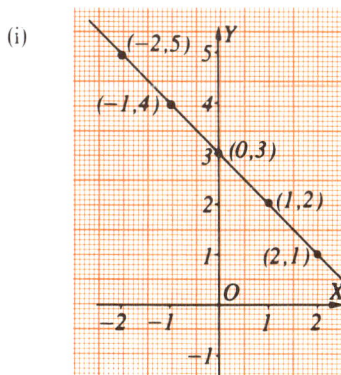

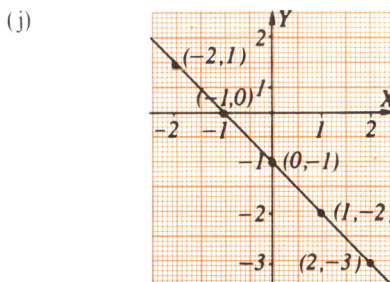

10.3.2   Plotting Linear Relationships

So far, we have been given a set of points from which we determined the rule describing their linear relationship. Now we shall work in reverse. Given the rule, we shall find a set of points belonging to the relationship and then plot these points on the Cartesian Plane.

Example 10.6

Plot the linear relationship $\{(x, y): y = 3x; \; -3 \leq x \leq 3\}$

Solution

Firstly, the set notation used to describe the relationship means:

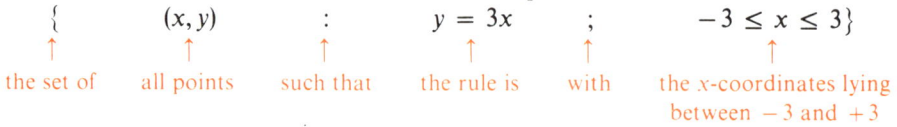

$$\{ \qquad (x,y) \qquad : \qquad y = 3x \qquad ; \qquad -3 \le x \le 3\}$$

the set of   all points   such that   the rule is   with   the x-coordinates lying between -3 and $+3$

The restriction on the x-coordinates $(-3 \le x \le 3)$ merely serves to indicate what part of the line $y = 3x$ is required.

To plot this straight line, we need only find and plot the coordinates of *some* points which belong to the given relationship and then join these points with a full line. Since the relationship states that we want x to lie between -3 and $+3$, the easiest points to find would be those whose x-coordinates are $-3, -2, -1, 0, 2$ and 3. In other words, we want to complete the following table of values using the given rule $y = 3x$:

x	-3	-2	-1	0	1	2	3
y							
(x,y)							

Table 10.1

To complete table 10.1 above, we merely substitute each x value into the rule $y = 3x$.

Thus, when $\quad x = -3, y = 3x$
$$= 3 \times -3$$
$$= -9$$

which gives us the point $(-3, -9)$.

Similarly, when $x = -2, y = 3x$
$$= 3 \times -2$$
$$= -6 \qquad \therefore \text{ point is } (-2, -6).$$

Also, when $\quad x = -1; y = 3x$
$$= 3 \times -1$$
$$= -3 \qquad \therefore \text{ point is } (-1, -3).$$

Continuing this way for $x = 0, 1, 2$ and 3 would give the points $(0,0)$ $(1,3)$ $(2,6)$ and $(3,9)$. The system of 'boxes' shown in table 10.1 can now be filled out using the above results.

	x	-3	-2	-1	0	1	2	3
$= 3x$	y	-9	-6	-3	0	3	6	9
	(x,y)	$(-3,-9)$	$(-2,-6)$	$(-1,-3)$	$(0,0)$	$(1,3)$	$(2,6)$	$(3,9)$

Table 10.2

Now, plotting the seven points found and joining them with a straight line, we have:

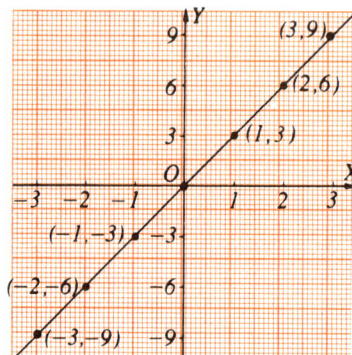

Fig 10.21

Example 10.7

Plot the linear relationships given by

(a) $\{(x, y): y = -2x; -3 \leq x \leq 3\}$
(b) $\{(x, y): y = x - 2; -3 \leq x \leq 3\}$
(c) $\{(x, y): y = x + 3; -3 \leq x \leq 3\}$

Solution

(a) $\{(x, y): y = -2x; -3 \leq x \leq 3\}$

	x	-3	-2	-1	0	1	2	3
$y = -2x$	y	$+6$	$+4$	$+2$	0	-2	-4	-6
	(x, y)	$(-3, 6)$	$(-2, 4)$	$(-1, 2)$	$(0, 0)$	$(1, -2)$	$(2, -4)$	$(3, -6)$

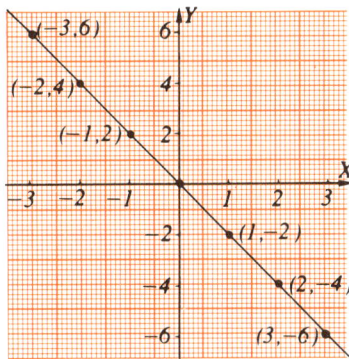

Fig 10.22

(b) $\{(x, y): y = x - 2; -3 \leq x \leq 3\}$

	x	-3	-2	-1	0	1	2	3
$y = x - 2$	y	-5	-4	-3	-2	-1	0	1
	(x, y)	$(-3, -5)$	$(-2, -4)$	$(-1, -3)$	$(0, -2)$	$(1, -1)$	$(2, 0)$	$(3, 1)$

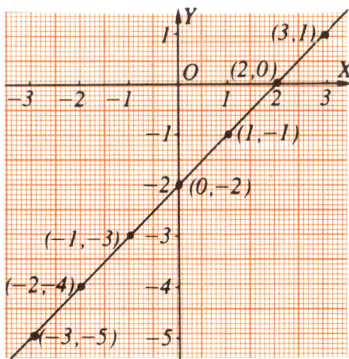

Fig 10.23

(c) $\{(x, y): y = x + 3; -3 \leq x \leq 3\}$

	x	-3	-2	-1	0	1	2	3
$y = x + 3$	y	0	1	2	3	4	5	6
	(x, y)	$(-3, 0)$	$(-2, 1)$	$(-1, 2)$	$(0, 3)$	$(1, 4)$	$(2, 5)$	$(3, 6)$

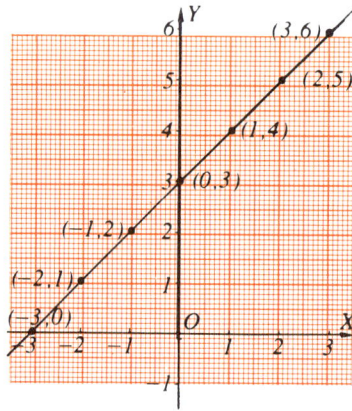

Fig 10.24

Exercise 10E

1. Copy and complete each of the following sets of boxes according to the rule given.

(a) $y = 5x$

x	-2	-1	0	1	2
y	-10			5	
(x, y)				$(1, 5)$	

(b) $y = 4x$

x	-2	-1	0	1	2
y		-4			8
(x, y)		$(-1, -4)$			

(c) $y = -x$

x	-3	-2	-1	0	1	2	3
y		2					
(x, y)							

(d) $y = -3x$

x	-2	-1	0	1	2
y	6				
(x, y)					

(e) $y = x + 2$

x	-3	-2	-1	0	1
y					
(x, y)					

(f) $y = x + 5$

x	-3	-2	-1	0	1
y					
(x, y)					

(g) $y = x - 1$

x	-2	-1	0	1	2
y					
(x, y)					

(h) $y = x - 3$

x	-1	0	1	2	3
y					
(x, y)					

2. Using your answers to question 1, plot each of the following linear relations on graph paper:
   (a) $\{(x, y): y = 5x; -2 \leq x \leq 2\}$
   (b) $\{(x, y): y = 4x; -2 \leq x \leq 2\}$
   (c) $\{(x, y): y = -x; -3 \leq x \leq 3\}$
   (d) $\{(x, y): y = -3x; -2 \leq x \leq 2\}$
   (e) $\{(x, y): y = x + 2; -3 \leq x \leq 1\}$
   (f) $\{(x, y): y = x + 5; -3 \leq x \leq 1\}$
   (g) $\{(x, y): y = x - 1; -2 \leq x \leq 2\}$
   (h) $\{(x, y): y = x - 3; -1 \leq x \leq 3\}$

3. Plot each of the following linear relations on a separate sheet of graph paper:
   (a) $\{(x, y): y = 2x; -3 \leq x \leq 3\}$
   (b) $\{(x, y): y = 6x; -2 \leq x \leq 2\}$
   (c) $\{(x, y): y = 7x; -2 \leq x \leq 2\}$
   (d) $\{(x, y): y = x; -3 \leq x \leq 3\}$
   (e) $\{(x, y): y = -4x; -2 \leq x \leq 2\}$
   (f) $\{(x, y): y = -5x; -2 \leq x \leq 2\}$
   (g) $\{(x, y): y = -8x; -2 \leq x \leq 2\}$
   (h) $\{(x, y): y = -6x; -2 \leq x \leq 2\}$

(i) $\{(x, y): y = x + 4; -5 \le x \le 1\}$
(k) $\{(x, y): y = x + 6; -5 \le x \le 1\}$
(m) $\{(x, y): y = x - 5; -1 \le x \le 5\}$
(o) $\{(x, y): y = x - 8; -1 \le x \le 5\}$

(j) $\{(x, y): y = x + 7; -5 \le x \le 1\}$
(l) $\{(x, y): y = x + 8; -5 \le x \le 1\}$
(n) $\{(x, y): y = x - 4; -1 \le x \le 5\}$
(p) $\{(x, y): y = x - 10; -1 \le x \le 5\}$

Consider the following arithmetic examples:

3×2 multiplication
$4 - 3$ subtraction

Both the above examples involve only one operation.
Consider now:

$3 \times 2 + 4$ multiplication *and* addition
$15 - 2 \times 6$ subtraction *and* multiplication

Both of these examples involve two operations. Any arithmetic problem that involves two or more operations must be dealt with in a certain order (see 1.1). Remember, the order of operations is:

---

1. Brackets (   )
2. Of, Division, Multiplication $\div$, $\times$
3. Addition, Subtraction $+$, $-$
Work the problem from left to right.

This is easily remembered using the word **BODMAS**.

These same principles also apply to algebraic expressions and equations.
So far, each of the linear relationships dealt with have required only one operation on x to obtain the corresponding y value. For example,

$y = 3x$ multiply x by 3 to get y.
$y = -2x$ multiply x by -2 to get y.
$y = x - 2$ subtract 2 from x to get y.
$y = x + 3$ add 3 to x to get y.

Many other linear relationships involve more than one operation on x in order to obtain y. For example,

1st operation   2nd operation
↑   ↑
$y = 2x + 1$ multiply x by 2   *and then*   add 1 to get y

When more than one operation on x is required it is essential to keep in mind the *order of operations* (BODMAS).

Example 10.8

Plot the linear relationships given by:
(a) $\{(x, y): y = 2x + 1; -2 \le x \le 2\}$
(b) $\{(x, y): y = -x + 3; -3 \le x \le 3\}$
(c) $\{(x, y): y = 2(x - 1); -2 \le x \le 2\}$

Solution

(a) $\{(x, y): y = 2x + 1; -2 \le x \le 2\}$
To get each y value we must first multiply x by 2 and then add 1.

	x	-2	-1	0	1	2
$y = 2x + 1$	y	$-4 + 1 = -3$	$-2 + 1 = -1$	$0 + 1 = 1$	$2 + 1 = 3$	$4 + 1 = 5$
	(x, y)	$(-2, -3)$	$(-1, -1)$	$(0, 1)$	$(1, 3)$	$(2, 5)$

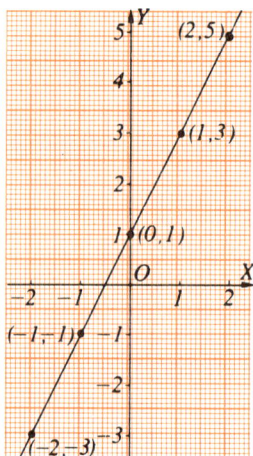

Fig 10.25

(b) $\{(x,y): y = -x + 3; -3 \le x \le 3\}$

To get each y value we must multiply x by -1 and then add 3.

$y = -x + 3$

x	-3	-2	-1	0	1	2	3
y	$-(-3) + 3 = 6$	$-(-2) + 3 = 5$	$-(-1) + 3 = 4$	$-0 + 3 = 3$	$-1 + 3 = 2$	$-2 + 3 = 1$	$-3 + 3 = 0$
(x,y)	$(-3, 6)$	$(-2, 5)$	$(-1, 4)$	$(0, 3)$	$(1, 2)$	$(2, 1)$	$(3, 0)$

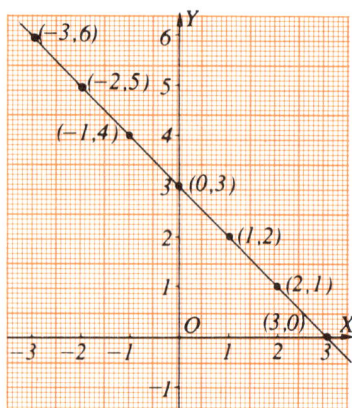

Fig 10.26

(c) $\{(x,y): y = 2(x - 1); -2 \le x \le 2\}$

Here, to get each y value we must *first* subtract 1 from x (brackets) and *then* multiply that result by 2.

	x	-2	-1	0	1	2
$y = 2(x - 1)$	y	$2(-2 - 1) = -6$	$2(-1 - 1) = -4$	$2(0 - 1) = -2$	$2(1 - 1) = 0$	$2(2 - 1) = 2$
	(x,y)	$(-2, -6)$	$(-1, -4)$	$(0, -2)$	$(1, 0)$	$(2, 2)$

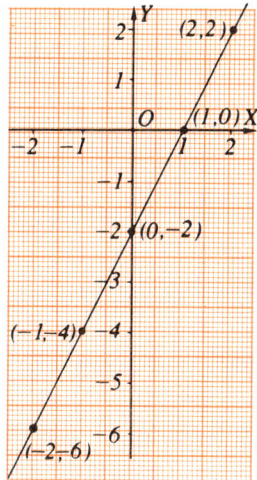

Fig 10.27

Example 10.9

Plot the linear relationship given by $\{(x, y): x + y = 4; -2 \le x \le 2\}$

Solution

Although the rule for this relationship, $x + y = 4$, is given in a different form from those already met in this chapter, it is still a linear relationship. In fact, the rule $x + y = 4$ is equivalent to

$$y = -x + 4 \quad \text{(subtracting } x \text{ from both sides)}$$

and this is the familiar form of a linear relationship.

$x + y = 4$ OR $y = -x + 4$

x	-2	-1	0	1	2
y	6	5	4	3	2
(x, y)	$(-2, 6)$	$(-1, 5)$	$(0, 4)$	$(1, 3)$	$(2, 2)$

(Notice that, as the rule $x + y = 4$ states, the sum of the two coordinates of each point is 4.)

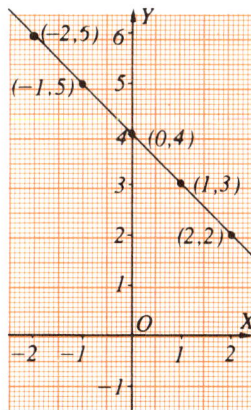

Fig 10.28

Exercise 10F

1. Write each of the following sentences as an algebraic expression:
   (Example: Double x and then subtract 5.
   Answer: $2x - 5$).
   (a) Multiply x by 3 and then subtract 1.
   (b) Multiply x by 7 and then add 4.
   (c) Add 2 to 4 times x.
   (d) Subtract 9 from 2 times x.
   (e) Multiply x by -2 and then add 3.

   (f)   Multiply x by -1 and then subtract 1.
   (g)   Add 7 to -3 times x.
   (h)   Subtract 4 from -4 times x.
   (i)   Add 1 to x and then multiply the result by 4.
   (j)   Add 5 to x and then multiply the result by 3.
   (k)   Subtract 3 from x and then multiply the result by 2.
   (l)   Subtract 2 from x and then multiply the result by -7.
   (m) Subtract x from 5 and then multiply the result by 2.
   (n)   Add x to -3 and then multiply the result by -5.

2. Describe the order of operations for each of the following algebraic expressions:
   (a) $2x - 2$ (b) $3x + 1$
   (c) $-4x + 3$ (d) $-x - 2$
   (e) $5x + 7$ (f) $2x + 6$
   (g) $4 - 2x$ (h) $5 - 3x$
   (i) $2(x - 3)$ (j) $3(x - 4)$
   (k) $5(x + 1)$ (l) $7(x + 2)$
   (m) $-4(x + 3)$ (n) $-2(x - 3)$

3. Copy and complete each of the following sets of boxes according to the rule given. Be careful to use the correct order of operations in each case.

   (a) $y = 3x - 2$

x	-3	-2	-1	0	1	2	3
y		-8			1		
(x, y)		$(-2, -8)$					

   (b) $y = 2x + 3$

x	-3	-2	-1	0	1	2	3
y			1				9
(x, y)			$(-1, 1)$				

   (c) $y = 4x + 5$

x	-3	-2	-1	0	1	2	3
y							
(x, y)							

   (d) $y = 3x - 4$

x	-3	-2	-1	0	1	2	3
y							
(x, y)							

   (e) $y = -x + 1$

x	-3	-2	-1	0	1	2	3
y							
(x, y)							

   (f) $y = -x - 3$

x	-3	-2	-1	0	1	2	3
y							
(x, y)							

   (g) $y = -2x - 5$

x	-3	-2	-1	0	1	2	3
y							
(x, y)							

   (h) $y = -3x + 7$

x	-3	-2	-1	0	1	2	3
y							
(x, y)							

   (i) $y = 3(x - 2)$

x	-3	-2	-1	0	1	2	3
y							
(x, y)							

   (j) $y = 2(x + 1)$

x	-3	-2	-1	0	1	2	3
y							
(x, y)							

(k) $y = -4(x + 1)$

x	-3	-2	-1	0	1	2	3
y							
(x, y)							

(l) $y = -3(x - 1)$

x	-3	-2	-1	0	1	2	3
y							
(x, y)							

4. Using your answers to question 3 plot each of the following linear relations for $-3 \leq x \leq 3$:
   (a) $\{(x, y): y = 3x - 2\}$ (b) $\{(x, y): y = 2x + 3\}$
   (c) $\{(x, y): y = 4x + 5\}$ (d) $\{(x, y): y = 3x - 4\}$
   (e) $\{(x, y): y = -x + 1\}$ (f) $\{(x, y): y = -x - 3\}$
   (g) $\{(x, y): y = -2x - 5\}$ (h) $\{(x, y): y = -3x + 7\}$
   (i) $\{(x, y): y = 3(x - 2)\}$ (j) $\{(x, y): y = 2(x + 1)\}$
   (k) $\{(x, y): y = -4(x + 1)\}$ (l) $\{(x, y): y = -3(x - 1)\}$

5. Plot each of the following linear relations for $-3 \leq x \leq 3$:
   (a) $\{(x, y): y = 2x + 5\}$ (b) $\{(x, y): y = 2x - 3\}$
   (c) $\{(x, y): y = 3x - 1\}$ (d) $\{(x, y): y = 3x + 2\}$
   (e) $\{(x, y): y = 5x - 3\}$ (f) $\{(x, y): y = 7x - 6\}$
   (g) $\{(x, y): y = -2x + 6\}$ (h) $\{(x, y): y = -3x + 8\}$
   (i) $\{(x, y): y = -3x - 2\}$ (j) $\{(x, y): y = -4x - 1\}$
   (k) $\{(x, y): y = -5x + 4\}$ (l) $\{(x, y): y = -2x - 6\}$
   (m) $\{(x, y): y = 3(x + 2)\}$ (n) $\{(x, y): y = 2(x - 3)\}$

6. Plot each of the following linear relations for $-3 \leq x \leq 3$:
   (a) $\{(x, y): x + y = 2\}$ (b) $\{(x, y): x + y = 5\}$
   (c) $\{(x, y): x + y = 1\}$ (d) $\{(x, y): x + y = 3\}$
   (e) $\{(x, y): x + y = -4\}$ (f) $\{(x, y): x + y = -1\}$
   (g) $\{(x, y): x + y = -5\}$ (h) $\{(x, y): x + y = -2\}$
   (i) $\{(x, y): x - y = 3\}$ (j) $\{(x, y): x - y = 1\}$
   (k) $\{(x, y): x - y = 7\}$ (l) $\{(x, y): x - y = 5\}$
   (m) $\{(x, y): x - y = -1\}$ (n) $\{(x, y): x - y = -2\}$

10.3.3 Horizontal and Vertical Linear Relationships

Consider the following group of points:

$$(-3, 3) \quad (-3, 1) \quad (-2, 2) \quad (-2, 3) \quad (-1, 3) \quad (0, 3)$$
$$(1, 2) \quad (1, 3) \quad (2, 0) \quad (2, 3) \quad (3, 2) \quad (3, 4)$$

Which of the above points conform to the rule $y = 3$? Notice that the rule places no restriction on what value x may take; however, the y value *must* be 3.

Reading across the first row of given points we find that the points $(-3, 3)$ $(-2, 3)$ $(-1, 3)$ and $(0, 3)$ all conform to $y = 3$ while the points $(-3, 1)$ and $(-2, 2)$ do not. In the second row only the points $(1, 3)$ and $(2, 3)$ conform to the rule $y = 3$.

Plotting the points which conform to the rule $y = 3$ we have,

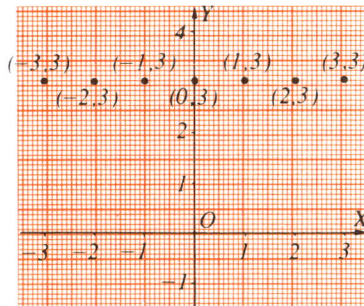

Fig 10.29

These points all lie in the same *horizontal* straight line. If a full line were drawn in to join up these points, the rule for the relationship would be

$$\{ \qquad (x, y) \qquad : \qquad y = 3\}$$

    ↑   ↑   ↑   ↑

the set of   all points   such that   the *y*-coordinate is *always* 3.

A linear relationship such as
$$\{(x, y): y = 3\}$$
gives rise to a *horizontal* straight line

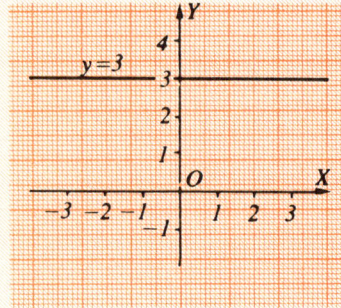

Fig. 10.30

Consider now the group of points:

$(-3, -3)$	$(1, -3)$	$(1, -2)$	$(0, -2)$	$(1, -1)$	$(1, 0)$
$(0, 1)$	$(2, 1)$	$(1, 2)$	$(3, 1)$	$(1, 3)$	$(2, 3)$

Which of the above points conform to the rule $x = 1$? Notice that the rule places no restriction on what value y may take; however, the x value *must* be 1.

From the first row we find the points $(1, -3)$ $(1, -2)$ $(1, -1)$ and $(1, 0)$ all conform to $x = 1$ while the points $(-3, -3)$ and $(0, -2)$ do not. From the second row, the points $(1, 2)$ and $(1, 3)$ both conform to the given rule.

Plotting the points which conform to the rule $x = 1$ we have,

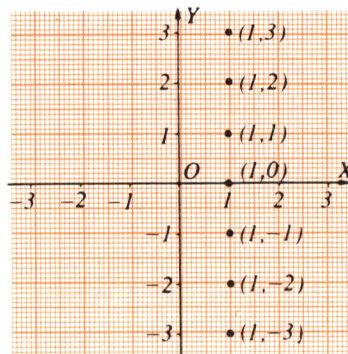

Fig. 10.31

These points all lie in the same *vertical* straight line. If a full line were drawn in to join up these points the rule for the relationship would be:

$$\{ \qquad (x, y) \qquad : \qquad x = 1\}$$

    ↑   ↑   ↑   ↑

the set of   all points   such that   the *x*-coordinate is *always* 1.

A linear relationship such as
$$\{(x, y): x = 1\}$$
gives rise to a *vertical* straight line

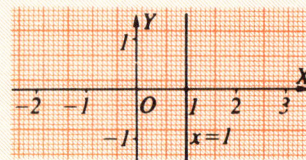

Fig 10.32

Exercise 10G

1. (a) Which of the points given below conform to the rule (i) $y = -2$ (ii) $x = 3$?

$(-3, -3)$ $(-3, -2)$ $(-3, -1)$ $(-3, 0)$ $(-3, 1)$ $(-3, 2)$ $(-3, 3)$
$(-2, -3)$ $(-2, -2)$ $(-2, -1)$ $(-2, 0)$ $(-2, 1)$ $(-2, 2)$ $(-2, 3)$
$(-1, -3)$ $(-1, -2)$ $(-1, -1)$ $(-1, 0)$ $(-1, 1)$ $(-1, 2)$ $(-1, 3)$
$(0, -3)$ $(0, -2)$ $(0, -1)$ $(0, 0)$ $(0, 1)$ $(0, 2)$ $(0, 3)$
$(1, -3)$ $(1, -2)$ $(1, -1)$ $(1, 0)$ $(1, 1)$ $(1, 2)$ $(1, 3)$
$(2, -3)$ $(2, -2)$ $(2, -1)$ $(2, 0)$ $(2, 1)$ $(2, 2)$ $(2, 3)$
$(3, -3)$ $(3, -2)$ $(3, -1)$ $(3, 0)$ $(3, 1)$ $(3, 2)$ $(3, 3)$

   (b) Plot the points found in (a).

2. (a) Which of the points given below conform to the rule (i) $y = 1$ (ii) $x = -1$?

$(-3, -3)$ $(-3, -2)$ $(-3, -1)$ $(-3, 0)$ $(-3, 1)$ $(-3, 2)$ $(-3, 3)$
$(-2, -3)$ $(-2, -2)$ $(-2, -1)$ $(-2, 0)$ $(-2, 1)$ $(-2, 2)$ $(-2, 3)$
$(-1, -3)$ $(-1, -2)$ $(-1, -1)$ $(-1, 0)$ $(-1, 1)$ $(-1, 2)$ $(-1, 3)$
$(0, -3)$ $(0, -2)$ $(0, -1)$ $(0, 0)$ $(0, 1)$ $(0, 2)$ $(0, 3)$
$(1, -3)$ $(1, -2)$ $(1, -1)$ $(1, 0)$ $(1, 1)$ $(1, 2)$ $(1, 3)$
$(2, -3)$ $(2, -2)$ $(2, -1)$ $(2, 0)$ $(2, 1)$ $(2, 2)$ $(2, 3)$
$(3, -3)$ $(3, -2)$ $(3, -1)$ $(3, 0)$ $(3, 1)$ $(3, 2)$ $(3, 3)$

   (b) Plot the points found in (a).

3. Plot the following linear relations for $-3 \le x \le 3$:

   (a) $\{(x, y): y = 2\}$ (b) $\{(x, y): y = 4\}$
   (c) $\{(x, y): y = 1\}$ (d) $\{(x, y): y = 5\}$
   (e) $\{(x, y): y = -3\}$ (f) $\{(x, y): y = -2\}$
   (g) $\{(x, y): y = -5\}$ (h) $\{(x, y): y = -1\}$

4. Plot the following linear relations:

   (a) $\{(x, y): x = 2\}$ (b) $\{(x, y): x = 3\}$
   (c) $\{(x, y): x = 4\}$ (d) $\{(x, y): x = 5\}$
   (e) $\{(x, y): x = -3\}$ (f) $\{(x, y): x = -1\}$
   (g) $\{(x, y): x = -5\}$ (h) $\{(x, y): x = -7\}$

5. Give the rule for each of the following linear relations:

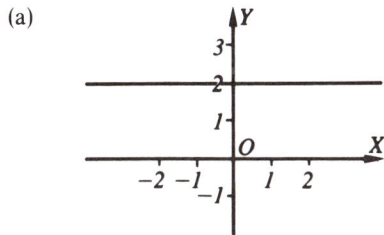

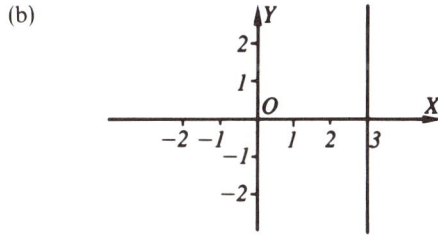

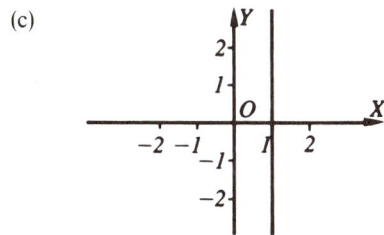

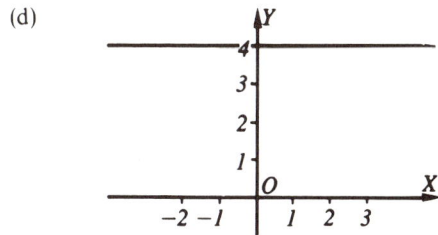

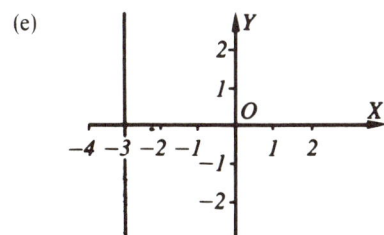

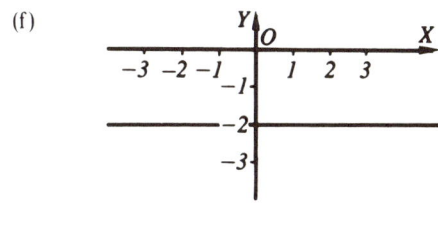

   (a)

   (b)

   (c)

   (d)

   (e)

   (f)

(g)

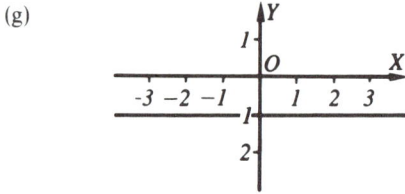

(h)

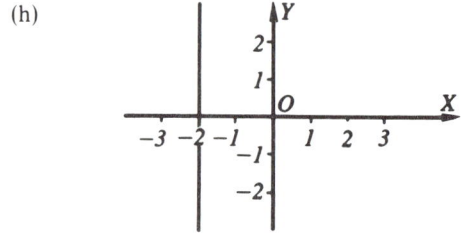

10.4   Intersection of Two Straight Lines

Example 10.10

(a) On the same set of axes plot the relations
   (i) $\{(x, y): y = 2x + 1; -3 \le x \le 3\}$
   (ii) $\{(x, y): y = -x + 4; -3 \le x \le 3\}$

(b) Using the graph in (a) find
   $\{(x, y): y = 2x + 1\} \cap \{(x, y): y = -x + 4\}$

Solution

(a) $y = 2x + 1$

x	-3	-2	-1	0	1	2	3
y	-5	-3	-1	1	3	5	7
(x, y)	$(-3, -5)$	$(-2, -3)$	$(-1, -1)$	$(0, 1)$	$(1, 3)$	$(2, 5)$	$(3, 7)$

$y = -x + 4$

x	-3	-2	-1	0	1	2	3
y	7	6	5	4	3	2	1
(x, y)	$(-3, 7)$	$(-2, 6)$	$(-1, 5)$	$(0, 4)$	$(1, 3)$	$(2, 2)$	$(3, 1)$

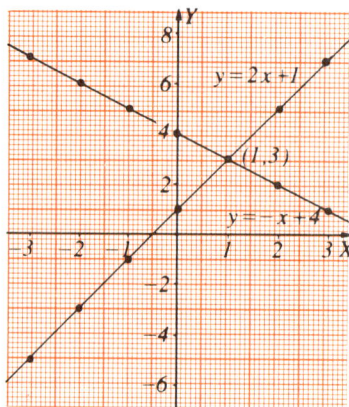

Fig. 10.33

(b) Remember, the symbol '$\cap$' means the intersection of two sets, i.e., the element or elements common to both sets. (see 2.2.1).

Thus, the notation

$$\{(x, y): y = 2x + 1\} \cap \{(x, y): y = -x + 4\}$$

means 'the element or elements common to the straight line $y = 2x + 1$ and the straight line $y = -x + 4$' or, alternatively, 'the intersection of the two straight lines.'

Looking at the graphs in (a) it can be seen that the two lines intersect at only one point, the point $(1, 3)$. Thus the point $(1, 3)$ is the only point that belongs to both straight lines. Hence,

$$\{(x, y): y = 2x + 1\} \cap \{(x, y): y = -x + 4\} \text{ gives the set } \{(1, 3)\}$$

If two different straight lines intersect, they intersect at only one point. The point of intersection can be found graphically by plotting both straight lines on the same set of axes.

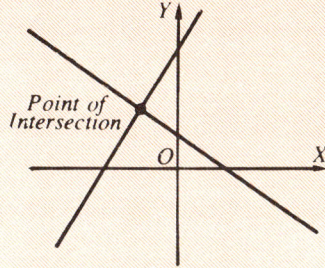

Fig. 10.34

Exercise 10H

1. (a) On the same set of axes plot
   (i) $\{(x, y): y = x + 1; -2 \le x \le 2\}$
   (ii) $\{(x, y): y = -x + 3; -2 \le x \le 2\}$
   (b) Using the graph in (a) find
   $\{(x, y): y = x + 1\} \cap \{(x, y): y = -x + 3\}$

2. (a) On the same set of axes plot
   (i) $\{(x, y): y = 2x - 3; -3 \le x \le 3\}$
   (ii) $\{(x, y): y = x - 1; -3 \le x \le 3\}$
   (b) Using the graph in (a) find
   $\{(x, y): y = 2x - 3\} \cap \{(x, y): y = x - 1\}$

3. Find, graphically, the intersection of each of the following pairs of straight lines whose rules are given below. Plot each pair of lines on graph paper for $-3 \le x \le 3$.
   (a) $y = 2x + 5$
   $y = -x + 2$
   (b) $y = 3x + 1$
   $y = 2x - 1$
   (c) $y = 4x$
   $y = -x + 5$
   (d) $y = -2x$
   $y = x - 6$

4. By plotting on graph paper for $-3 \le x \le 3$, find
   (a) $\{(x, y): y = x - 7\} \cap \{(x, y): y = -2x - 4\}$
   (b) $\{(x, y): y = 2x + 3\} \cap \{(x, y): y = -x - 3\}$
   (c) $\{(x, y): y = 3x + 2\} \cap \{(x, y): y = -5x + 2\}$
   (d) $\{(x, y): y = x + 1\} \cap \{(x, y): y = -2x - 2\}$
   (e) $\{(x, y): y = 3\} \cap \{(x, y): y = 4x - 1\}$
   (f) $\{(x, y): x = 2\} \cap \{(x, y): y = 4x - 5\}$
   (g) $\{(x, y): y = 3x - 2\} \cap \{(x, y): y = -x + 4\}$
   (h) $\{(x, y): y = 5x + 1\} \cap \{(x, y): y = -x - 2\}$
   (i) $\{(x, y): x + y = 3\} \cap \{(x, y): x - y = -1\}$
   (j) $\{(x, y): x + y = 1\} \cap \{(x, y): x - y = -5\}$

11 Ratio and Percentage

11.1 Ratios

11.1.1 Writing Ratios

We often need to compare two numbers or quantities.
Considering the pictures below, it is possible to make the following comparisons:

Here we have 3 cats and 1 dog.

Fig 11.1

We say that the number of cats to dogs is 3 to 1, which can be written as the ratio 3 : 1, where the symbol : means *to*.
The ratio 3 : 1 tells us that there are 3 cats for every 1 dog.
The order in which a comparison or ratio is expressed is important. The quantity which is mentioned first in the comparison must be written first in the ratio.

Here we have 2 cars and 1 bicycle.

Fig 11.2

Thus the ratio of the number of cars to bicycles is 2 : 1.
However the ratio of the number of bicycles to cars is 1 : 2.
Both ratios tell us that there are 2 cars for every 1 bicycle.

> A ratio is a comparison of two quantities, say *a* and *b*, given in a particular order.
> The ratio is written $a : b$, where the symbol : means *to*.

Example 11.1
Write the following as a ratio in the order given:

(a)

mugs to cups

Fig 11.3

(b) A bowl contains five oranges to nine apples.

Solution
(a) Order given is mugs to cups.
There are 4 mugs and 3 cups.
Therefore the ratio of the number of mugs to cups is 4 : 3.
(b) Order given is oranges to apples.
There are 5 oranges and 9 apples.
Therefore the ratio of the number of oranges to apples is 5 : 9.

Exercise 11A
1. Express each of the following as a ratio in the order given:
    (a)   knives to forks

    (b) mice to cheese

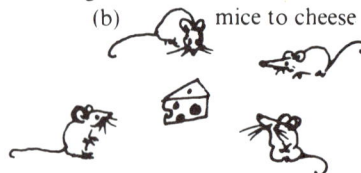

(c)   clothes to pegs

(d)   birds to trees

(e)   doors to windows

(f)   drums to whistles

(g)   cars to trucks

(h)   rabbits to ducks

(i)   notes to coins

(j)   clouds to birds

(k)   birds to kites

(l)   pens to books

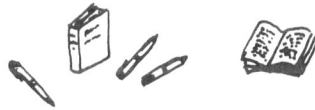

(m)   giraffes to emus

(n)   triangles to squares

(o)   bats to balls

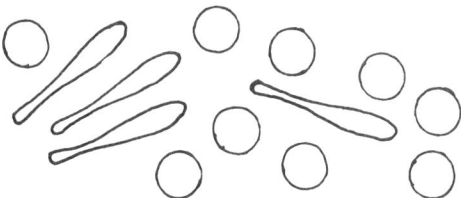

(p)   apples to pears

(q)   guitars to drums

(r)   spades to buckets

(s)   eggs to cartons

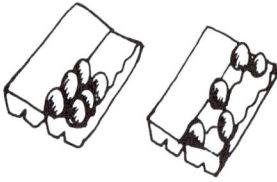

(t)   milk cartons to milk bottles

2. Write down the ratio of the following in the order given:
   (a)   Nine girls to ten boys.
   (b)   Three adults to five children.
   (c)   Five indians to one cowboy.
   (d)   Four soldiers to one sailor.
   (e)   Three orange trees to two lemon trees.
   (f)   Seven red apples to five green apples.
   (g)   Four classical records to eleven pop records.
   (h)   Nine cardigans to five jumpers.
   (i)   Three ties to eight shirts.
   (j)   Two scarves to five hats.
   (k)   Three cats to eleven kittens.
   (l)   Thirteen puppies to two dogs.
   (m)   Two cups to three saucers.
   (n)   Eight chairs to three tables.
   (o)   Six basketballs to seven footballs.
   (p)   Eleven golf balls to ten tennis balls.

3. The Cochrane family have four daughters and the Lloyd family three sons. Write down the ratio of:
   (a)   girls to boys in the two families;
   (b)   boys to girls in the two families.

4. Scott had two puppies and Mrs. Smith had three kittens. Write down the ratio of:
   (a)   kittens to puppies;
   (b)   puppies to kittens.

5. In a particular Year 2 class, nine students elect to do basketball and seventeen elect to do football as their sport. Write down the ratio of:
   (a)   basketball players to football players;
   (b)   football players to basketball players;
   (c)   basketball players to students in the class;
   (d)   football players to students in the class.

6. A swimming club has nineteen members. If there are nine junior members and ten senior members, write down the ratio of:
   (a)   junior members to senior members;
   (b)   senior members to junior members;
   (c)   junior members to club members;
   (d)   senior members to club members.

11.1.2   Writing Ratios as Fractions

Another way to write a ratio is as a fraction.
Here we have 3 candles and 5 matches.

Fig 11.4

The ratio of the number of candles to matches is 3 : 5.
Written as a fraction the ratio of the number of candles to matches is

$$\frac{\text{number of candles}}{\text{number of matches}} = \frac{3}{5},$$

$$\text{So } 3 : 5 = \frac{3}{5}$$

Remember that the order in which a ratio is expressed is important.

The quantity which is mentioned *first* in the ratio must be written in the *numerator* of the fraction. Thus the ratio of matches to candles is $\frac{5}{3}$. This is not changed to a mixed number.

The ratio $a:b$ may be written as a fraction $\dfrac{a}{b}$.

Example 11.2

Express the following comparisons as   (i) a ratio   (ii) a fraction in the order given.

(a)

Fig 11.5

engines to carriages

(b) A vegetable garden contains eleven lettuces and six cabbages.

Solution

(a) Order given is engines to carriages

There is 1 engine and 4 carriages

(i)   Ratio of the number of engines to carriages is 1 : 4

(ii) Written as a fraction 1 : 4 = $\dfrac{1}{4}$

(b) Order given is lettuces to cabbages

There are 11 lettuces and 6 cabbages

(i)   Ratio of the number of lettuces to cabbages is 11 : 6

(ii) Written as a fraction 11 : 6 = $\dfrac{11}{6}$

Exercise 11B

1. Write the following ratios as fractions:

   (a)   4 : 1
   (b)   1 : 3
   (c)   3 : 4
   (d)   6 : 7
   (e)   1 : 6
   (f)   4 : 3
   (g)   2 : 5
   (h)   5 : 1
   (i)   3 : 7
   (j)   7 : 2
   (k)   6 : 5
   (l)   9 : 10
   (m) 11 : 7
   (n)   3 : 2
   (o)   20 : 21
   (p) 24 : 25

2. Write each of the following expressions as
   (i) a ratio
   (ii) a fraction in the order given:

   (a)   drawing pins to paper clips
   (b)   brushes to combs

   (c)   planes to yachts
   (d)   buttons to cotton reels

   (e)   flags to hats
   (f)   ticks to crosses

(g)   numbers to letters

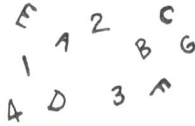

(h)   sweets to ice creams

(i)   gingerbread men to cakes

(j)   buttons to cardigans

3. Write each of the following as
   (i) a ratio   (ii) a fraction in the order given:
   (a) Eight slides to three prints.
   (b) Four ducks to five swans.
   (c) Two pears to five bananas.
   (d) Six carrots to five parsnips.
   (e) Three women to seven men.
   (f) Four lions to three tigers.
   (g) Twelve sweets to seven chocolates.
   (h) Eight biscuits to nine cakes.
   (i) Four blondes to nine brunettes.
   (j) Three giants to two dwarfs.

4. Twins Lucy and Euan have three identical T-shirts and five different T-shirts. Write as a *fraction* the ratio of the number of:
   (a) identical T-shirts to different T-shirts;
   (b) different T-shirts to identical T-shirts.

5. John tossed a coin twenty times and a head appeared eleven times. Write as a fraction the ratio of the number of:
   (a) heads to tails;
   (b) tails to heads.

6. Kate goes to kindergarten three days a week and play-school one day a week. Write as a *fraction* the ratio of the number of days:
   (a) at kindergarten to play-school;
   (b) at play-school to kindergarten.

7. Two dogs Rufus and Cindy required the following food each week: Rufus ate seven tins of dog food, while Cindy ate only five. Write as a *fraction* the ratio of the number of tins of food required for:
   (a) Rufus to Cindy;
   (b) Cindy to Rufus.

11.1.3   Simplifying Ratios

Consider the picture below:

Fig 11.6

Here we have 2 hammers and 8 nails. Thus the ratio of hammers to nails is 2 : 8 or, written as a fraction, $\frac{2}{8}$.

We know that when using fractions our answers should be expressed in simplest form.

So $\frac{2}{8} = \frac{1}{4}$ in simplest form.

Therefore 2 : 8 = 1 : 4 in simplest form.

Thus the *simplest ratio* of hammers to nails is 1 : 4.

The ratio 1 : 4 means that for every one hammer there are four nails.

Fig 11.7

So the 2 hammers and 8 nails could be arranged to show the ratio 1 hammer for every 4 nails.

Since to simplify a fraction we cancel by the highest common factor:

$$\frac{\cancel{12}^4}{\cancel{15}_5} = \frac{4}{5} \text{ cancelling by 3}$$

To simplify a ratio we must divide the numbers by the highest common factor

$$12 : 15$$
$$\div 3 \Big(\qquad \Big) \div 3 \qquad \text{dividing by 3}$$
$$4 : 5$$

> To express a ratio in simplest form divide both quantities by their highest common factor.

Example 11.3

Write the following as a ratio in simplest form:

(a) 9 : 3   (b) 24 : 36

Solution

(a)
$$9 : 3$$
$$\div 3 \Big(\qquad \Big) \div 3$$
$$3 : 1$$

3 is a factor of both 3 and 9 so divide both numbers by 3.

(b)
$$24 : 36$$
$$\div 4 \Big(\qquad \Big) \div 4$$
$$6 : 9$$

4 is a factor of both 24 and 36, so divide by 4

$$\div 3 \Big(\qquad \Big) \div 3$$
$$2 : 3$$

3 is a factor of both 6 and 9, so divide by 3.

Alternatively,
$$24 : 36$$
$$\div 12 \Big(\qquad \Big) \div 12$$
$$2 : 3$$

12 is the highest common factor of 24 and 36, so divide by 12.

Exercise 11C

1. Express each of the following ratios in simplest form:

(a) 3 : 3   (b) 5 : 5
(c) 4 : 8   (d) 2 : 6
(e) 12 : 4   (f) 10 : 5
(g) 32 : 8   (h) 24 : 4
(i) 7 : 70   (j) 10 : 40
(k) 5 : 25   (l) 8 : 40
(m) 18 : 2   (n) 48 : 6
(o) 60 : 3   (p) 45 : 3
(q) 5 : 90   (r) 5 : 150
(s) 100 : 4   (t) 96 : 6

2. Express each of the following ratios in simplest form:

(a) 6 : 4   (b) 15 : 6
(c) 12 : 16   (d) 6 : 14
(e) 15 : 21   (f) 24 : 18
(g) 10 : 25   (h) 20 : 24
(i) 40 : 90   (j) 100 : 30
(k) 36 : 42   (l) 81 : 72
(m) 60 : 24   (n) 10 : 15
(o) 80 : 180   (p) 96 : 24
(q) 150 : 45   (r) 240 : 90
(s) 288 : 108   (t) 108 : 126

3. A shop sells sixty pasties, forty-five pies and eighty sausage rolls. Write down the following ratios in simplest form:

(a) number of sausage rolls to pies;
(b) number of pasties to sausage rolls;
(c) number of pasties to pies.

4. For her twins birthday party, Shelly buys fifteen hats, thirty-six balloons and twelve prizes. Write down the following ratios in simplest form:
   (a) number of balloons to prizes;
   (b) number of prizes to hats;
   (c) number of hats to balloons.

5. Margie's jigsaw has 3000 pieces and Judy's jigsaw has 2500 pieces. Write down the ratio of the number of pieces of:
   (a) the larger jigsaw to the smaller jigsaw;
   (b) the smaller jigsaw to the larger jigsaw.

6. Melbney has a population of 4000 people, Sydbourne has a population of 3600. Write down the ratio of:
   (a) Melbney's population to Sydbourne's population;
   (b) Sydbourne's population to Melbney's population.

7. Sixty-four out of eighty cars successfully completed the car rally. Write down the following ratios in simplest form:
   (a) number of successful entrants to total entrants;
   (b) number of unsuccessful entrants to total entrants;
   (c) number of successful entrants to unsuccessful entrants.

8. In a Year 1 class of thirty-six students, twenty are under twelve years of age. Write down the following ratios in simplest form:
   (a) number of students under 12 to total number of students;
   (b) number of students over 12 to total number of students;
   (c) number of students under 12 to number of students over 12.

11.1.4   Ratio in Measurement

Ratio can be used to compare measurements.

Consider the heights of the children below.

Fig 11.8

The ratio of the height of the smaller child to that of the taller child is 120 : 150

$$120 : 150$$
$$12 : 15$$
$$4 : 5$$

In simplest form, the ratio of heights is 4 : 5.

Notice that even though the measurement has units, a *ratio* has *no unit* at all.

However, when comparing quantities we must ensure that they have the same units.

Consider the example below:

Fig 11.9

Fiona has 2 *blocks* of chocolate and Jimmy has 5 *pieces* of chocolate.

It is incorrect to say that the ratio of Fiona's chocolate to Jimmy's chocolate is 2 : 5 as this ratio shows incorrectly that Fiona has less chocolate than Jimmy.

We must express both quantities in terms of the *same units*.

Fiona has 2 *blocks* of chocolate or 10 *pieces* of chocolate.

Jimmy has 5 *pieces* of chocolate.

The ratio of Fiona's chocolate to Jimmy's chocolate (in pieces) is 10 : 5.

But 10 : 5 = 2 : 1.

Therefore, the ratio of Fiona's to Jimmy's chocolate in simplest form is 2 : 1. This means that Fiona has two pieces to every one piece of Jimmy's.

To compare two quantities they must have the same units.
A ratio has no units.

Example 11.4

A cake requires 1 hour of cooking time compared with scones which require 15 minutes' cooking time. Write down the ratio of cooking time for the cake to that for the scones.

Solution

Cake requires 1 hour or 60 *minutes*.
Scones require 15 *minutes*.
Ratio of cooking time for the cake to that for the scones in *minutes* is 60 : 15.

But

$$
\begin{array}{c}
60 : 15 \\
= _{\div 15} \quad 12 : 3 \\
4 : 1
\end{array}
\begin{array}{l}
{\scriptstyle \div 5} \\
{\scriptstyle \div 3}
\end{array}
$$

The simplest ratio of cooking time for the cake to that for the scones is 4 : 1.

Exercise 11D

1. Write the following ratios from left to right:

(a) ratio of weights of blocks

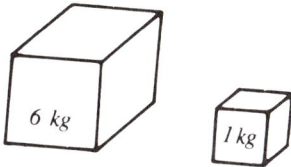

6 kg   1 kg

(b) ratio of heights of books

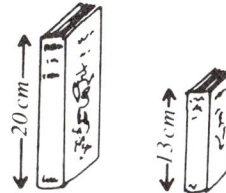

20 cm   13 cm

(c) ratio of costs of toys

£4   £5

(d) ratio of amounts of water

3 litres   2 litres

(e) ratio of lengths of pencils

12 cm   7 cm

(f) ratio of weights of boys

50 kgs   37 kgs

(g) ratio of amounts of milk

1 litres   2 litres

(h) ratio of costs of toys

£8   £3

2. Write the following ratios in simplest form in the order given:

(a) 5 minutes to 10 minutes
(b) 6 days to 12 days
(c) 8 days to 4 days
(d) 15 litres to 3 litres
(e) 60 km/h to 30 km/h
(f) 25°C to 50°C
(g) 12 litres to 18 litres
(h) 36 cents to 24 cents
(i) 40°C to 28°C
(j) 16 years to 18 years
(k) 27 years to 45 years
(l) 48 seconds to 32 seconds
(m) 120 m to 36 m
(n) 10 months to 8 months

(o) 9 months to 6 months   (p) 32 cm to 44 cm
(q) 54 seconds to 48 seconds   (r) 24 km/h to 60 km/h
(s) 84 pence to 60 pence   (t) 50 minutes to 48 minutes

3. Write the following ratios in simplest form in the order given:
   (a) 12 minutes to 1 hour   (b) 21 days to 1 week
   (c) 8 months to 2 years   (d) 50 pence to £1
   (e) 10 days to 4 weeks   (f) 20 seconds to 2 minutes
   (g) 60 pence to £5   (h) 6 months to 4 years
   (i) 40 seconds to 3 minutes   (j) 30 minutes to 5 hours
   (k) 2 hours to 40 minutes   (l) 2 weeks to 10 days
   (m) 3 years to 10 months   (n) £2.40 to 16 pence
   (o) 3 weeks to 42 days   (p) 4 minutes to 20 seconds
   (q) £1.80 to 96 pence   (r) 5 years to 8 months
   (s) 5 minutes to 50 seconds   (t) 3 hours to 54 minutes

4. Lorraine spends £10 a month on fruit and £25 a month on meat. Write down the following ratios in simplest form:
   (a) spending each month on fruit to meat;
   (b) spending each month on meat to fruit.

5. Sue and Simon are renovating their house. If they spend £40 on paint and £84 on wallpaper, find the ratio in simplest form of:
   (a) spending on paint to wallpaper;
   (b) spending on wallpaper to paint.

6. A photograph is 10 cm wide and 12 cm long. Write the following ratios in simplest form:
   (a) length to width;
   (b) width to length.

7. Pam has 3 blocks of chocolate and Guy has 15 pieces left. If each block of chocolate has eighteen pieces, find the ratio of pieces in simplest form in the order given:
   (a) Pam to Guy;
   (b) Guy to Pam.

8. A car travels 200 km from Birmingham to Swansea and uses 20 litres of petrol. On the return journey it travels 240 km using 25 litres of petrol. Write down the following ratios in simplest form in the order given:
   (a) distance travelled between Birmingham and Swansea to the return journey;
   (b) petrol used on the return journey to the outward journey.

9. A rectangular metal sheet of area 48 cm² weighs 26 gm and a circular metal sheet of area 36 cm² weighs 20 gm. Write down the following ratios in simplest form in the order given:
   (a) area of circular metal sheet to rectangular metal sheet;
   (b) weight of rectangular metal sheet to circular metal sheet.

11.2   Percentages

11.2.1   Writing Percentages as Fractions

| 10% discount on all items | 25% deposit balance 60 days | Earn 9% interest on all investments |

Advertisements like those above can be seen every day in newspapers, on television and are used frequently in conversation.

The word *percent* comes from the Latin words *per centum* which mean *out of one hundred*. The symbol % is used to indicate a percentage. Perhaps this symbol reminds you of the digits in 100 and arose through repeated writing of /100 eventually leading to the abbreviation %. Thus 9% is read nine percent and means *nine out of one hundred*.

Since 9% means 9 parts out of 100 parts, 9% can be written as $\dfrac{9}{100}$.

Similarly 10% means 10 out of 100 and can be written as the fraction $\dfrac{10}{100}$.

However the fraction $\dfrac{10}{100}$ can be simplified.

$$\dfrac{10}{100} = \dfrac{1}{10} \quad \text{cancelling by 10}$$

Therefore $10\% = \dfrac{10}{100} = \dfrac{1}{10}$

A percentage is indicated by the symbol %, and means a fraction out of one hundred.

E.g. $11\% = \dfrac{11}{100}$

The fraction should be written in simplest form.

E.g. $5\% = \dfrac{5}{100} = \dfrac{1}{20}$

Example 11.5

Express the following percentages as fractions in simplest form:

(a) 17%   (b) 60%

(c) 37.5%   (d) $7\frac{1}{2}\%$

Solution

(a) $17\% = \dfrac{17}{100}$ 17 out of 100   (b) $60\% = \dfrac{60}{100}$ 60 out of 100

$$= \dfrac{3}{5}$$

(c) $37.5\% = \dfrac{37.5}{100}$ (d) $7\frac{1}{2}\% = \dfrac{7\frac{1}{2}}{100}$

$$= \dfrac{37.5 \times 10}{100 \times 10} \qquad\qquad\qquad = \dfrac{7\frac{1}{2} \times 2}{100 \times 2}$$

$$= \dfrac{375}{1000} \qquad\qquad\qquad\qquad = \dfrac{15}{200}$$

$$= \dfrac{3}{8} \qquad\qquad\qquad\qquad\quad = \dfrac{3}{40}$$

Exercise 11E

Express the following percentages as fractions in simplest form:

1. 1%   2. 7%
3. 19%   4. 27%
5. 83%   6. 99%
7. 8%   8. 15%
9. 20%   10. 36%
11. 48%   12. 50%
13. 75%   14. 100%
15. 2.5%   16. 12.5%
17. 62.5%   18. 87.5%
19. $2\frac{1}{2}\%$ 20. $33\frac{1}{3}\%$
21. $66\frac{2}{3}\%$ 22. $37\frac{1}{2}\%$
23. $12\frac{1}{2}\%$ 24. $62\frac{1}{2}\%$

11.2.2   Writing Fractions as Percentages

We have learnt that any percentage can be written as a fraction with a denominator of 100.

E.g. $7\% = \dfrac{7}{100}$

Therefore any fraction with a denominator of 100 can be written as a percentage.

E.g. $\dfrac{7}{100} = 7\%$

Equivalent fractions enable us to write fractions with a required denominator.
For percentages we need to write the fraction with a denominator of 100.
For example

$$\dfrac{9}{10} = \dfrac{\square}{100}$$

If the fraction is multiplied by $\dfrac{100}{1}$ the percentage can also be found:

$$\dfrac{9}{10} = \dfrac{9}{{}_1 10} \times \dfrac{\overset{10}{100}}{1}\%$$
$$= 90\%$$

$$\therefore \dfrac{9}{10} = \dfrac{\boxed{90}}{100} = 90\%$$

Similarly

$$\dfrac{3}{4} = \dfrac{75}{100} = 75\% \quad \text{Also} \quad \dfrac{3}{4} = \dfrac{3}{{}_{}4_1} \times \dfrac{\overset{25}{100}}{1}\% = 75\%$$

To change a fraction to a percentage multiply by $\dfrac{100}{1}$ and include the % sign.

Example 11.6

Express the following fractions as percentages:

(a) $\dfrac{3}{20}$

(b) $\dfrac{11}{300}$

Solution

(a) $\dfrac{3}{20} = \dfrac{3}{{}_1 20} \times \dfrac{\overset{5}{100}}{1}\%$ multiplying by $\dfrac{100}{1}$

$\qquad = \dfrac{15}{1}\%$ cancelling by 20

$\qquad = 15\%$

(b) $\dfrac{11}{300} = \dfrac{11}{{}_3 300} \times \dfrac{\overset{1}{100}}{1}\%$ multiplying by $\dfrac{100}{1}$

$\qquad = \dfrac{11}{3}\%$ cancelling by 100

$\qquad = 3\tfrac{2}{3}\%$ expressing the answer as a mixed number.

Exercise 11F

1. Write the following as percentages:

(a) $\dfrac{7}{100}$

(b) $\dfrac{1}{100}$

(c) $\dfrac{3}{100}$

(d) $\dfrac{9}{100}$

(e) $\dfrac{21}{100}$

(f) $\dfrac{17}{100}$

(g) $\dfrac{39}{100}$

(h) $\dfrac{43}{100}$

(i) $\dfrac{71}{100}$

(j) $\dfrac{69}{100}$

(k) $\dfrac{89}{100}$

(l) $\dfrac{77}{100}$

(m) $\dfrac{93}{100}$

(n) $\dfrac{83}{100}$

2. Express the following as percentages:

(a) $\frac{1}{10}$ (b) $\frac{7}{10}$

(c) $\frac{9}{20}$ (d) $\frac{11}{20}$

(e) $\frac{21}{50}$ (f) $\frac{33}{50}$

(g) $\frac{17}{20}$ (h) $\frac{3}{20}$

(i) $\frac{39}{50}$ (j) $\frac{27}{50}$

(k) $\frac{1}{2}$ (l) $\frac{1}{4}$

(m) $\frac{2}{5}$ (n) $\frac{3}{5}$

(o) $\frac{12}{25}$ (p) $\frac{18}{25}$

(q) $\frac{3}{4}$ (r) $\frac{4}{5}$

(s) $\frac{24}{25}$ (t) $\frac{14}{25}$

3. Express the following fractions as percentages:

(a) $\frac{9}{200}$ (b) $\frac{21}{200}$

(c) $\frac{11}{500}$ (d) $\frac{69}{700}$

(e) $\frac{1}{3}$ (f) $\frac{2}{3}$

(g) $\frac{5}{8}$ (h) $\frac{3}{8}$

11.2.3   Percentage Parts of a Whole

We have found that $100\% = \frac{100}{100} = 1$, so 100% represents one whole.
This square represents one whole (i.e. 100%).
The shaded part has 30 squares out
of 100 squares, so represents 30%.

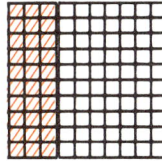

Fig 11.10

30% shaded

To find the percentage unshaded we must find $100\% - 30\%$

$$
\begin{aligned}
100\% - 30\% &= \frac{100}{100} - \frac{30}{100} \\
&= \frac{100 - 30}{100} \\
&= \frac{70}{100} \\
&= 70\%
\end{aligned}
$$

Now   shaded part + unshaded part = whole

      i.e. 30% + 70% = 100%

Thus to calculate the unshaded part we could have subtracted the percentage shaded from the whole.

    I.e. percentage unshaded $= 100\%$ − percentage shaded

                       $= 100\% - 30\%$

                       $= 70\%$

Now consider the block of chocolate below:

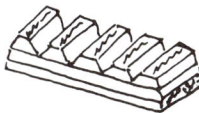

Fig 11.11

If Heather ate 3 pieces of chocolate, she has eaten $\frac{3}{5}$ of the whole. Since $\frac{3}{5} = \frac{3}{5} \times \frac{100}{1}\% = 60\%$, Heather has eaten 60% of the chocolate.

Therefore the percentage remaining $= 100\% -$ percentage eaten
$$= 100\% - 60\%$$
$$= 40\%$$

Exercise 11G

1. Find
   (a) $100\% - 60\%$
   (b) $100\% - 70\%$
   (c) $100\% - 25\%$
   (d) $100\% - 35\%$
   (e) $100\% - 48\%$
   (f) $100\% - 22\%$
   (g) $100\% - 83\%$
   (h) $100\% - 81\%$
   (i) $100\% - 32\%$
   (j) $100\% - 78\%$

2. Jackie answered 80% of the test questions correctly. What percentage did she answer incorrectly?

3. David's cricket team has won 90% of its matches this year. What percentage of matches has the team lost?

4. 42% of babies born on a particular day were male. What percentage were female?

5. 26% of students in Year 2 are the youngest in their family. What percentage are not the youngest?

6. Celia sells 86% of her sea shells. What percentage does she have left?

7. Sue waters 50% of her pot-plants. What percentage are not watered?

8. In the shapes below
   (i) express the fraction shaded as a percentage;
   (ii) therefore find the percentage unshaded.

   (a)    (b)

   (c)    (d)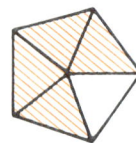

   (e)    (f)

   (g) 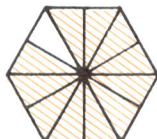   (h)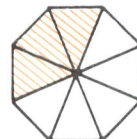

11.2.4   Percentages Greater than 100%

Percentages represent a fraction out of 100.

Thus 50% represents 50 out of 100, written $\frac{50}{100} = \frac{1}{2}$, and 100% represents 100 out of 100, written $\frac{100}{100} = 1$.

Consider a percentage greater than 100%, say 150%.

150% means 150 out of 100, written $\frac{150}{100} = \frac{3}{2} = 1\frac{1}{2}$.

> A percentage greater than 100% will have its numerator greater than the denominator when written as a fraction, and therefore represents a number greater than one.

Example 11.7

(a) Express 140% as a fraction.

(b) Express $1\frac{1}{4}$ as a percentage.

Solution

(a) $140\% = \frac{140}{100}$

$$= \frac{14^7}{10_5}$$

$$= \frac{7}{5}$$

$$= 1\frac{2}{5}$$

(b) $1\frac{1}{4} = 1\frac{1}{4} \times \frac{100}{1}\%$

$$= \frac{5}{_1 4} \times \frac{100^{25}}{1}\%$$

$$= 125\%$$

Exercise 11H

1. Express the following percentages as fractions:

   (a) 120%
   (b) 180%
   (c) 150%
   (d) 160%
   (e) 125%
   (f) 175%
   (g) 275%
   (h) 225%
   (i) 104%
   (j) 128%
   (k) 172%
   (l) 148%
   (m) 196%
   (n) 164%
   (o) 200%
   (p) 300%
   (q) 800%
   (r) 400%
   (s) 500%
   (t) 900%

2. Express the following as percentages:

   (a) $1\frac{1}{2}$
   (b) $1\frac{1}{5}$
   (c) $1\frac{3}{4}$
   (d) $2\frac{1}{4}$
   (e) $1\frac{1}{10}$
   (f) $1\frac{3}{10}$
   (g) $2\frac{2}{5}$
   (h) $4\frac{1}{2}$

11.2.5   Expressing One Quantity as a Percentage of Another

A percentage can be used to compare two quantities.

Consider a swimming club which has 60 members.

If 15 members are girls, we could express the number of girls as a percentage of the total club membership.

First, we must express the number of girls as a fraction of the club membership.

Fraction of members which are girls $= \dfrac{\text{number of girls}}{\text{number of members}} = \dfrac{15}{60}.$

To change a fraction to a percentage multiply by $\frac{100}{1}$.

$\therefore$ Percentage of members which are girls $= \dfrac{15^5}{60_{\mathcal{X}_1}} \times \dfrac{100^5}{1}\%$

$$= 25\%$$

Thus 25% of club members are girls.

To express one quantity as a percentage of another:

(a) write one quantity as a fraction of the other;

(b) multiply by $\frac{100}{1}$ and include the percentage symbol.

E.g.   40 out of $50 = \dfrac{40}{50} = \dfrac{40}{50_1} \times \dfrac{100^2}{1}\% = 80\%$

Example 11.8
What percentage is 24 of 40?

Solution

Writing as a fraction, 24 out of 40 $= \frac{24}{40}$

Thus, expressing as a percentage $\frac{24}{40} = \frac{24^{\,6}}{_{1}40} \times \frac{100}{1}\%$

$$= 60\%$$

Therefore 24 is 60% of 40.

Exercise 11I
1. What percentage is the first number of the second number?

   (a) 16, 100   (b) 18, 100
   (c) 47, 100   (d) 33, 100
   (e) 10, 20   (f) 10, 50
   (g) 24, 60   (h) 20, 80
   (i) 21, 30   (j) 39, 75
   (k) 36, 48   (l) 32, 40
   (m) 120, 200   (n) 150, 300
   (o) 2, 16   (p) 21, 56
   (q) 66, 99   (r) 12, 90
   (s) 24, 45   (t) 56, 96

2. Kevin gained the following results in his final exams. Find his percentage mark for each subject.

   (a) English: 90 out of 100   (b) French: 70 out of 100
   (c) Mathematics: 64 out of 80   (d) Geography: 60 out of 75
   (e) Science: 78 out of 120   (f) History: 77 out of 140

3. Helen played 12 games of squash, winning 9 of these. What percentage of games did Helen win?

4. In a spelling test Pamela correctly spelt 19 out of 20 words. What percentage did she spell correctly?

5. In the school cross-country race 72 out of 80 students completed the course. What percentage of students completed the course? What percentage failed to complete the course?

6. A market gardener found that 63 out of 90 lettuces he planted grew. What percentage of lettuces grew? What percentage did not grow?

7. The twenty-five members of a football club voted for captain.
   (a) If 18 members voted for Greg, what percentage of the members voted for him?
   (b) What percentage voted for the other nominee, Glenn?

8. Bert the odd job man earns £25 per week. He earns £20 for delivering papers and £5 for mowing lawns.
   (a) What percentage of his earnings comes from delivering papers?
   (b) What percentage of his earnings comes from mowing lawns?

11.2.6   Finding a Given Percentage of an Amount
A survey shows that of 50 people interviewed, 8% were unemployed.
To find the number who were unemployed we must find 8% of 50.
Since 8% can be written as the fraction $\frac{8}{100}$,

$$8\% \text{ of } 50 = \tfrac{8}{100} \text{ of } 50$$

Remembering that the word *of* means multiply

$$8\% \text{ of } 50 = \frac{8}{100} \text{ of } 50 = \frac{8^{\,4}}{100_{\,z_1}} \times 50^{\,1}$$
$$= 4$$

Therefore 4 people were unemployed.

> To find a given percentage of an amount:
> (a) write the percentage as a fraction;
> (b) multiply by that fraction.
>
> E.g. 25% of $80 = \dfrac{25}{100} \times 80 = 20$

Example 11.9
Find 12% of £250.

Solution

$$12\% \text{ of £250} = \frac{12}{100} \text{ of £250}$$

$$= £\frac{12}{100} \times \frac{250}{1}$$

$$= £\frac{12}{10} \times \frac{25}{1}$$

$$= £\frac{12}{2} \times \frac{5}{1}$$

$$= £30$$

Exercise 11J

1. Find

   (a) 20% of 500
   (b) 30% of 200
   (c) 75% of 80
   (d) 25% of 60
   (e) 50% of 126
   (f) 40% of 220
   (g) 24% of 50
   (h) 36% of 50
   (i) 8% of 25
   (j) 25% of 8
   (k) 80% of 75
   (l) 70% of 50
   (m) 35% of 120
   (n) 45% of 980
   (o) 64% of 150
   (p) 88% of 350

2. Find

   (a) 6.5% of 200
   (b) 4.5% of 400
   (c) 12.5% of 600
   (d) 24.5% of 200
   (e) 37.5% of 5000
   (f) 36.5% of 5000
   (g) 64.5% of 1000
   (h) 87.5% of 1000
   (i) 8.4% of 4500
   (j) 9.2% of 8000
   (k) 10.8% of 3500
   (l) 12.6% of 500
   (m) 18.6% of 500
   (n) 24.8% of 3500
   (o) 32.2% of 9500
   (p) 38.4% of 6000

3. Find

   (a) $1\frac{1}{2}\%$ of 400
   (b) $4\frac{1}{2}\%$ of 200
   (c) $2\frac{1}{5}\%$ of 1000
   (d) $2\frac{1}{4}\%$ of 2800
   (e) $33\frac{1}{3}\%$ of 3600
   (f) $66\frac{2}{3}\%$ of 1200
   (g) $37\frac{1}{2}\%$ of 1200
   (h) $62\frac{1}{2}\%$ of 400

4. Find

(a) 100% of 82
(b) 100% of 27
(c) 100% of 15
(d) 100% of 140
(e) 200% of 50
(f) 300% of 20
(g) 300% of 60
(h) 400% of 80
(i) 500% of 140
(j) 200% of 150
(k) 120% of 400
(l) 125% of 480
(m) 145% of 600
(n) 150% of 840
(o) 180% of 2500
(p) 175% of 600

5. 84% of a class of 25 can swim. How many in the class can swim?

6. 40% of a class of 30 are left-handed. How many in the class are left-handed?

7. 4% of packets of a certain brand of cereal were underweight. If 450 packets were examined, how many packets were underweight?

8. A water tank holds 5800 litres. If the tank is 55% full, how much water is in the tank?

9. Marion receives $8\frac{1}{2}\%$ interest on the £650 she has in the bank. How much interest does she receive?

10. In studying a rock which weighed 15gm, a scientist found $37\frac{1}{2}\%$ was copper. What was the actual weight of copper in the rock?

11. A newsagent orders 500 papers per day. Due to a mechanical failure, production is reduced and therefore all newsagents receive 82.4% of their daily requirements. How many papers will this newsagent receive?

12. A manufacturer finds that his production this month has increased by 15.6%. If he produced 2000 containers last month, by how much has his production increased this month?

13. A car firm requires 15% deposit when buying a new car. If Mr. White bought a car valued at £8000, how much deposit is required?

14. Mrs. White needs to pay 18% as a deposit on a lounge suite. If the lounge suite costs £900, how much deposit must she pay?

Measurement

In this chapter we will look at various types of measurement in the metric system of units. Measurement has developed through the ages because people needed to answer questions such as, 'how far?,' 'how much?,' 'how long?,' 'how heavy?,' etc.

For example,

'How far is it from one town to another?'—*length*.

'How much land do I have?'—*area*.

'How much grain does one grain bin hold?'—*capacity*.

'How much space does the grain bin take up?'—*volume*.

'How heavy is that sack?'—*mass*.

'How long did it take to hoe the land?'—*time*.

Although all these questions could be answered according to the units of measure used in that town, every town varied and thus the same questions resulted in different answers in different towns. It was the French who first made an effort to establish a standardised system of measurements in the late 18th century. They had found a great need for this since within one province there were one hundred and ten different measures for corn and twenty-one variations of the pound weight.

By the end of the 18th century they had published the metric system of weights and measures which is now in operation in many countries.

12.1   Length

The units of length in the Metric System are based on the *metre*; hence the name 'metric'. The metre is about the distance between your fingertips when your arms are outstretched. We abbreviate metre to just m.

Fig 12.1

The metre is sometimes too large to measure certain objects and therefore it is divided into smaller units.

(a) The *decimetre* is about the distance between the tip of your middle finger and the tip of your index finger when they are stretched apart. We abbreviate decimetre to dm.

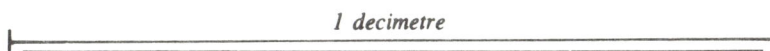

1 decimetre   Fig 12.2

The prefix 'deci' means $\frac{1}{10}$ so a decimetre is $\frac{1}{10}$ of a metre.

(b) The *centimetre* is about the width of your fingernail. We abbreviate centimetre to cm.

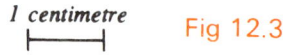

1 centimetre

Fig 12.3

The prefix 'centi' means $\frac{1}{100}$ so a centimetre is $\frac{1}{100}$ of a metre.

(c) The *millimetre* is about the width of a pinhead.
We abbreviate millimetre to mm.

H

1 millimetre   Fig 12.4

The prefix 'milli' means $\frac{1}{1000}$ so a millimetre is $\frac{1}{1000}$ of a metre.
The metre is also sometimes too small to measure certain distances and therefore a longer measure is needed.

(d) The *kilometre* is two and a half laps of a running track. We abbreviate kilometre to km.
The prefix 'kilo' means 1000 so a kilometre is 1000 metres.

Exercise 12A

1. What unit of length would you use to measure the following:
    (a) distance from London to Manchester; (b) the length of a sports ground;
    (c) the width of a finger nail; (d) the length of a car;
    (e) the height of a tennis net; (f) the distance of the flight from Prestwick to Heathrow;
    (g) the width of the nib of a pen; (h) the length of your hair;
    (i) the height of a tree; (j) the width of a building;
    (k) the size of your waist; (l) the thickness of your shoelace;
    (m) the distance around a swimming pool?

2. Estimate the length (in cm) of the listed objects, then measure their length and find the difference between your estimate and the actual length.

	Object	*Estimation*	*Measurement*	*Difference*
(a)	length of book			
(b)	width of book			
(c)	height of desk			
(d)	length of book			
(e)	width of book			
(f)	your hand span			
(g)	length of your foot			
(h)	distance around your waist			
(i)	length of your arm			
(j)	length of your leg			
(k)	height of door			
(l)	height of window			
(m)	length of classroom			
(n)	width of classroom			
(o)	length of blackboard			

12.1.1 Perimeter

> The perimeter of a plane figure is the distance around the border of that figure. The units are mm, cm, m, km, etc.

For example, finding the perimeter of a square of side length 3 cm,

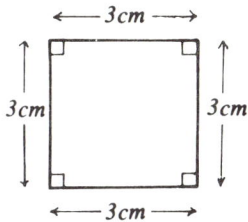

Perimeter = the sum of the length of all sides
= 3 + 3 + 3 + 3 cm
= 12 cm.

Fig 12.5

Example 12.1

Find the perimeter of each of the following:

(a)

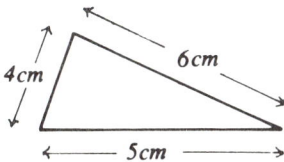

Fig 12.6 (a)

(b)

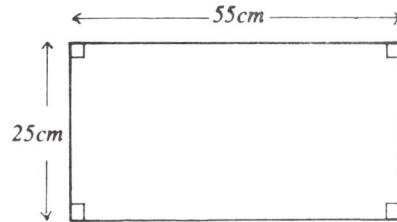

Fig 12.6 (b)

(c) A rectangular paddock which is 30 m in length and 22 m in width.

Solution

(a)

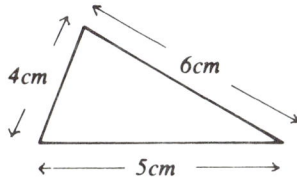

Perimeter = 4 cm + 6 cm + 5 cm
= 15 cm.

(b)

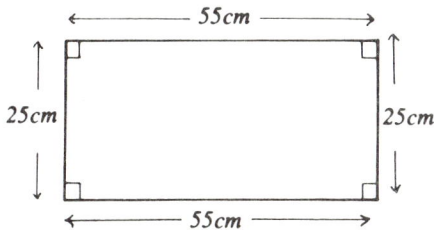

As we have seen in Chapter 9 the opposite sides of a rectangle are equal.
∴ Perimeter = 55 cm + 25 cm + 55 cm + 25 cm
= 160 cm.

(c) The paddock is rectangular with a length of 30 m and a width of 22 m. The diagram therefore is as shown in fig 12.7.

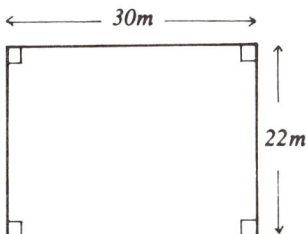

Remember that the opposite sides of a rectangle are equal.
∴ Perimeter = 30 m + 22 m + 30 m + 22 m
= 104 m.

Fig 12.7

Exercise 12B

Find the perimeters of each of the following:

1. (a)

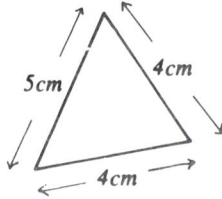

(b)

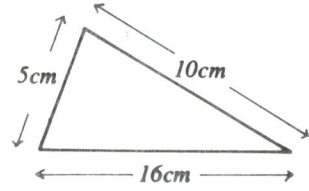

(c)

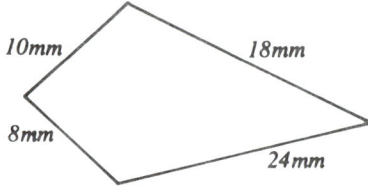

(d)

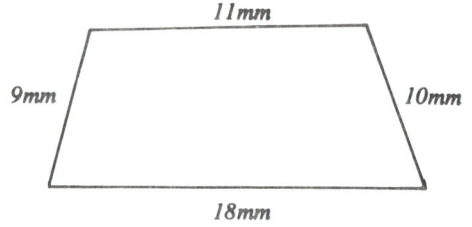

(e)

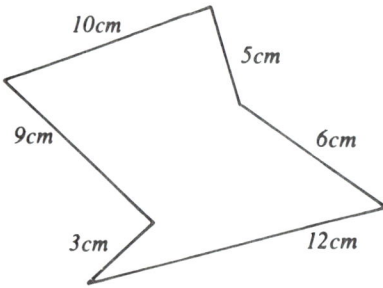

(f)

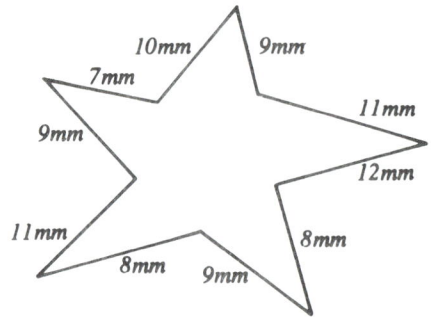

(g)

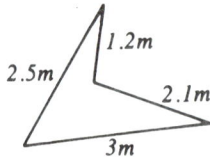

(h)

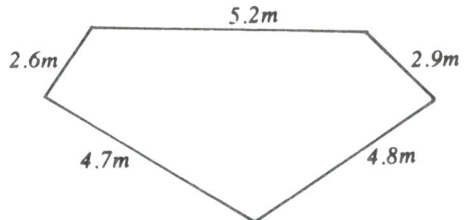

(i)

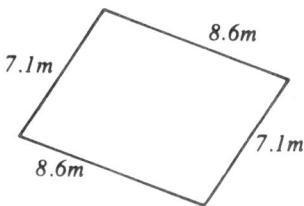

(j)

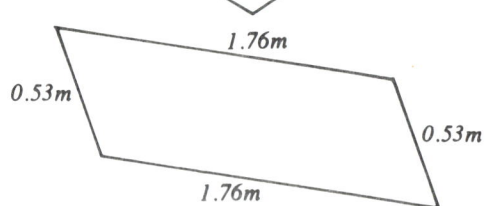

(k)

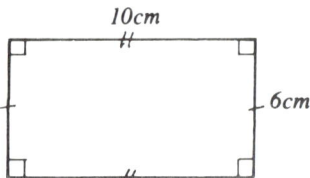

(l)

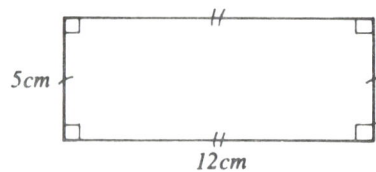

(m)

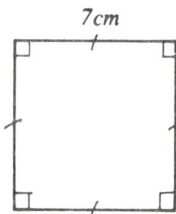

(n)

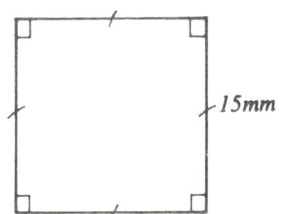

(o)

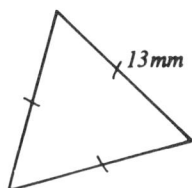

13mm

(p)

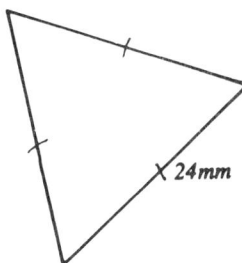

24mm

(q)

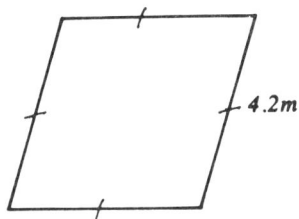

4.2m

(r)

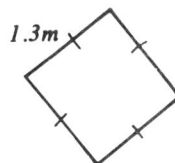

1.3m

(s)

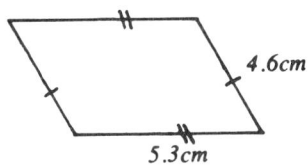

4.6cm

5.3cm

(t)

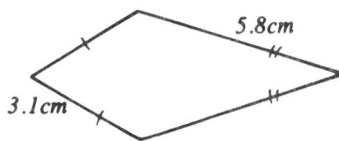

5.8cm

3.1cm

2. Find the perimeter of each of the following by first measuring the length of each side in the unit stated.

(a)

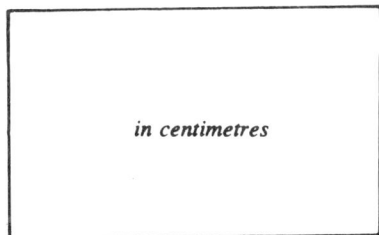

in centimetres

(b)

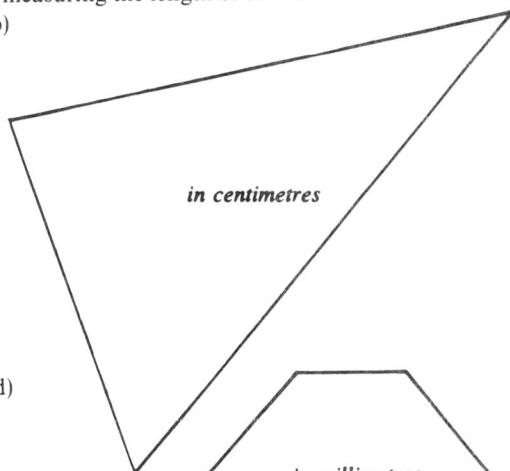

in centimetres

(c)

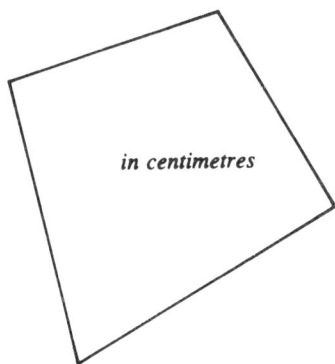

in centimetres

(d)

in millimetres

(e)

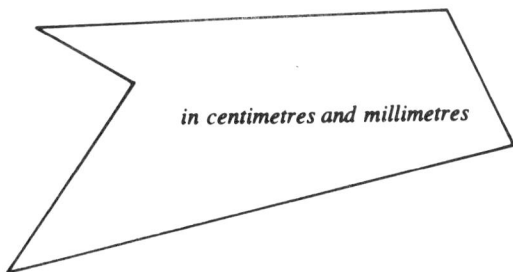

in centimetres and millimetres

(f)

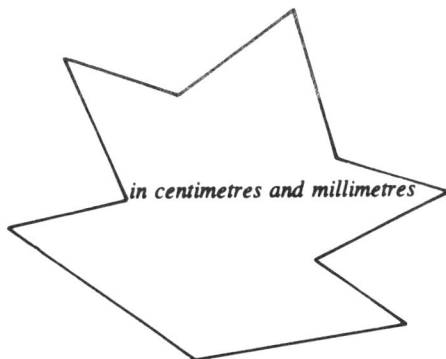

in centimetres and millimetres

(g)

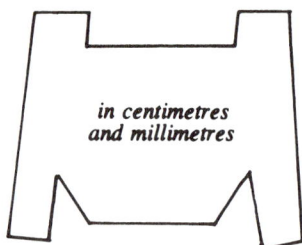

in centimetres
and millimetres

(h)

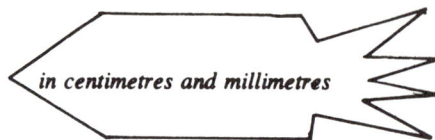

in centimetres and millimetres

3. (a) The length of a rectangular paddock is 16 m. If the width is 9 m, find the perimeter.

(b) The width of a rectangular car park is 72 m and the length is 85 m. Find the perimeter.

(c) The length of a rectangular city block is 108 m and the width is 88 m. Find the perimeter.

(d) The length of a rectangular rug is 128 cm and the width is 65 cm. What is the perimeter of the rug?

(e) A square table has a length of 2 m. What is the perimeter of the top?

(f) A square room is 8.2 m in length. Find the perimeter of the room.

(g) During football training the team runs round the edge of a rectangular field which is 175 m long and 85 m wide. How far do they run if they go round the field

(i) once   (ii) three times?

(h) A bed has a length of 190 cm and a width of 150 cm. Find the length of lace needed to go around the base of the bed.

(i) A man walks his dog round a block each night. If the block is 125 metres in length and 96 metres wide, find the distance the man and his dog walk.

(j) A sheep station in central Australia needs new fencing around the border of the property. If the length of the property is 60 km and the width is 36 km, find the length of the fencing needed.

(k) A gardener looks after a certain plot of grass in the park. If the plot is 270 m long and 182 m wide, find the distance he must walk when he is trimming the edges.

(l) A rectangular shaped room needs new skirting boards around the edge. Find the length of skirting board needed if the room is 12.6 metres long and 8.4 metres wide.

12.1.2   Conversion of Units

The metric system is being gradually introduced to take the place of the Imperial system of units of measurement for length, which were the inch, foot, yard and mile.

When we compare the two systems, we realize the enormous advantage that the Metric System has over the Imperial System.

For example, to change from one unit to another in the Imperial System, you first had to remember a complicated series of conversion numbers and then do a long calculation.

However, in the metric system everything is based on using the factors 10, 100, 1000, etc., and these factors are built into the name of each unit by means of the prefix.

For example, 1 *centi*metre $= \frac{1}{100}$ of a metre

or   1 metre $= 100$ centimetres

Note that the prefix is related to the metre only.

Look at 1 centimetre on your ruler:

Fig 12.8

You should be able to see that it is divided into 10 equal parts. Each part is 1 mm, so

1 cm = 10 mm

or 1 mm = 1 cm ÷ 10   (conversion factor of 10).

On a 1 metre ruler you should be able to see centimetres marked. If you were to count the number of centimetres you would find there are 100 cm on a 1 metre ruler.

∴ 1 m = 100 cm

or 1 cm = 1 m ÷ 100   (conversion factor of 100)

Similarly, you may find millimetres marked on a 1 metre ruler. Counting the number of millimetres on a 1 metre ruler you would find 1000.

∴ 1 m = 1000 mm

or 1 mm = 1 m ÷ 1000   (conversion factor of 1000)

The other common conversion is from kilometres to metres or visa versa.

We have already seen from the use of prefixes that 'kilo' means 1000.

Thus 1 km = 1000 metres

or 1 m = 1 km ÷ 1000   (conversion factor of 1000)

So far we have covered the four most common conversions and found that

(a) . . . mm   =   . . . cm

(b) . . . cm   =   . . . m

(c) . . . mm   =   . . . m

(d) . . . m   =   . . . km

Notice that in each of the above instances we are changing a *smaller unit to a larger unit*, and therefore there are less of a larger unit so we *divide* by the conversion factor.

Example 12.2

(a) Convert 55 mm to cm.

(b) Convert 4200 m to km.

(c) How many metres is 423 cm equivalent to?

Solution

(a) To convert mm to cm, we need to *divide* as we are converting to a *larger* unit. The conversion factor is 10.

Thus   55 mm = . . . cm

55 ÷ 10 = 5.5

∴ 55 mm = 5.5 cm

(b) To convert m to km, we need to *divide* as we are converting to a *larger* unit. The conversion factor is 100.

Thus   4200 m = . . . km

4200 ÷ 1000 = 4.2

∴ 4200 m = 4.2 km

(c) The conversion factor for cm to m is 100.

Thus $\overset{\div\,100}{\overgroup{423\text{ cm}}} = \ldots\text{ m}$

$423 \div 100 = 4.23$

$\therefore 423\text{ cm} = 4.23\text{ m}$

Now consider changing from a larger unit to a smaller unit. For example, converting centimetres to millimetres. Looking at your ruler again find a length of 4 cm. If you were to count the number of millimetres in that section you would find there are 40 mm.

Thus $1\text{ cm} = 10\text{ mm}$

or $1\text{ cm} = 1\text{ mm} \times 10$ (conversion factor of 10)

Note that the conversion factor remains the same.

Fig 12.9

If we were to continue to convert *larger units to smaller units* we would see the same pattern developing as before, except that we would be *multiplying* the larger unit by the conversion factor to give the smaller unit.

(a) $\ldots\text{ cm}$ $=$ $\ldots\text{ mm}$

(b) $\ldots\text{ m}$ $=$ $\ldots\text{ cm}$

(c) $\ldots\text{ m}$ $=$ $\ldots\text{ mm}$

(d) $\ldots\text{ km}$ $=$ $\ldots\text{ m}$

Thus the completed pattern of the 4 basic conversions is:

(a) mm $\overset{\div 10}{\underset{\times 10}{=}}$ cm	(b) cm $\overset{\div 100}{\underset{\times 100}{=}}$ m
(c) mm $\overset{\div 1000}{\underset{\times 1000}{=}}$ m	(d) m $\overset{\div 1000}{\underset{\times 1000}{=}}$ km

When converting from a smaller unit to a larger unit divide by the conversion unit.
When converting from a larger unit to a smaller unit multiply by the conversion unit.

These basic conversions enable us to derive other conversions, for example, mm to km and cm to km, etc.

Example 12.3

(a) Convert 520 cm to metres.
(b) Convert 2.4 km to metres.
(c) How many centimetres in 2 km?
(d) 1 000 000 mm is equal to . . . km.

Solution

(a) To convert centimetres to metres, we need to *divide* as we are converting to a larger unit. The conversion factor is 100.

Thus   520 cm   = ... m

520 ÷ 100 = 5.20

∴ 520 cm = 5.20 m

(b) To convert kilometres to metres we need to *multiply* as we are converting to a smaller unit. The conversion factor is 1000.

Thus   2.4 km   = ... m

2.4 × 1000 = 2400

∴ 2.4 km = 2400 m

(c) To convert km to cm we need to *multiply* as we are converting to a smaller unit.

As we do not know the conversion factor we will need to do the calculation in two steps.

Step 1:   2 km   = 2000 metres

Step 2:   2000 m   = ... cm

2000 × 100 = 200 000

2 km = 2000 m = 200 000 cm

(d) 1 000 000 mm = ... km

As we are converting to a larger unit we will need to *divide* but we do not know the conversion factor from mm to km and so we will do it in two steps.

Step 1:   1 000 000 mm = 1 000 metres

Step 2:   1000 metres   = ... km

1000 ÷ 1000   = 1

∴ 1 000 000 mm = 1000 m = 1 km

Exercise 12C

1. State whether (i) you would multiply or divide and (ii) by what factor, when doing the following conversions:
   (a) cm to mm
   (b) cm to m
   (c) m to mm
   (d) mm to cm
   (e) mm to m
   (f) km to m
   (g) km to cm
   (h) m to km
   (i) cm to km
   (j) m to cm

2. Copy and complete the following:
   (a) 2 m = ... × 100 cm = ... cm
   (b) 6 m = ... × 1000 mm = ... mm
   (c) 850 cm = ... ÷ 100 m = ... m
   (d) 2500 mm = ... ÷ 1000 mm = ... mm
   (e) 4.5 cm = ... × 10 mm = ... mm
   (f) 1.34 km = ... × 1000 m = ... m
   (g) 7500 mm = ... ÷ 1000 m = ... m
   (h) 4620 m = ... ÷ 1000 km = ... km
   (i) 12.6 m = ... × 100 cm = ... cm
   (j) 0.38 cm = ... × 10 mm = ... mm
   (k) 4820 m = ... ÷ 1000 km = ... km
   (l) 72.6 mm = ... ÷ 1000 m = ... m

3. Copy and complete the following:
   (a) 12 m = 12 × ... cm = ... cm
   (b) 45 m = 45 × ... mm = ... mm
   (c) 250 m = 250 ÷ ... km = ... km
   (d) 74 mm = 74 ÷ ... cm = ... cm
   (e) 2.2 km = 2.2 × ... m = ... m
   (f) 38 cm = 38 × ... mm = ... mm
   (g) 450 mm = 450 ÷ ... cm = ... cm
   (h) 305 m = 305 ÷ ... km = ... km

4. Convert the following measurements to centimetres:
   (a) 200 mm
   (b) 560 mm
   (c) 32 mm
   (d) 8 mm
   (e) 3 m
   (f) 6 m
   (g) 7.2 m
   (h) 0.4 m
   (i) 1 km
   (j) 1.5 km

5. Convert the following measurements to millimetres:

(a) 4 cm   (b) 12 cm
(c) 9.6 cm   (d) 5.23 cm
(e) 7.4 m   (f) 3.7 m
(g) 0.25 m   (h) 0.71 m
(i) 2 km   (j) 0.6 km

6. Convert the following measurements to metres:

(a) 700 cm   (b) 300 cm
(c) 560 cm   (d) 890 cm
(e) 2385 cm   (f) 7207 cm
(g) 62 cm   (h) 80 cm
(i) 5000 mm   (j) 8000 mm
(k) 1800 mm   (l) 2040 mm
(m) 3619 mm   (n) 8423 mm
(o) 430 mm   (p) 102 mm
(q) 5 km   (r) 7 km
(s) 2.34 km   (t) 1.78 km
(u) 0.8 km   (v) 0.57 km

7. Copy and complete the following by inserting $<$, $>$ or $=$ signs:

(a) 4 m . . . 23 cm   (b) 9.2 m . . . 450 mm
(c) 760 cm . . . 4.3 m   (d) 124 cm . . . 1240 mm
(e) 65 km . . . 725 m   (f) 30 000 mm . . . 46 m
(g) 92 cm . . . 4 m   (h) 35 km . . . 35 000 m
(i) 95 mm . . . 950 cm   (j) 640 mm . . . 6.4 m
(k) 7.52 m . . . 810 cm   (l) 39 mm . . . 5.2 cm
(m) 76 000 m . . . 76 km   (n) 320 mm . . . 0.32 m
(o) 0.24 cm . . . 0.024 m   (p) 6.5 km . . . 650 000 cm

Example 12.4

(a) Find the sum of 4 m, 78 cm, and 16 mm and express your answer in metres.

(b) Find the perimeter of a rectangular city block if its length is 1.2 km and its width is 840 metres.

Solution

(a) 4 m + 78 cm + 16 mm cannot be calculated until all the measurements are in the same unit. As the answer is to be expressed in metres, we should express each measurement in metres first. For the two measurements *not* given in metres,

$$78 \text{ cm} = \ldots \text{m}. \quad \therefore 78 \div 100 \text{ m} = 0.78 \text{ m}$$

$$16 \text{ mm} = \ldots \text{m}. \quad \therefore 16 \div 1000 \text{ m} = 0.016 \text{ m}$$

Thus
$$
\begin{aligned}
& 4.0 \quad \text{m} \\
& 0.78 \quad \text{m} \\
+\; & 0.016 \text{ m} \\
\hline
=\; & 4.796 \text{ m}
\end{aligned}
$$

$\therefore$ 4 m + 78 cm + 16 mm = 4.796 metres

(b)

Perimeter = 1.2 km + 840 m + 1.2 km + 840 m

Fig 12.10

Before we can calculate the perimeter we need to express the measurements in the same unit, either kilometres or metres.

(i) In kilometres: or   (ii) In metres:

$\overset{\div 1000}{\overbrace{\qquad}}$

$\overset{\times 1000}{\overbrace{\qquad}}$

840 m   = ... km   1.2 km   = ... m
∴ 840 ÷ 1000 km = 0.84 km   1.2 × 1000 m = 1200 m

Perimeter =   1.2   Perimeter =   1200
                0.84   840
                1.2   1200
              + 0.84   +   840
              ─────────   ─────────
                4.08 km   4080 m

Thus the perimeter of the city block is either 4.08 km or 4080 metres.

> Measurements must be expressed in the same units before any calculations can be performed.

Exercise 12D

1. Find the sum of each of the following measurements expressing your answer in metres:
   (a) 2 km + 250 m + 0.6 km   (b) 72 cm + 8 m + 137 cm
   (c) 45 mm + 1.3 m + 562 mm   (d) 12 cm + 512 mm + 0.3 m
   (e) 2.8 m + 125 cm + 320 mm   (f) 16 cm + 10 mm + 0.8 m
   (g) 652 cm + 605 mm + 6 m   (h) 4.6 m + 46 cm + 460 mm
   (i) 7.35 km + 368 m + 0.76 km   (j) 69 m + 0.381 km + 2.053 km

2. Find the perimeters of the following, expressing your answers in the unit stated:

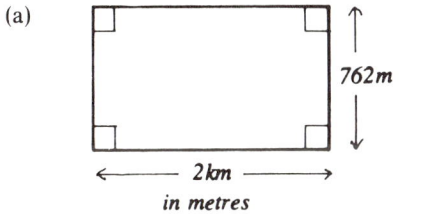

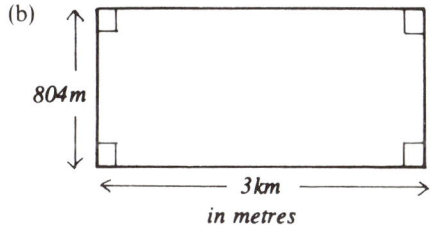

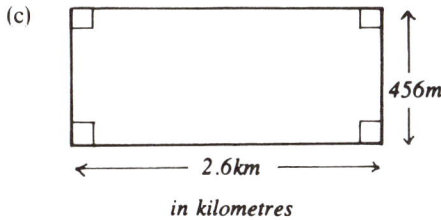

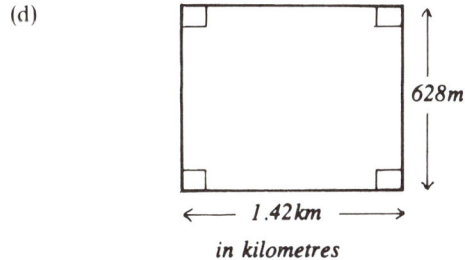

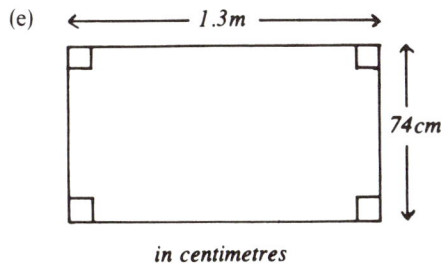

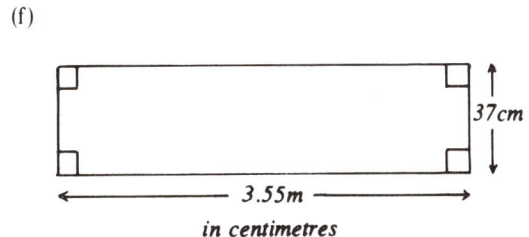

(a)

762 m

2 km
in metres

(b)

804 m

3 km
in metres

(c)

456 m

2.6 km
in kilometres

(d)

628 m

1.42 km
in kilometres

(e)

1.3 m

74 cm

in centimetres

(f)

37 cm

3.55 m
in centimetres

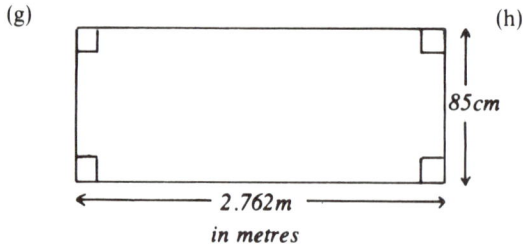

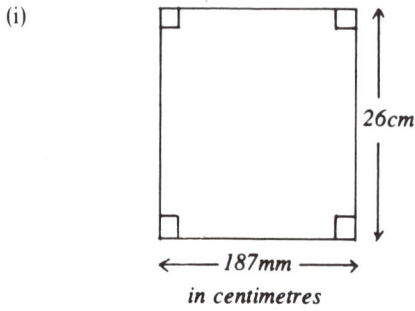

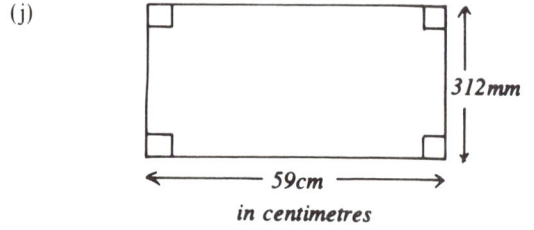

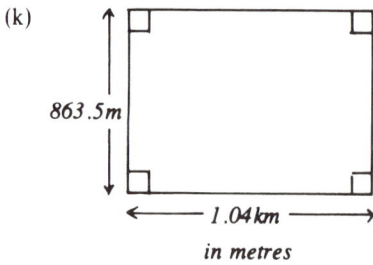

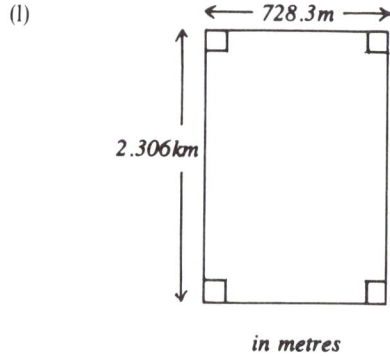

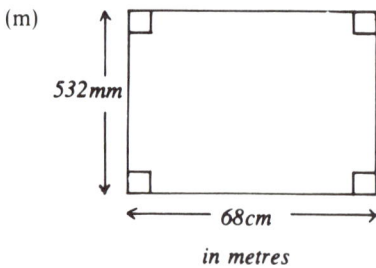

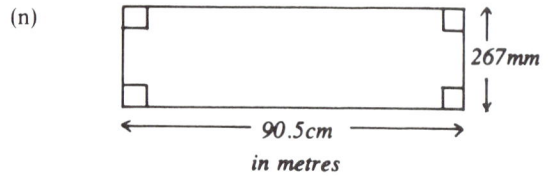

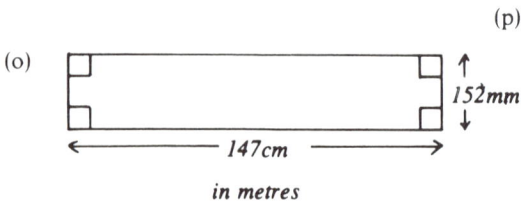

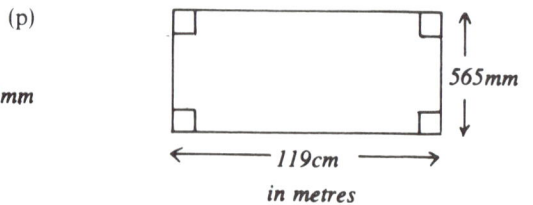

(g)

85cm

2.762m

in metres

(h)

94cm

2.425m

in metres

(i)

26cm

187mm

in centimetres

(j)

312mm

59cm

in centimetres

(k)

863.5m

1.04km

in metres

(l)

728.3m

2.306km

in metres

(m)

532mm

68cm

in metres

(n)

267mm

90.5cm

in metres

(o)

152mm

147cm

in metres

(p)

565mm

119cm

in metres

12.2   Area

The measurement 'area' is used to find the amount of space within the bounds of a plane figure; for example, the area of land bounded by a fence, the area of a table top bounded by its edges etc. Thus, to measure area we need to use a basic shape of known area and then count the number of these shapes which fit in the figure. For example, the area of the rectangle, using the square as the unit of measurement, is 18 squares.

Fig 12.11

The area of the parallelogram, using the triangle as the unit of measurement is 16 triangles.

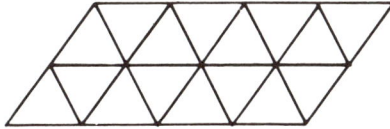

Fig 12.12

However, in each of the above examples the measurement is not precise as we do not know the area of the square or triangle. Therefore we need a unit of measurement with a specific area. We can use a square of side 1 cm.

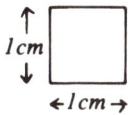

This is '1 square centimetre' or has a area of '1 cm^2' (1 centimetre squared).

There are several units of measurement of area in the metric system.

(a) The *square millimetre* is about the area of a pinhead.

□   1 mm^2 (1 millimetre squared)

(b) The *square centimetre* is about the area of your fingernail.

1 cm^2 (1 centimetre squared)

(c) The *square metre* is about the area of the roof of a 'Mini' motor car.

1 m^2 (1 metre squared)

(d) The *square kilometre*.

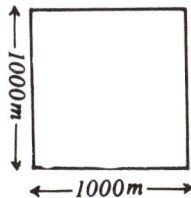

1 km^2 (1 kilometre squared)

(e) The *hectare* is the unit of measurement for land. For example, the size of a farmer's property is measured in hectares.

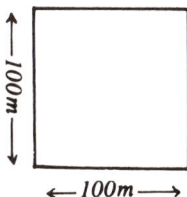

1 hectare

Example 12.5

Find the area of each of the following:

(a)

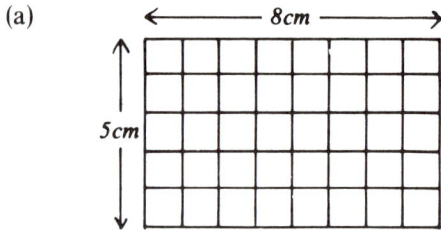

Fig 12.13 (a)

(b)

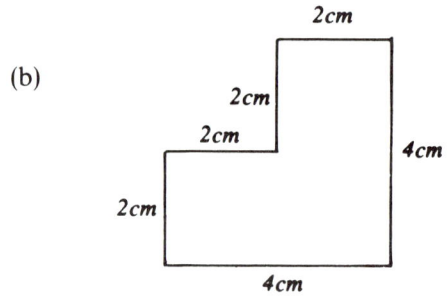

Fig 12.13 (b)

Solution

(a)

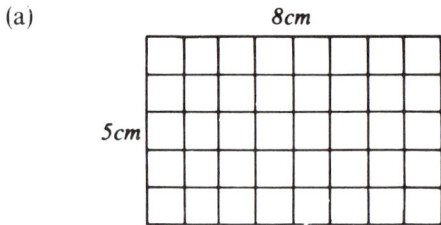

By counting the number of square centimetres in the rectangle we can find the area.
There are 8 cm^2 in each row and 5 rows.
Thus the area is 40 cm^2.

(b)

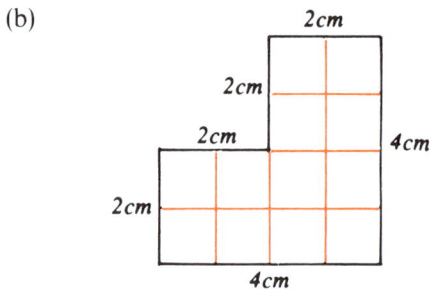

We have been given the lengths of each side, so it is possible to draw in the square centimetres. Now by counting the number of square centimetres we find the area to be 12 cm^2.

Exercise 12E

1. Find the area of each of the following:

   (a)

   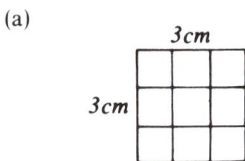

   (b)

   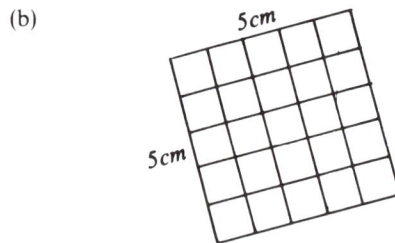

   (c)

   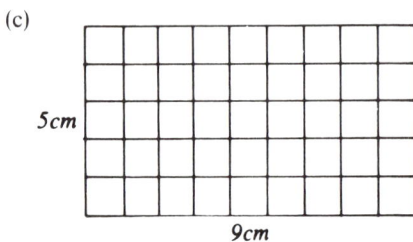

   (d)

   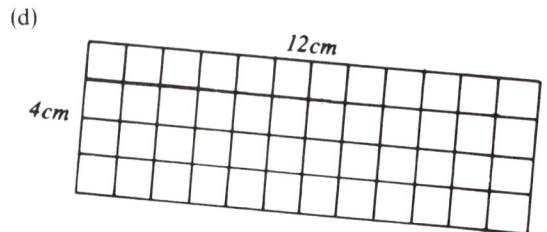

(e)

5cm

11cm

(f)

8cm

3cm

(g)

5cm

3cm

10cm

4cm

(h)

3cm

3cm

3cm

6cm

5cm

(i)

9cm

6cm

6cm

3cm

(j)

4cm

3cm

6cm

7cm

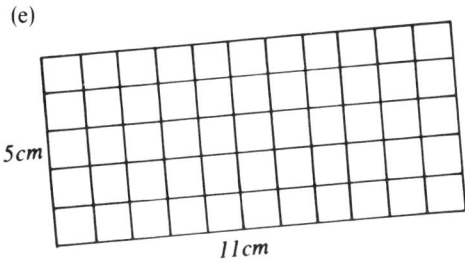

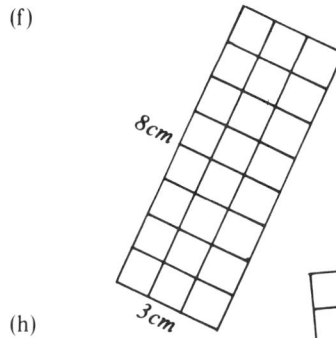

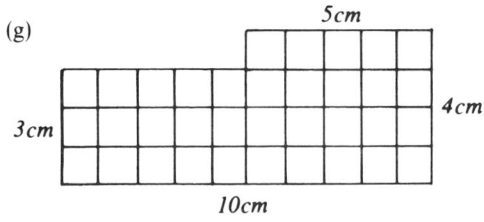

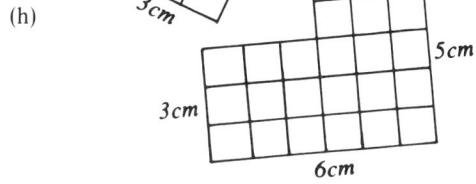

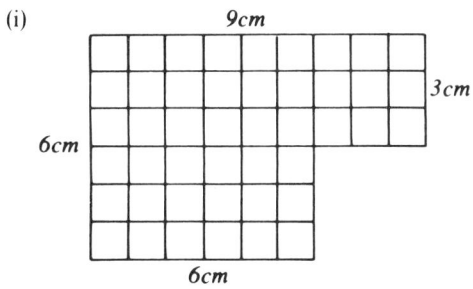

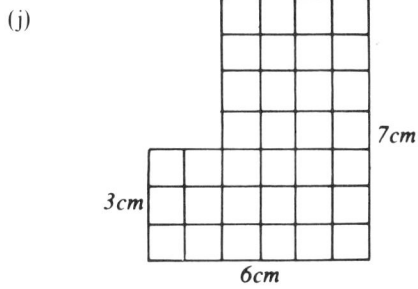

2. Find the area of each of the following:

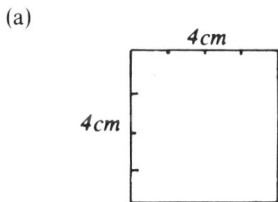

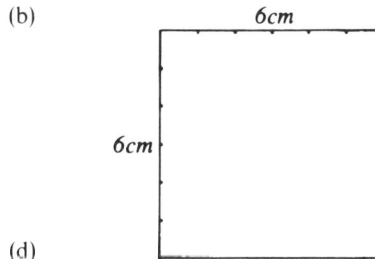

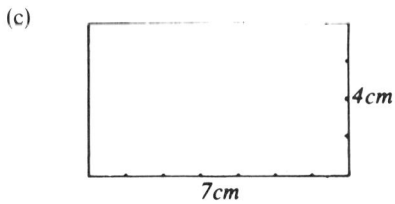

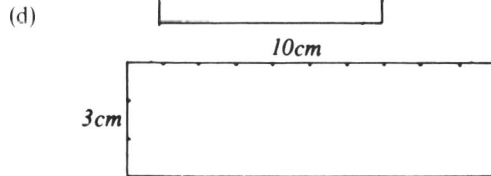

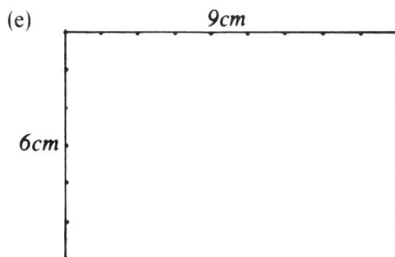

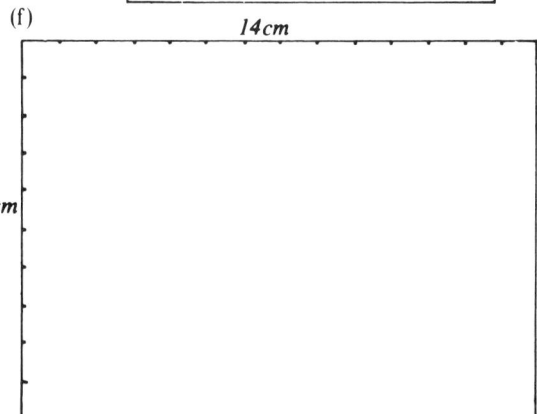

(a)

4cm

4cm

(b)

6cm

6cm

(c)

7cm

4cm

(d)

10cm

3cm

(e)

9cm

6cm

(f)

14cm

10cm

(g) 3cm 3cm 3cm 3cm 6cm

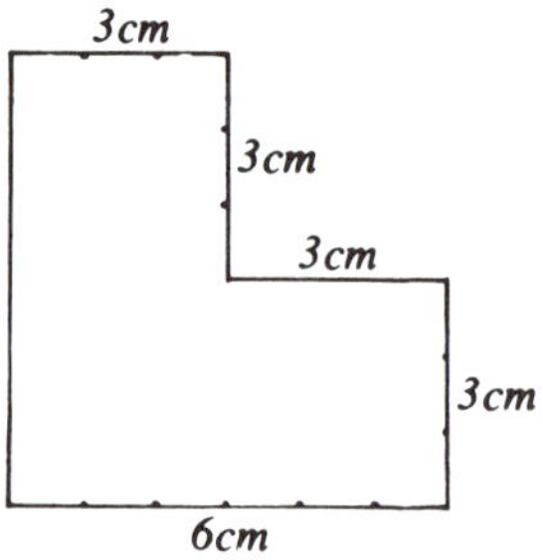

(h) 4cm 2cm 2cm 2cm 4cm

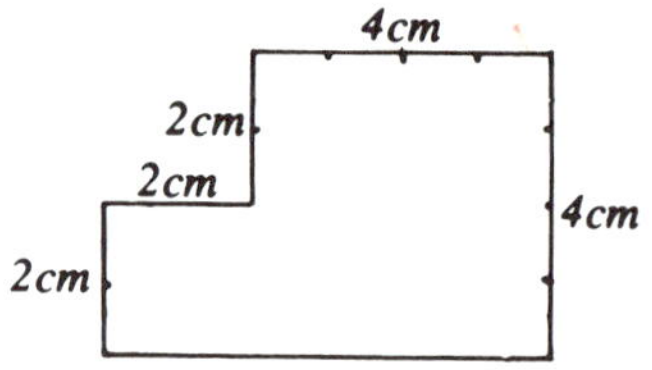

(i) 5cm 2cm 4cm 8cm

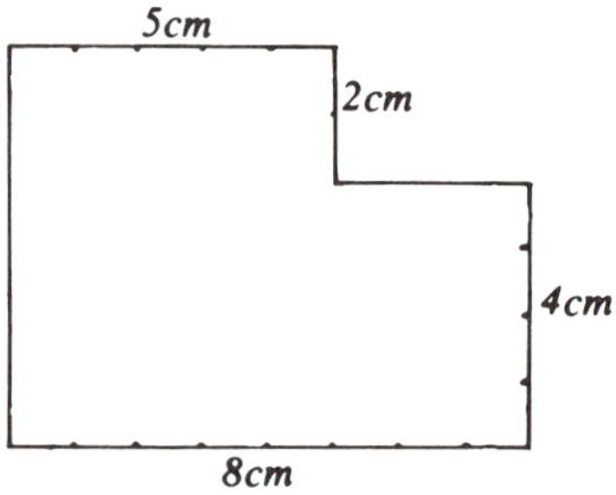

(j) 2cm 2cm 7cm 7cm

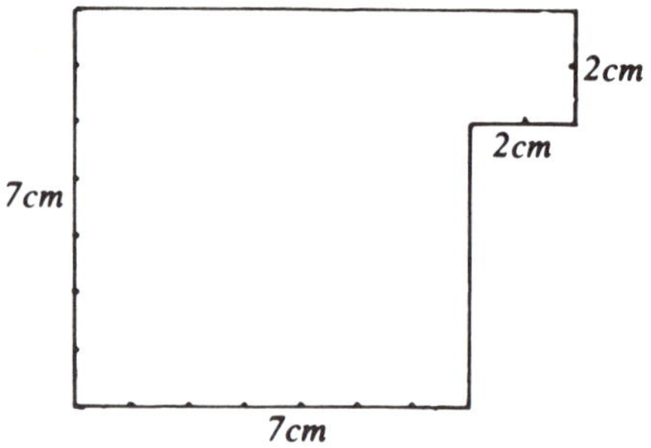

(k) 5cm 4cm 9cm 7cm

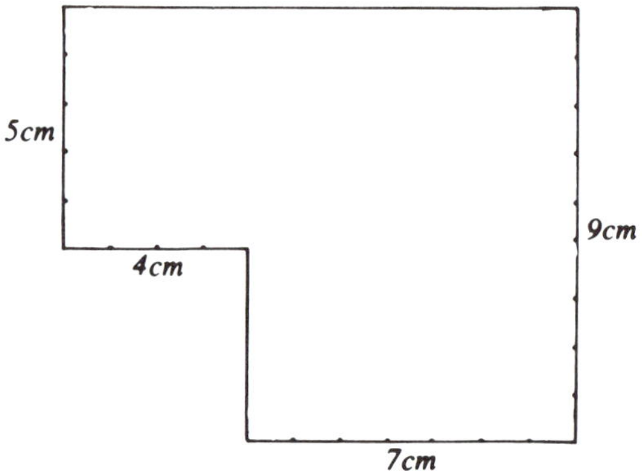

(l) 9cm 3cm 3cm 8cm

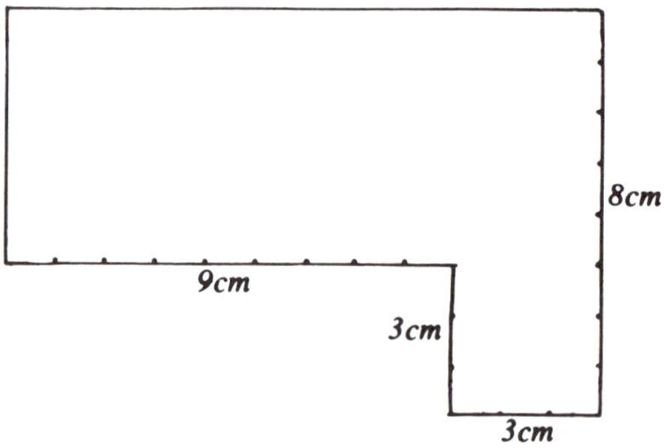

12.2.1   Area of a Rectangle

Consider the square and rectangle below.

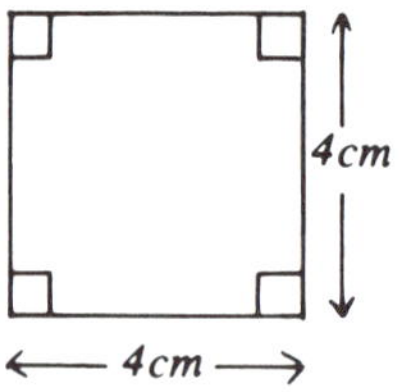

4cm 4cm

Fig 12.14

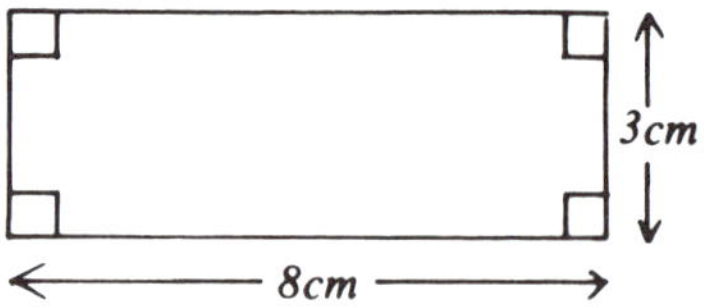

3cm 8cm

Fig 12.15

We have seen that by dividing the above into square centimetres we can calculate the area.

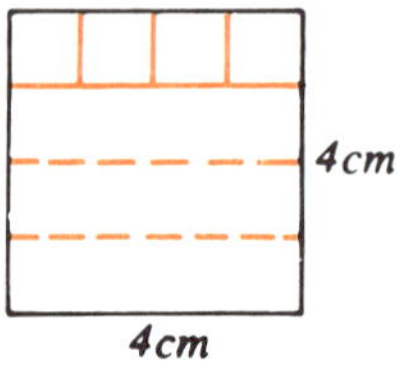

4cm 4cm

In the square there are 4 cm² in a row and there are 4 rows, therefore if we multiply the number of squares in a row by the number of rows, we find the area to be 16 cm².

We can check this by counting the number of squares.

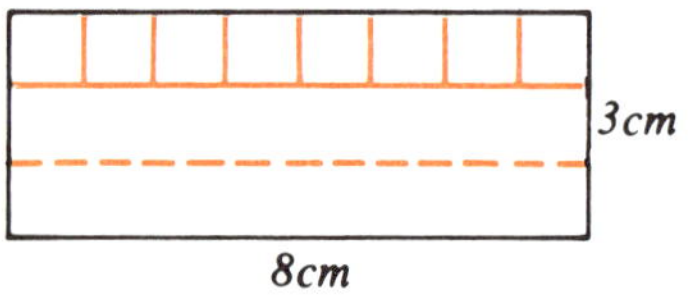

3cm 8cm

In the rectangle there are 8 cm² in each row and there are 3 rows, therefore if we multiply the number of squares in a row by the number of rows we find the area to be 24 cm².

This can also be checked by counting the squares.

In both the above instances the *number of squares in a row* corresponds to the *length* of the square or rectangle and the *number of rows* corresponds to the *width* of the square or rectangle. Thus, we should find that by multiplying the length by the width of a square or rectangle gives us the area.

Example 12.6

Find the area of the following rectangle:

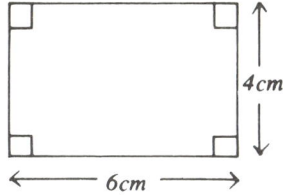

Fig 12.16

Solution

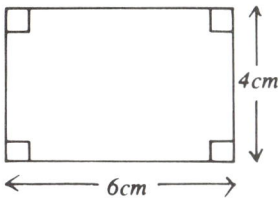

Area of the rectangle = the length multiplied by the width.

So,   Area = 6 cm × 4 cm
              = 24 cm^2

Checking by counting squares:

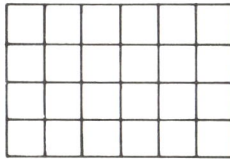

we find there are 24 square centimetres.

> In general, the area of a rectangle is equal to the length multiplied by the width and is expressed in square units.
>
> $$A = L \times W$$

Example 12.7

Find the area of each of the following:

(a)

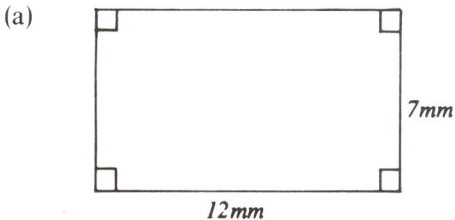

Fig 12.17 (a)

(b)

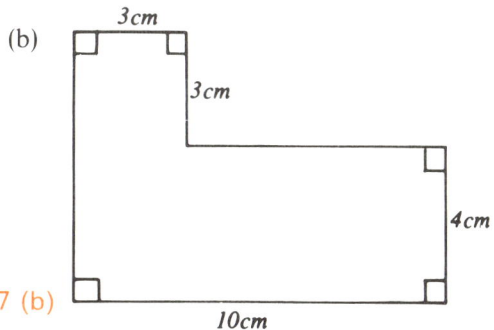

Fig 12.17 (b)

(c) An estate agent was required to find the area of land on a property, not including that taken up by the house. He measured the length of the property and found it to be 120 m and the width to be 70 m. Then he measured the length and width of the house which he found to be square and measured 40 m along each side. What was the area of land?

Solution

(a)

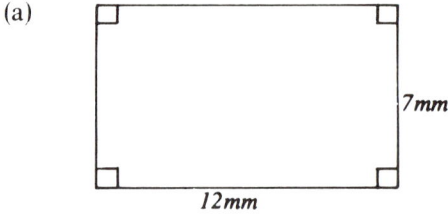

Area of a rectangle $= L \times W$
$= 12 \text{ mm} \times 7 \text{ mm}$
$= 84 \text{ mm}^2$

(b)

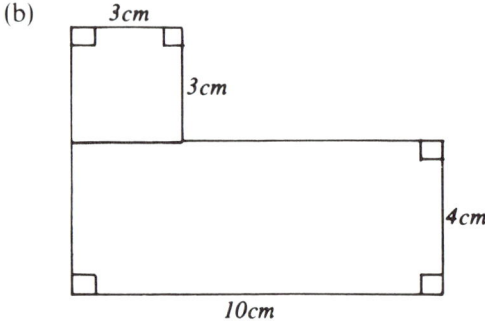

To find the area of the entire shape, it is easier to divide it into two seperate rectangles, find the area of each and then add the seperate areas together to give the total area of the shape.

Thus

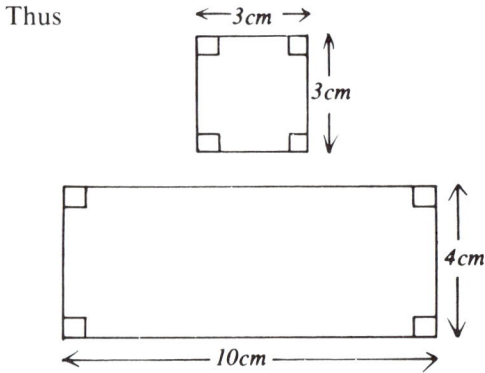

$A = L \times W$
$= 3 \text{ cm} \times 3 \text{ cm}$
$= 9 \text{ cm}^2$

$A = L \times W$
$= 10 \text{ cm} \times 4 \text{ cm}$
$= 40 \text{ cm}^2$

∴ Total area of the shape $= 9 \text{ cm}^2 + 40 \text{ cm}^2$
$= 49 \text{ cm}^2$

(c) The property the estate agent measured was rectangular and the house on it was square, as shown on the diagram.

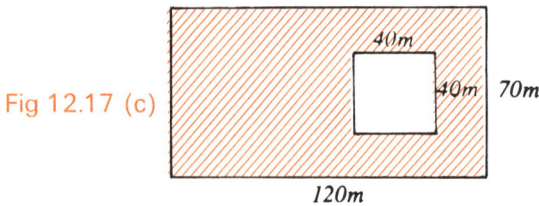

Fig 12.17 (c)

The area he needed to find is shaded.
The shaded area is the area of the property
− area of the house
$= 120 \text{ m} \times 70 \text{ m} − 40 \text{ m} \times 40 \text{ m}$
$= 8400 \text{ m}^2 − 1600 \text{ m}^2$
$= 6800 \text{ m}^2.$

Exercise 12F

1. Find the area of each of the following:

   (a)

   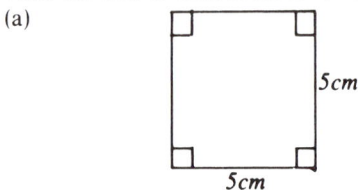

   (b)

   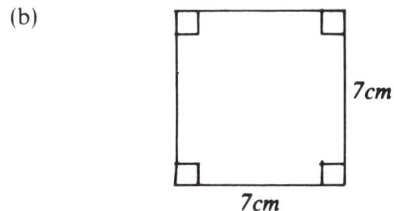

(c)

3mm

7mm

(d)

7mm

9mm

(e)

8m

12m

(f)

4m

15m

(g)

6cm

14cm

(h)

5cm

16cm

(i)

10mm

18mm

(j)

8mm

25mm

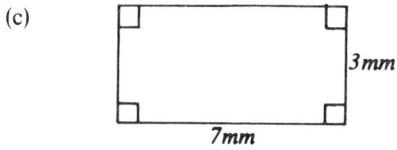

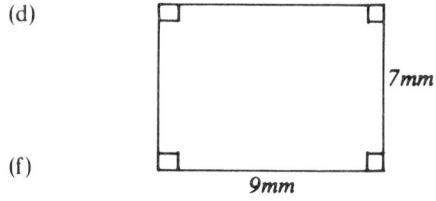

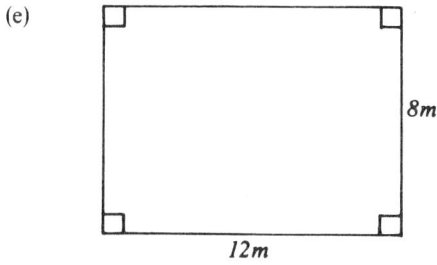

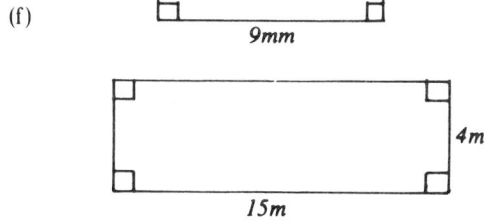

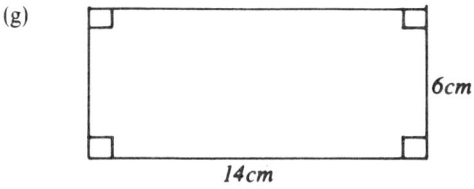

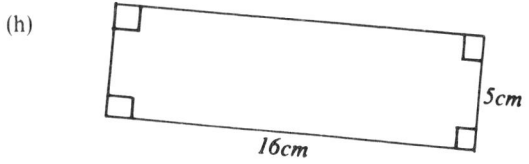

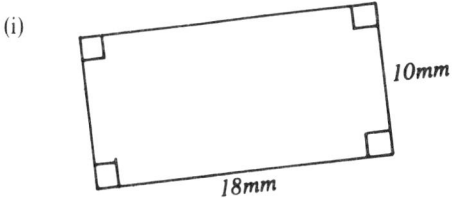

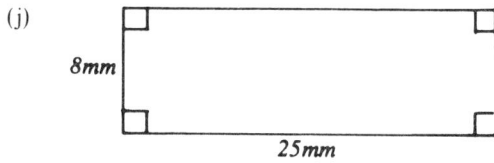

2. Find the total area of each of the following:

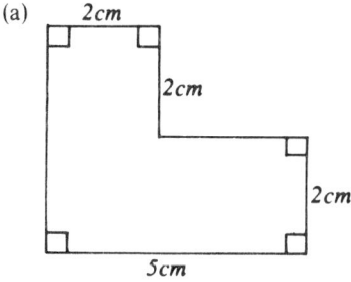

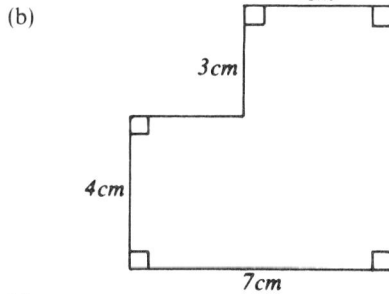

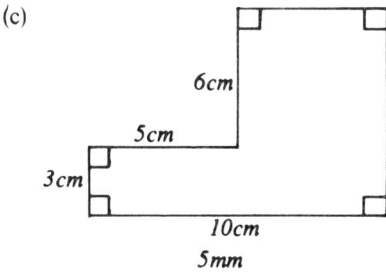

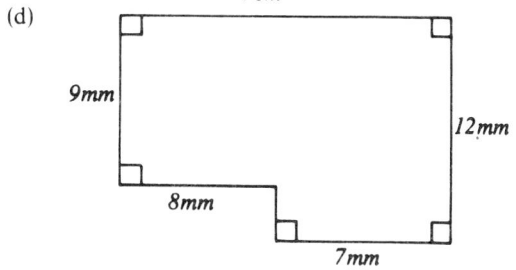

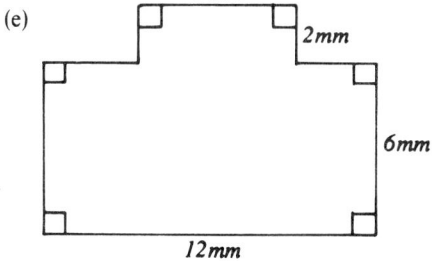

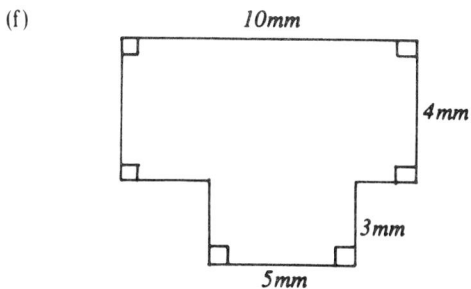

(a)

2cm

2cm

2cm

5cm

(b)

4cm

3cm

4cm

7cm

(c)

6cm

5cm

3cm

10cm

(d)

9mm

12mm

8mm

7mm

(e)

5mm

2mm

6mm

12mm

(f)

10mm

4mm

3mm

5mm

(g)

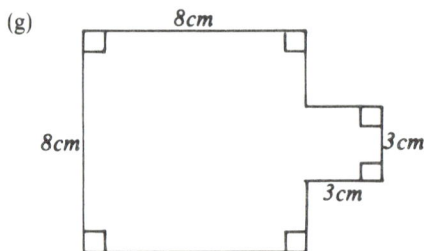

(h)

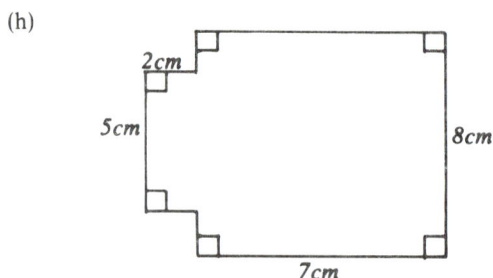

(i)

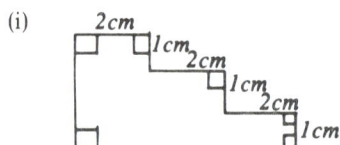

(j)

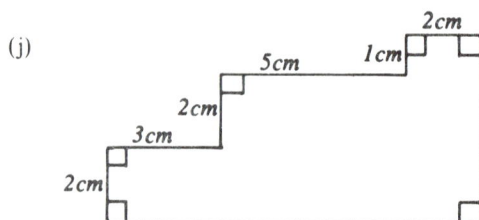

3. Draw a diagram for each of the following and find the area required.

(a) A square has a side length of 9 cm. What is its area?

(b) A rectangle is 16 cm long and 10 cm wide. Find its area.

(c) If the square in part (a) was drawn inside the rectangle in part (b); find the area outside the square but inside the rectangle.

(d) A square has a side length of 15 m. Find its area.

(e) A rectangle has a length of 25 m and a width of 20 m. Find its area.

(f) Draw the square in part (d) inside the rectangle in part (e) and find the area outside the square but inside the rectangle.

(g) The Smith family have a back garden which is 20 m long and 12 m wide. They put in a swimming pool which was 10 m long and 4 m wide. What is the area of the remaining back garden?

(h) At the Comprehensive School they have a rectangular playing field which is 150 m in length and 60 m in width. During the cricket season a certain area is fenced off to protect the cricket pitch. If that area is 40 m long and 8 m wide, find the remaining area which can still be used.

(i) A small farm consists of two paddocks, one which has a house and a few sheds on it and the other which is for cattle. If the paddock for the cattle is 500 m long and 320 m wide and the hayshed which is also on the paddock is 45 m long and 30 m wide, find the area which the cattle have to graze on.

(j) The Southvale Tennis club have an area of land on which they are going to build two tennis courts. If the land available is 80 metres in length and 22 metres in width and each tennis court is 28 metres long and 12 metres wide, find the area of land occupied by (i) each tennis court (ii) both tennis courts and (iii) the region round the tennis courts.

12.2.2   Area of a Triangle

Consider the right-angled triangle below.

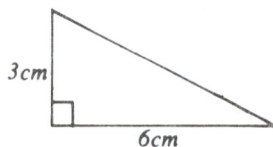

Fig 12.18

Dividing it into square centimetres, we find that it is difficult to count the number of squares exactly.

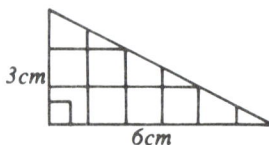

Fig 12.19

However it is easy to see that a right-angled triangle is exactly half a rectangle of the same dimensions.

(c)

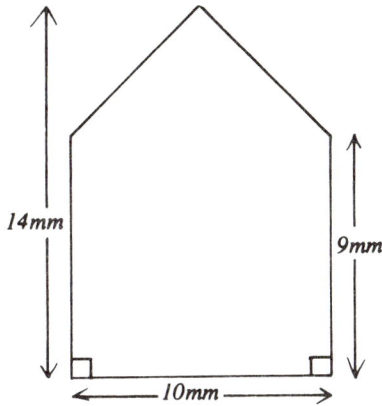

Fig 12.29 (c)

Solution

(a)

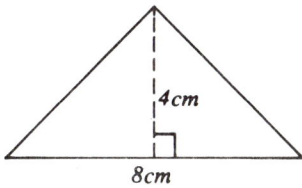

Area of triangle $= \frac{1}{2} \times b \times h$
$\qquad = \frac{1}{2} \times 8$ cm $\times 4$ cm
$\qquad = \frac{1}{2} \times 32$ cm^2
$\qquad = 16$ cm^2

(b)

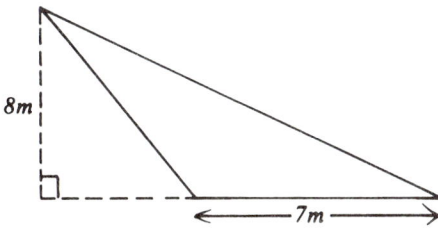

Note: the height of any triangle is the distance from the base (or level of the base) to the vertex opposite the base. In this case the height falls outside the triangle.

$\therefore$ Area $= \frac{1}{2} \times b \times h$
$\qquad = \frac{1}{2} \times 7$ m $\times 8$ m
$\qquad = \frac{1}{2} \times 56$ m^2
$\qquad = 28$ m^2

(c)

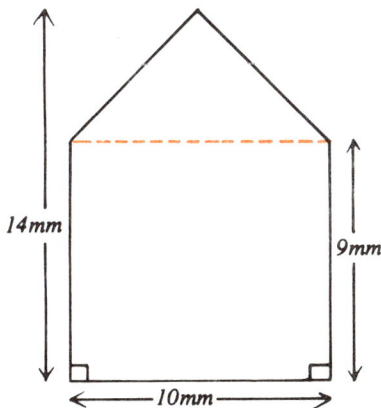

To be able to find the total area of the shape, it will be necessary to divide it into a rectangle and a triangle.

Thus, the area of the rectangle
$\qquad = L \times W$
$\qquad = 10$ mm $\times 9$ mm
$\qquad = 90$ mm^2

Now, to calculate the area of the triangle we need its height. It can be seen that the distance from the top of the triangle to the base of the rectangle is 14 mm, and the width of the rectangle is 9 mm.

Therefore, the height of the triangle $= 14$ mm $- 9$ mm
$\qquad\qquad\qquad\qquad\qquad\qquad = 5$ mm

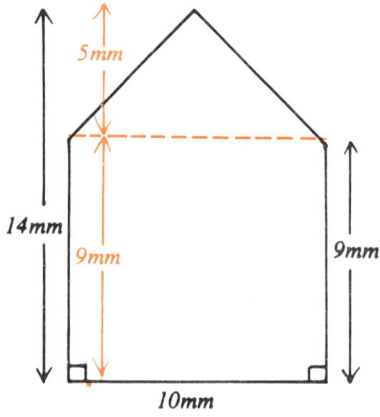

$$\therefore \text{ Area of triangle } = \tfrac{1}{2} \times b \times h$$
$$= \tfrac{1}{2} \times 10 \text{ mm} \times 5 \text{ mm}$$
$$= \tfrac{1}{2} \times 50 \text{ mm}^2$$
$$= 25 \text{ mm}^2$$

Fig 12.30

$$\text{Total area} = \text{area of rectangle } + \text{ area of triangle}$$
$$= \quad 90 \text{ mm}^2 \quad + \quad 25 \text{ mm}^2$$
$$= \quad 115 \text{ mm}^2$$

Exercise 12G

1. Find the area of each of the triangles below:

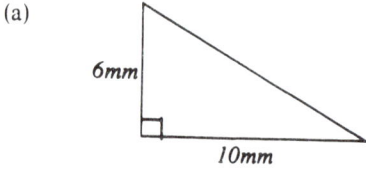

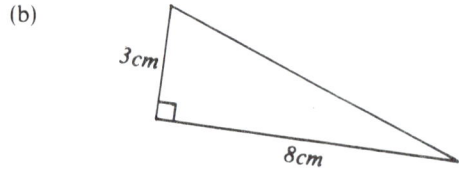

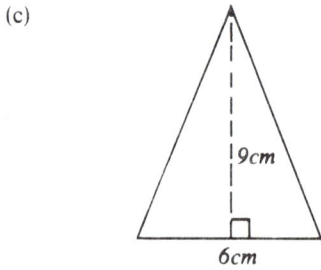

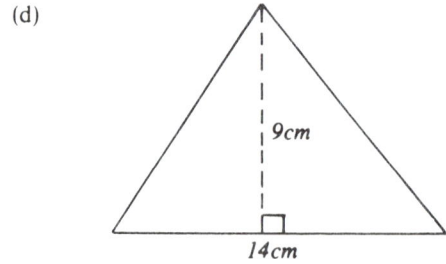

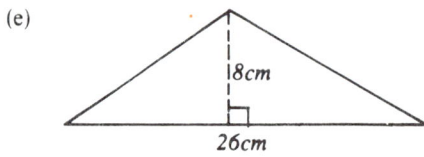

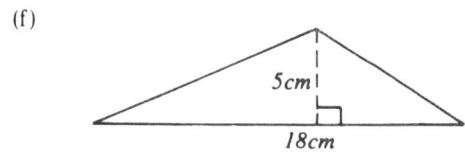

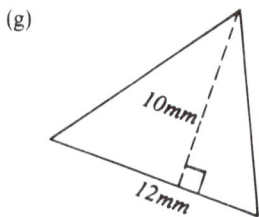

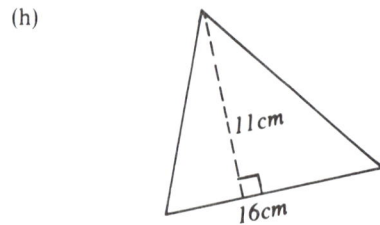

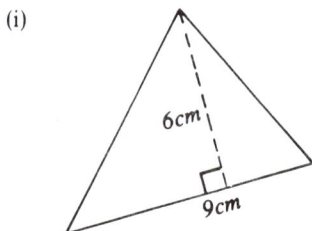

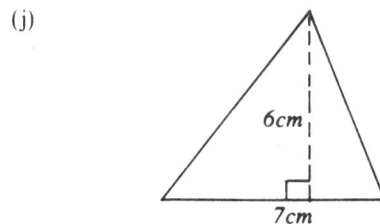

   (a)

   6mm
   10mm

   (b)

   3cm
   8cm

   (c)

   9cm
   6cm

   (d)

   9cm
   14cm

   (e)

   8cm
   26cm

   (f)

   5cm
   18cm

   (g)

   10mm
   12mm

   (h)

   11cm
   16cm

   (i)

   6cm
   9cm

   (j)

   6cm
   7cm

(k)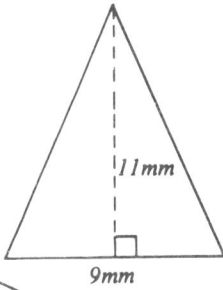

11mm

9mm

(l)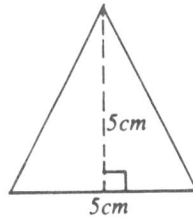

5cm

5cm

(m)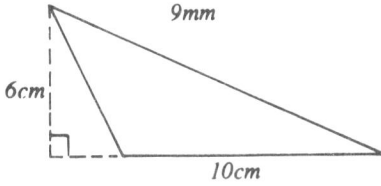

6cm

10cm

(n)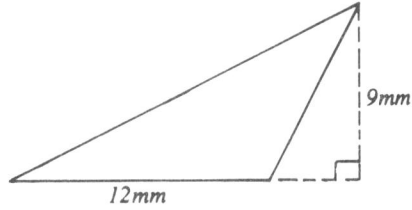

9mm

12mm

(o)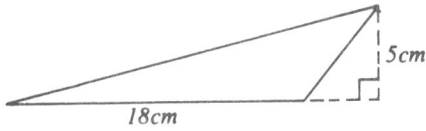

5cm

18cm

(p)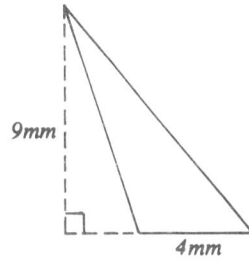

9mm

4mm

(q)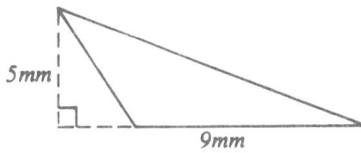

5mm

9mm

(r)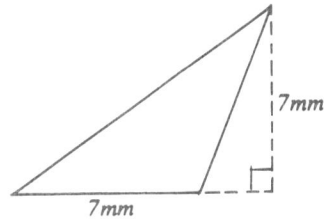

7mm

7mm

2. Find the total area of each of the following shapes:

(a)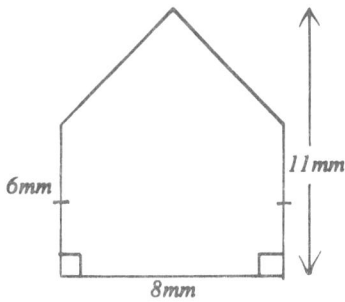

6mm

11mm

8mm

(b)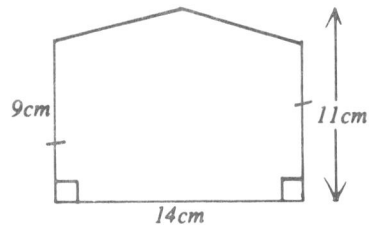

9cm

11cm

14cm

(c)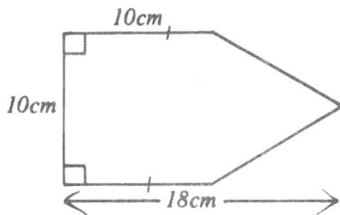

10cm

10cm

18cm

(d)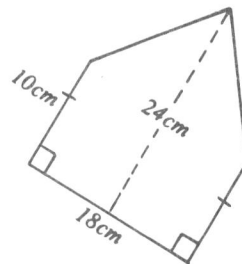

10cm

24cm

18cm

(e)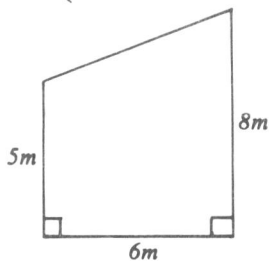

8m

5m

6m

(f)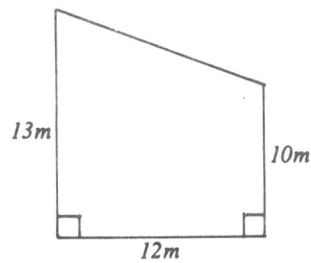

13m

10m

12m

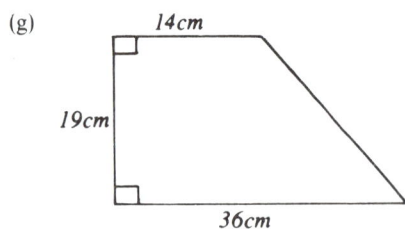

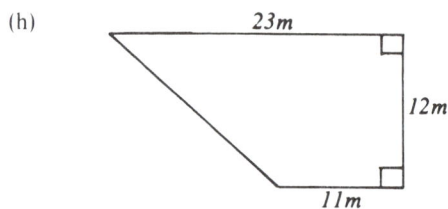

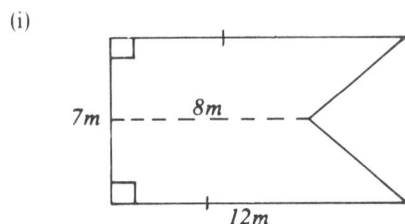

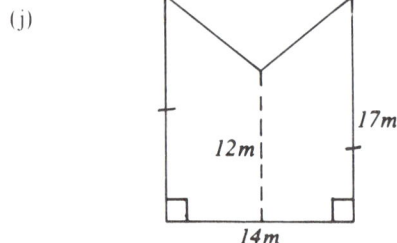

(g)

14cm

19cm

36cm

(h)

23m

12m

11m

(i)

7m

8m

12m

(j)

12m

17m

14m

12.3   Volume

The measurement 'volume' is related to 3-dimensional objects (i.e. objects with length, width and height) and the amount of space they occupy. For example, the amount of space a cardboard box occupies is called its volume, the amount of space a house occupies is called the volume of that house etc.

Just as the unit of measurement for area needs a basic shape of known area, the unit of measurement for volume will need to be a basic shape which has a known volume and the number of those shapes that fit exactly into an object will be the volume of that object.

The basic shape for area is the square unit, that is, the square centimetre (cm^2), the square metre (m^2) etc.

The basic shape for volume is easiest if it is also based on the square but it must be 3-dimensional. A 3-dimensional square is called a cube.

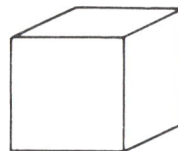

A cube

Fig 12.31

Thus, the units of measurement for volume are:

(a) the *cubic millimetre* which would occupy about the same volume as the end of your pencil lead.

1 mm^3 (1 millimetre cubed)

(b) the *cubic centimetre* which occupies about the same space as a die.

1 cm^3 (1 centimetre cubed)

(c) the *cubic metre*, which is used to measure larger volumes, is written 1 m^3 (1 metre cubed).

As stated earlier, volume is related to 3-dimensional objects. However, these objects can take many different shapes and forms.
For example,

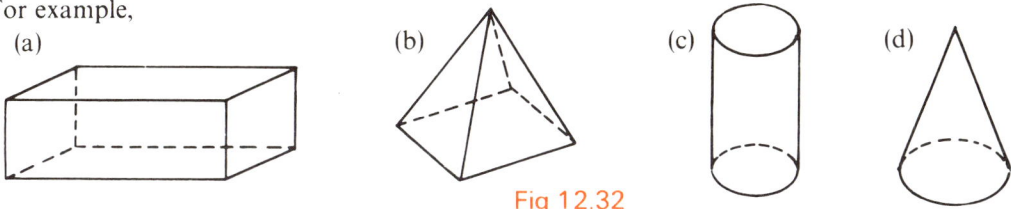

(a)   (b)   (c)   (d)

Fig 12.32

Although all of the above are 3-dimensional objects we will be dealing only with those which have a *uniform cross-section*. When such objects are cut through, parallel to an outside face, the face of the 'cut' will look and measure exactly the same, no matter where the cut is made. The objects (a) and (c) above have this property. They are called *prisms*.

> A prism is a 3-dimensional figure which has a uniform cross-section.

12.3.1   Volume of Rectangular Prisms

A rectangular prism is a prism with a uniform rectangular cross-section.

Example 12.9

Find the volume of each of the following prisms by counting the number of cubic centimetres.

(a)

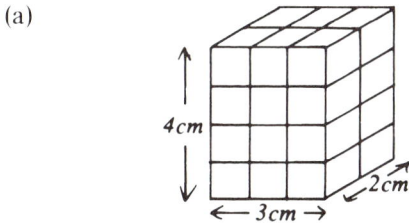

(b)

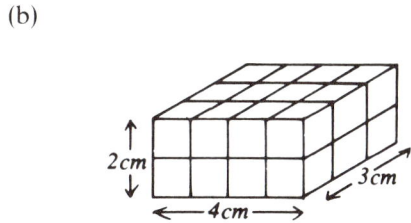

Solution

(a)

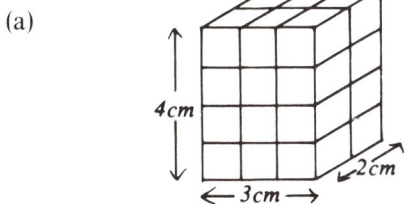

The diagram does not show us *all* the cubic centimetres contained within the prism. Therefore, we need to account for those not seen to find the total number of cubic centimetres.

To do this it is easiest to count the number of cm^3 in each layer and then the number of layers. Since there are 6 cubic centimetres in the top layer and there are 4 layers each with 6 cm^3, the volume of the prism = 4 layers × 6 cm^3

= 24 cm^3

(b)

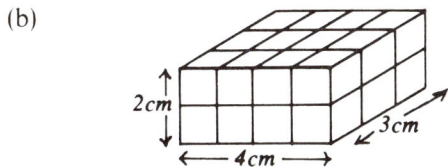

Similarly, the easiest approach is to count the cubic centimetres in a layer and the number of layers.

Thus, volume of the prism = 2 layers × 12 cm^3
= 24 cm^3.

N.B. In both the above examples the volume was 24 cm^3, even though the shapes differed. It is an interesting fact that objects can differ in shape but still have the same volume.

Exercise 12H

Count the number of cubic centimetres in the prisms below and thus find their volumes.

1.
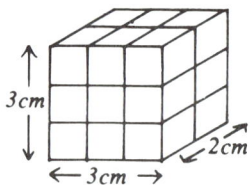

2.
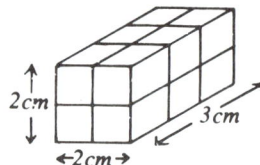

3.
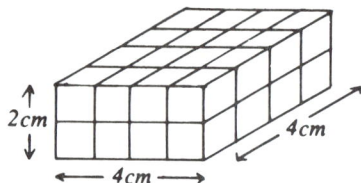

4.
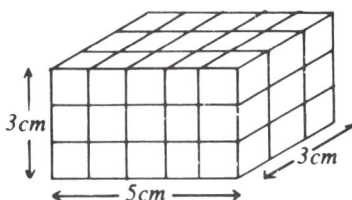

5.
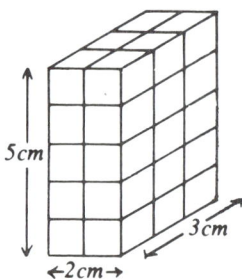

6.
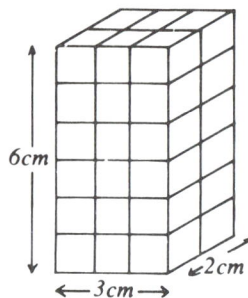

7.
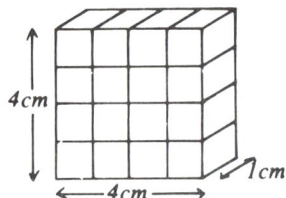

8.
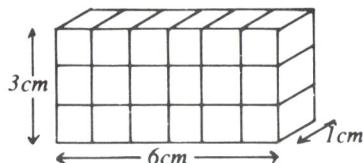

9.
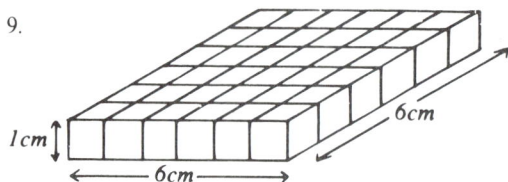

10.
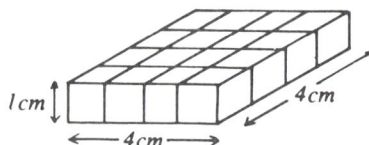

11.
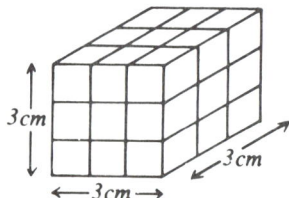

12.
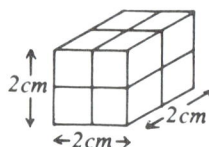

13.
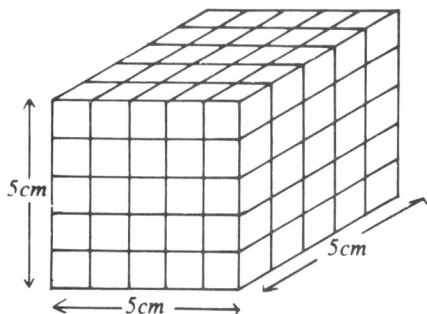

14.
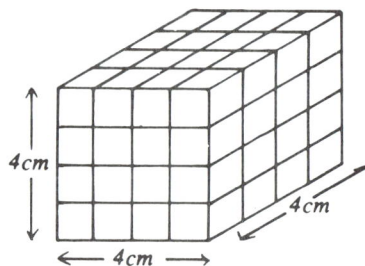

15.

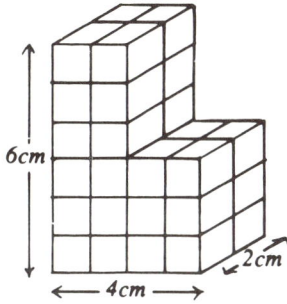

6cm

4cm

2cm

16.

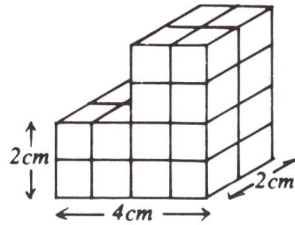

2cm

4cm

2cm

17.

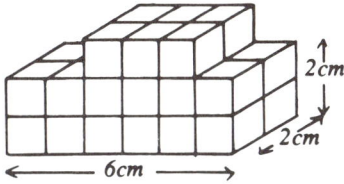

2cm

2cm

6cm

18.

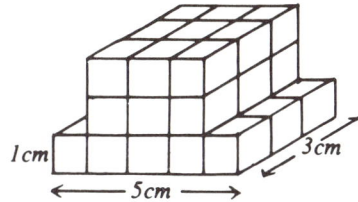

1cm

5cm

3cm

19.

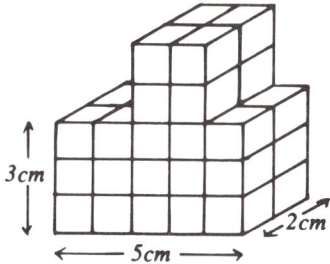

3cm

5cm

2cm

20.

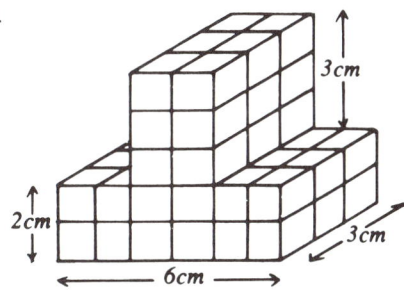

3cm

2cm

6cm

3cm

Consider the prism below.

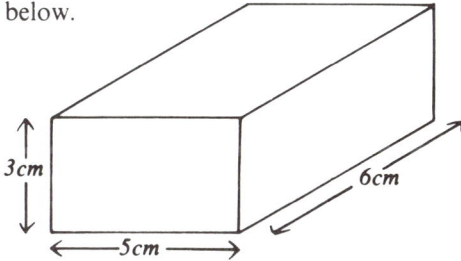

3cm

6cm

5cm

Fig 12.33

To find its volume we could draw in the cubic centimetres and count them but as we have seen in the previous examples, we do not actually count each cube within the prism.

More simply if we knew the number of cubes within one layer, the height of the prism (3 cm) would tell us the number of layers.

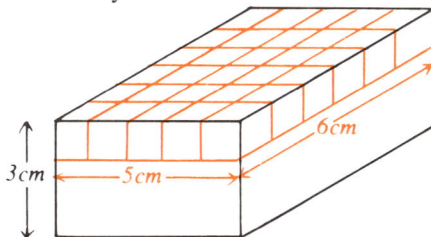

6cm

3cm

5cm

Fig 12.34

Thus, the diagram now shows us the cubic centimetres in one layer; that is, 6 rows of 5 cm² or 30 cm².

In fact what we have found is the area of the top or *base* of the prism.

I.e.   Area = length × width
              = 5 cm × 6 cm
              = 30 cm²

Now, by multiplying the number of cubes in one layer by the number of layers or height, we will have the volume.

I.e.   Volume = 30 cm³ × 3 layers
                      = 90 cm³

More simply again, to find the volume of a prism,

         Volume = number of cm³ in 1 layer × number of layers
                = Area of base × height.

In general,

> Volume of a prism = Area of base × height and is measured in cubic units.

Example 12.10

Find the volume of each of the following prisms:

(a)

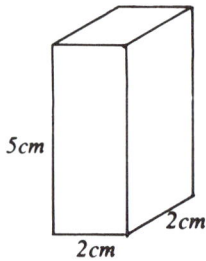

(b)

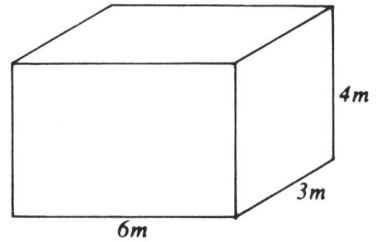

Solution

(a)

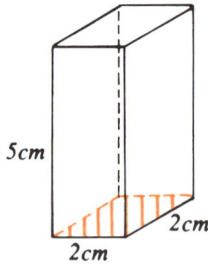

Volume = Area of base × height
       = (L × W) × height
       = (2 cm × 2 cm) × height
       = 4 cm² × 5 cm
       = 20 cm³

(b)

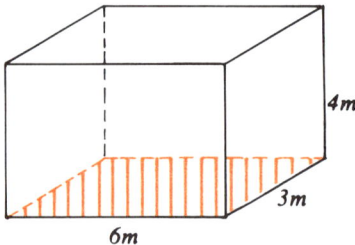

Volume = Area of base × height
       = (L × W) × height
       = 6 m × 3 m × height
       = 18 m² × 4 m
       = 72 m³

Exercise 12I

Find the volume of each of the following prisms:

1.

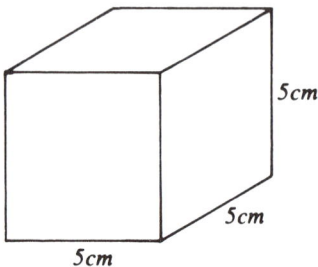

2.

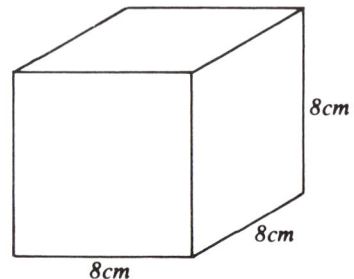

3.

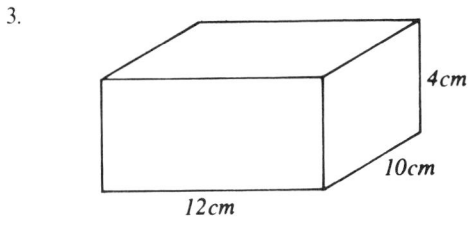

4cm
10cm
12cm

4.

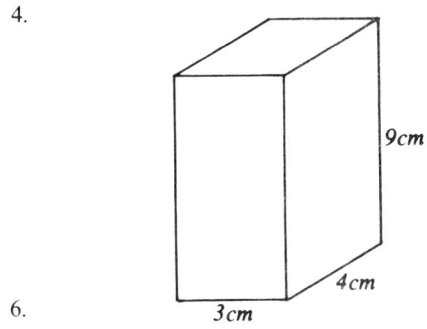

9cm
4cm
3cm

5.

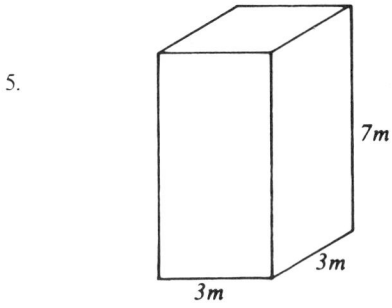

7m
3m
3m

6.

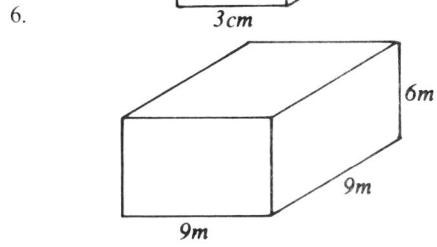

6m
9m
9m

7.

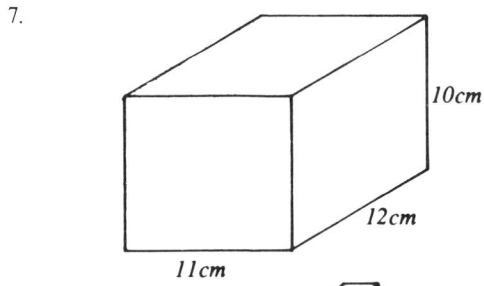

10cm
12cm
11cm

8.

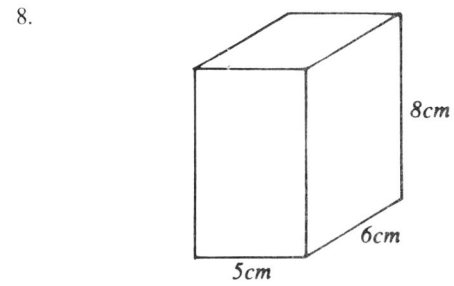

8cm
6cm
5cm

9.

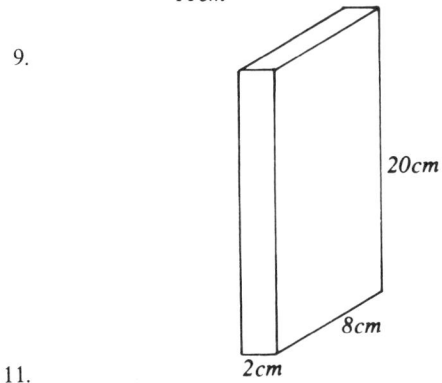

20cm
8cm
2cm

10.

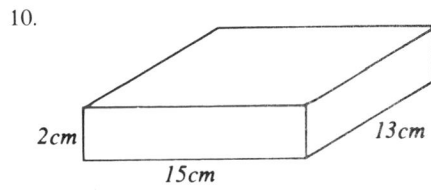

2cm
13cm
15cm

11.

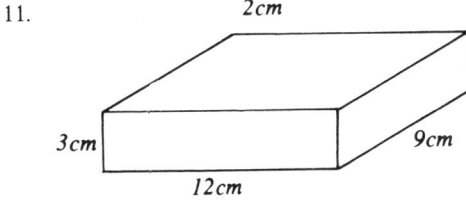

3cm
9cm
12cm

12.

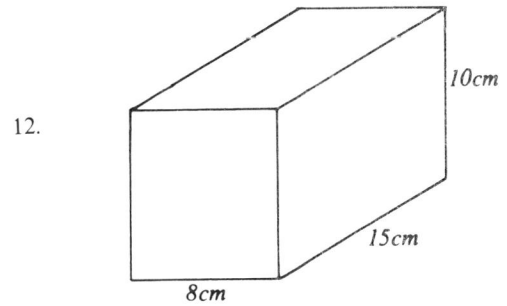

10cm
15cm
8cm

13.

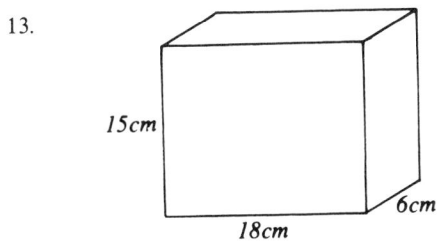

15cm
6cm
18cm

14.

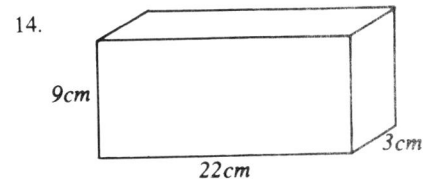

9cm
3cm
22cm

15.
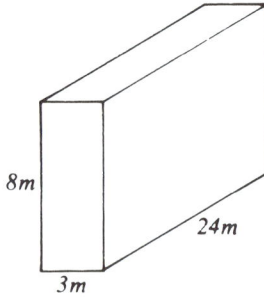

16.
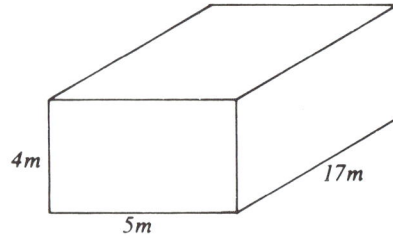

12.3.2   Capacity

Capacity is a measurement similar to volume, in that they both deal with 3-dimensional objects. Whereas volume measures the amount of space an object occupies, capacity measures the amount of space an object contains.

Capacity is related mainly to liquid measurements and the basic unit of capacity is the *litre*. The units of capacity most commonly used are

(a) the *millilitre* which is about the same as 2 drops from an eye-dropper.

   1 millilitre

We abbreviate millilitre to ml.

(b) the *litre* which is generally the amount of soft drink in a large bottle.

A capacity of 1 litre

We abbreviate litre to *l*.

Similarly to the other metric units of measurement, 'milli' means $\frac{1}{1000}$

∴ 1 millilitre $= \frac{1}{1000}$ of a litre

or 1000 millilitres = 1 litre.

Exercise 12J

1. Which unit of capacity would you measure the following in?
    (a) The capacity of a can of soft drink;   (b) The capacity of a swimming pool;
    (c) The capacity of a tablespoon;   (d) The capacity of a teacup;
    (e) The amount of petrol in a car;   (f) The amount of beer in a keg;
    (g) The amount of medicine to be taken;   (h) The amount of orange juice in a bottle;
    (i) The amount of vinegar put on fish and chips;
    (j) The amount of water pumped on to a fire.

2. (a) If a bottle contains 1 litre of soft drink, how many millilitres does it contain?
    (b) If a milk carton holds 2 litres of milk, how many millilitres of milk are there?
    (c) If a glass holds $\frac{1}{2}$ litre of cordial, how many millilitres does it contain?
    (d) If a teacup contains $\frac{1}{4}$ litre of tea, how many millilitres does it contain?
    (e) If a bucket contains $8\frac{1}{2}$ litres of water, how many millilitres does it contain?

(f) If a jug holds $1\frac{3}{4}$ litres of milk, how many millilitres does it contain?

(g) If a can of paint holds 800 ml, how many litres does it hold?

(h) If a bottle holds 600 ml of detergent, how many litres does it hold?

(i) If a kettle holds 950 millilitres of water, how many litres does it hold?

(j) If a saucer holds 320 millilitres of milk, how many litres does it hold?

(k) If an egg cup holds 80 millilitres of water, how many litres does it hold?

(l) If a can contains 485 ml of soup, how many litres does it contain?

(m) If a test-tube contains 74 millilitres of acid, how many litres does it contain?

(n) If a thermos flask contains 1375 ml of coffee, how many litres does it contain?

(o) If a container holds 2455 millilitres of ice-cream, how many litres does it contain?

(p) If a jug contains 1842 millilitres of orange juice, how many litres does it contain?

12.4   Mass

The mass of an object tells us how hard it is to push that object. For example, a truck would have a greater mass than a car, as it is harder to push.

Usually we speak about the 'weight' of an object rather than its mass, but strictly speaking mass and weight are *different* and have different units of measure. The distinction between mass and weight is likely to be discussed in your science lessons.

The units of measurement for mass most commonly used are,

(a) the *gram* which is about the mass of water contained in $\frac{1}{2}$ a thimble.

1 gram

We abbreviate gram to g.

(b) the *kilogram,* which is about the mass of water contained in a jug.

1 kilogram

We abbreviate kilogram to kg.

Just as with the other metric units of measurement, the prefix *kilo* means 1000. Thus a *kilogram* is 1000 grams or a gram is $\frac{1}{1000}$ of a kilogram.

There is a larger unit of measure than that with a prefix of 'kilo'. This is the *tonne*, which is about the mass of a small car.

1 tonne

A *tonne* is equivalent to *1000 kilograms* or a kilogram is $\frac{1}{1000}$ of a tonne.

Exercise 12K

1. State the unit of mass used to measure each of the following:

   (a) the mass of a bag of apples;   (b) the mass of a bag of mushrooms;

   (c) the mass of a person;   (d) the mass of a truck;

   (e) the mass of a sparrow;   (f) the mass of a ship;

   (g) the mass of a box of matches;   (h) the mass of a chair;

   (i) the mass of a train;   (j) the mass of a tin of paint;

   (k) the mass of a packet of biscuits;

   (l) the mass of water that falls over a waterfall each day;

   (m) the mass of a letter;

   (n) the mass of a beam of wood;

   (o) the mass of coal mined each day.

2. Express each of the following measurements in grams:

   (a) 1 kilogram   (b) 2 kilograms

   (c) 5 kilograms   (d) 3 kilograms

   (e) $\frac{1}{2}$ kilogram   (f) $\frac{1}{4}$ kilogram

   (g) $\frac{3}{4}$ kilogram   (h) $\frac{1}{10}$ kilogram

(i) $1\frac{1}{4}$ kilograms   (j) $1\frac{1}{2}$ kilograms
(k) $2\frac{3}{10}$ kilograms   (l) $3\frac{9}{10}$ kilograms
(m)   0.5 kilogram   (n)   0.2 kilograms
(o)   0.25 kilogram   (p)   0.75 kilograms
(q)   0.473 kilograms   (r)   0.205 kilograms
(s)   2.39 kilograms   (t)   4.62 kilograms
(u)   5.071 kilograms   (v)   9.027 kilograms

3. Express each of the following measurements in kilograms:
   (a)   1000 g   (b)   4000 g
   (c)   250 g   (d)   500 g
   (e)   750 g   (f)   200 g
   (g)   2540 g   (h)   3760 g
   (i)   820 g   (j)   365 g
   (k)   57 g   (l)   89 g
   (m)   125.8 g   (n)   268.7 g
   (o)   16.32 g   (p)   19.58 g

4. Express each of the following measurements in kilograms:
   (a)   1 tonne   (b)   3 tonne
   (c) $\frac{1}{2}$ tonne   (d) $\frac{1}{4}$ tonne
   (e) $\frac{7}{10}$ tonne   (f) $\frac{1}{10}$ tonne
   (g) $1\frac{3}{4}$ tonne   (h) $2\frac{9}{10}$ tonne
   (i)   0.6 tonne   (j)   0.8 tonne
   (k)   3.54 tonne   (l)   8.73 tonne
   (m)   12.017 tonne   (n)   23.704 tonne

12.5   Time

Time is the only measurement which does not use the metric system of units.

It is, in fact, based on the system of units developed by the Babylonians, who measured everything with a base 60 rather than base 10 (metric). Thus the most notable difference between the metric units and units of time is the conversion factors. There is no apparent pattern between units.

I.e.

60 seconds (s)	= 1 minute (min)
60 minutes	= 1 hour (h)
24 hours	= 1 day
7 days	= 1 week
28–31 days	= 1 month
$365\frac{1}{4}$ days	= 1 year
52 weeks	= 1 year
12 months	= 1 year
366 days	= leap year
10 years	= 1 decade
100 years	= 1 century

Table 12.1

Table 12.1 shows all the common units of measurement of time and the relationship between those units.

It should be noted that the number of days in a month varies. To help in remembering the number of days in each month there is a short but simple rhyme:

> 30 days has September,
> April, June and November,
> All the rest have 31,
> Excepting February alone,
> which has 28 days clear
> and 29 in each leap year.

It can be seen from the table that 1 day is equivalent to 24 hours. However, from our knowledge of time we know that watches and clocks often show only 12 hours. This is because 1 day is divided into 2 periods, *a.m.* (ante meridiem = before midday), which is from 12 midnight to 12 noon and *p.m.* (post meridiem = after midday), which is from 12 noon to 12 midnight.

Thus, 3 a.m. is 3 hours after midnight

whereas 3 p.m. is 3 hours after noon (midday)

and from 3 a.m. to 3 p.m. is a period of 12 hours.

Example 12.11

(a) Find the sum of 3 h 27 mins 18 s and 5 h 43 mins 52 s.

(b) Find the period of time which elapses between 3.42 p.m. and 7.20 a.m.

Solution

(a)

	3 h	27 mins	18 s
+	5 h	43 mins	52 s
			70 s
=		1 mins	10 s
		71 mins	
=	1 h	11 mins	
=	9 h		
total	9 h	11 mins	10 s

Add seconds column first, then minutes and hours

(b) 3.42 p.m. to 7.20 a.m.:

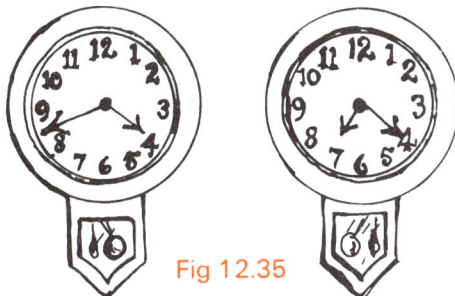

Fig 12.35

As the first clock is p.m. (before midnight) and the second clock is a.m. (after midnight), the simplest way to find the time difference is in relation to midnight.

3.42 p.m. tells us that it is 18 minutes before 4.00 p.m.; and 4.00 p.m. is 8 hours before midnight.

∴ 3.42 is 8 hours and 18 minutes before midnight.

7.20 a.m. tells us more simply that it is 7 hours and 20 minutes past midnight.

∴ time elapsed between 3.42 p.m. and 7.20 a.m. is

	8 h	18 mins
+	7 h	20 mins
	15 h	38 mins

Exercise 12L

1. State the unit of time used to measure each of the following:
   (a) time taken to bake a cake;
   (b) time spent on school holidays;
   (c) time taken to travel by plane to New Zealand;
   (d) time taken to run 100 metres;
   (e) time spent in gaol for murder;
   (f) time taken to travel from London to Australia in 1778;
   (g) time taken for Easter holidays;
   (h) time taken to blink;
   (i) time taken to build a house;
   (j) time spent sleeping each day;
   (k) time taken to boil an egg;

    (l) time the average person is alive;

    (m) time taken to drive from Lands End to John O'Groats by car;

    (n) time taken to play a football match.

2. Using table 12.1, find the number of :

    (a) seconds in an hour;   (b) hours in a week;

    (c) days in a decade;   (d) weeks in a century;

    (e) months in a century;   (f) minutes in a day;

    (g) seconds in a day;   (h) months in a decade;

    (i) weeks in a decade;   (j) decades in a century;

3. (a) Find the number of minutes in each of the following:

       (i) $\frac{1}{2}$ hour   (ii) $\frac{3}{4}$ hour

       (iii) $2\frac{1}{3}$ hours   (iv) $4\frac{1}{4}$ hours

       (v) $5\frac{5}{6}$ hours   (vi) $7\frac{2}{3}$ hours

       (vii) $15\frac{1}{4}$ hours   (viii) $20\frac{1}{6}$ hours

       (ix) 1 day and 4 hours   (x)   1 day and 3 hours

    (b) Find the number of hours in each of the following:

       (i) $\frac{1}{2}$ day   (ii) $\frac{3}{4}$ day

       (iii) $2\frac{1}{3}$ days   (iv) $4\frac{1}{4}$ days

       (v) $5\frac{5}{6}$ days   (vi) $7\frac{2}{3}$ days

       (vii) 1 week, 3 days   (viii) 1 week, 6 days

       (ix) 1 week $4\frac{1}{2}$ days   (x)   1 week $2\frac{1}{4}$ days

    (c) Find the number of days in each of the following (using 1 month = 30 days):

       (i) $\frac{1}{2}$ a week   (ii) $2\frac{1}{2}$ weeks

       (iii) $4\frac{6}{7}$ weeks   (iv) $8\frac{3}{7}$ weeks

       (v)   1 month 2 weeks   (vi) 1 month 3 weeks

       (vii) $1\frac{1}{3}$ months   (viii) $2\frac{1}{2}$ months

       (ix) $4\frac{1}{5}$ months   (x) $3\frac{5}{6}$ months

4. Evaluate each of the following:

    (a) 4 h 7 mins 5 s + 2 h 24 mins 18 s   (b) 6 h 37 mins 45 s + 5 h 12 mins 3 s

    (c) 1 h 48 mins 23 s + 2 h 8 mins 46 s   (d) 3 h 23 mins 16 s + 4 h 7 mins 53 s

    (e) 7 h 12 mins 47 s + 3 h 52 mins 8 s   (f) 2 h 39 mins 11 s + 8 h 43 mins 29 s

    (g) 6 h 56 mins 28 s + 2 h 31 mins 48 s   (h) 3 h 19 mins 46 s + 1 h 46 mins 19 s

    (i) 2 h 42 mins 18 s + 2 h 17 mins 42 s

5. Find the period of time which elapses between

    (a) 8.00 a.m. and 2.00 p.m.   (b) 6.00 a.m. and 4.00 p.m.

    (c) 7.00 p.m. and 3.00 a.m.   (d) 11.00 p.m. and 5.00 a.m.

    (e) 4.00 a.m. and 7.00 p.m.   (f) 3.00 a.m. and 11.00 p.m.

    (g) 2.00 p.m. and 9.00 a.m.   (h) 6.00 p.m. and 10.00 a.m.

    (i) 1.30 a.m. and 12.30 p.m.   (j) 8.30 a.m. and 7.30 p.m.

    (k) 6.20 p.m. and 3.20 a.m.   (l) 2.15 p.m. and 8.15 a.m.

    (m) 11.45 a.m. and 10.45 p.m.   (n) 12.50 a.m. and 9.50 p.m.

    (o) 6.30 p.m. and 5.45 a.m.   (p) 9.45 p.m. and 3.30 a.m.

    (q) 2.25 a.m. and 3.50 p.m.   (r) 6.40 a.m. and 11.25 p.m.

    (s) 8.06 p.m. and 4.24 p.m.   (t) 10.28 p.m. and 8.44 p.m.

    (u) 1.16 a.m. and 5.09 p.m.   (v) 2.19 a.m. and 6.26 p.m.

    (w) 12.43 p.m. and 6.18 a.m.   (x) 8.52 p.m. and 11.18 a.m.

    (y) 7.53 a.m. and 2.36 p.m.   (z) 3.34 a.m. and 9.43 p.m.

12.5.1 The 24-Hour Clock

The 24-hour clock, as the name implies, deals with a 24-hour period beginning at 12 (midnight). Rather than referring to a specific time as 9.00 a.m. or 9 o'clock in the morning, one would refer to it as 09 00, 'o nine hundred hours'.

This system is generally used for airline, train or bus timetables and other official purposes, although some countries have adopted the system for general use.

Example 12.12

Find the equivalent time in a.m. or p.m. of each of the following:

(a) 07 15 (b) 18 00

Solution

(a) 07 15 is before 12 noon so it must be equivalent to 7.15 a.m.

(b) 18 00 (18 hundred hours) is after 12 noon so it must be 6 hours after 12 noon and equivalent to 6.00 p.m.

Exercise 12M

1. Find the equivalent time in a.m. or p.m. of the following

   (a) 02 30   (b) 07 40
   (c) 00 15   (d) 00 52
   (e) 10 10   (f) 11 12
   (g) 12 00   (h) 10 50
   (i) 15 00   (j) 19 00
   (k) 13 42   (l) 14 18
   (m) 20 20   (n) 22 32
   (o) 23 10   (p) 20 30

2. Rewrite each of the following times using the 24-hour clock:

   (a) 8.50 a.m.   (b) 3.20 a.m.
   (c) 11.22 a.m.   (d) 9.16 a.m.
   (e) 12.42 a.m.   (f) 12.18 a.m.
   (g) 1.05 a.m.   (h) 2.53 a.m.
   (i) 2.00 p.m.   (j) 5.00 p.m.
   (k) 4.15 p.m.   (l) 6.52 p.m.
   (m) 7.38 p.m.   (n) 3.12 p.m.
   (o) 8.08 p.m.   (p) 9.43 p.m.

12.6   Three figure Bearings

When an air traffic controller is telling the pilot of an aeroplane the direction in which he must fly, it is not enough for him to say, 'Fly a course just a little to the right of north'. The pilot must know exactly in which direction to fly.

To do this the controller will give the pilot his course as a 3-figure bearing, such as 125°. This bearing is always the angle measured clockwise from north.

(Remember that there are 360° in a full turn.)

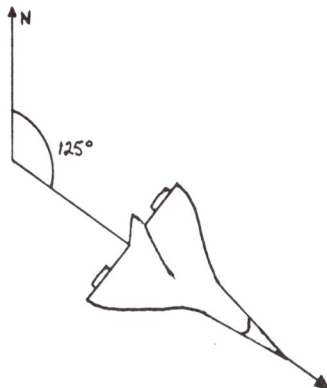

Fig 12.36

As it is a *three*-figure bearing, an angle of 82° clockwise from north is written as 082°.

Fig 12.37

How would an angle of 7° clockwise from north be written?

Example 12.13

Show, on a diagram, three figure bearings of 009° and 300°.

Solution

009° is 9° clockwise from north.
300° is 300° clockwise from north. (i.e. a reflex angle.)

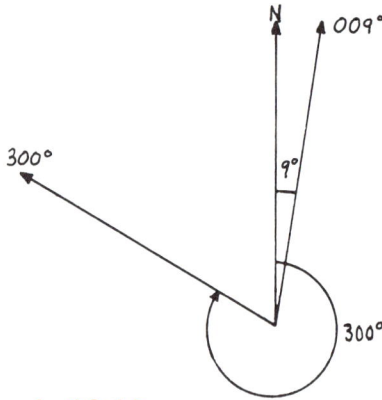

N.B. Always show the direction of north on diagrams which use bearings.

Example 12.14

Draw a diagram to represent the path of a boat which leaves port sailing on a bearing of 030° for 10 km and then changes course to a bearing of 116°.

Solution

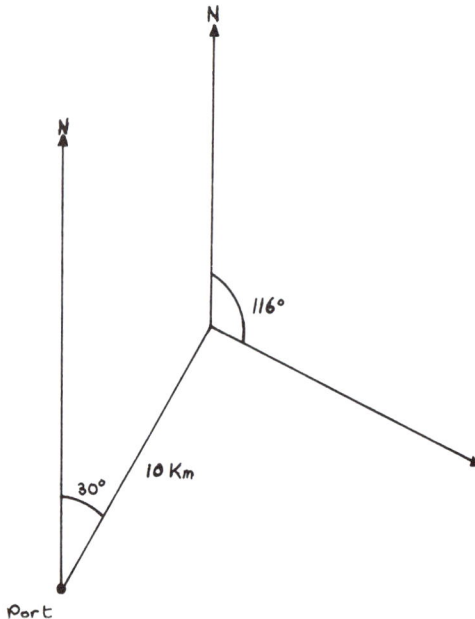

Exercise 12N

1. Draw diagrams to show the following three-figure bearings:
   (a) 100°   (b) 120°
   (c) 090°   (d) 270°
   (e) 008°   (f) 355°
   (g) 180°   (h) 012°

2. What are the names for the directions shown in your diagram for parts c, d and g above?

3. Draw diagrams to represent the following courses:
   (a) A boat leaves harbour sailing on a course of 172° for 5 km and then changes course to 204°.
   (b) An explorer walks for 2 km on a course of 292° and then walks due west for 1 km. To get back to his camp he then has to walk on a bearing of 197° for 3 km.

4. A man walks on a bearing of 136° and then changes course to 176°. Through what angle did he turn?

5. Stephen walks on a bearing of 120° and then turns anti-clockwise through 15°. In which direction is he now facing?

6. A plane flies on a course of 003° and then changes to 357°. Through what angle did the plane turn?

7. An explorer leaves camp and walks due north. In what direction is his camp from him?

Revision Exercises for Chapters 9–12

1. Construct angles of the following magnitude:
   (a) 35°   (b) 55°
   (c) 105°   (d) 165°
   (e) 138°   (f) 95°
   (g) 200°   (h) 240°
   (i) 310°   (j) 340°

2. Name the types of angles below:
   (a)   (b)

      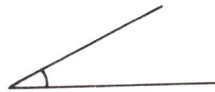

   (c)   (d)

(e)

(f)

(g)

(h)

(i)

(j)

(k)

(l)

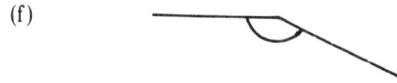

3. (a) Find the complement of an angle of magnitude 16°;
   (b) Find the supplement of an angle of magnitude 56°;
   (c) Find the measure of an angle with a complement of 71°;
   (d) Find the measure of an angle with a supplement of 108°;
   (e) Find the angle whose measure is the same as its complement;
   (f) Find the angle which has a supplement of the same magnitude.

4. Find the value of each of the pronumerals in the following, stating a reason for your answer:

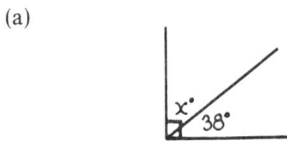

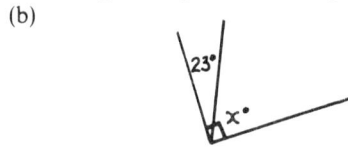

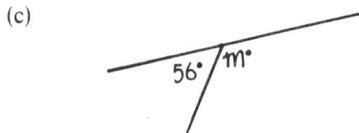

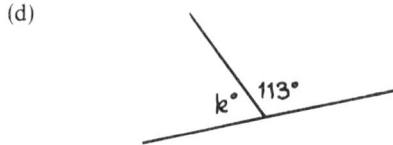

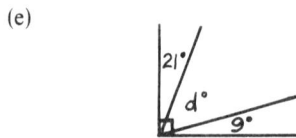

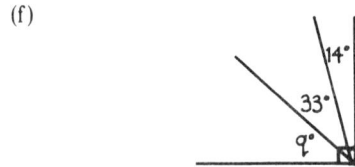

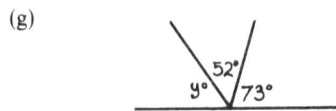

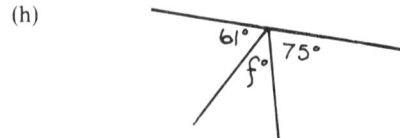

(a)

$x°$ $38°$

(b)

$23°$ $x°$

(c)

$56°$ $m°$

(d)

$k°$ $113°$

(e)

$21°$ $d°$ $9°$

(f)

$14°$ $33°$ $q°$

(g)

$52°$ $y°$ $73°$

(h)

$61°$ $f°$ $75°$

(i)

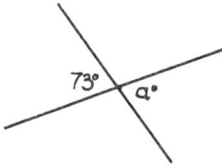

(j)

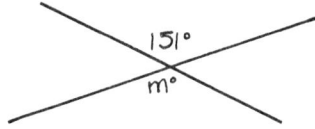

(k)

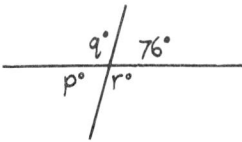

(l)

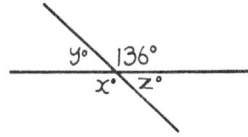

5. Find the value of the pronumerals in each of the following and state a reason for your answer:

(a)

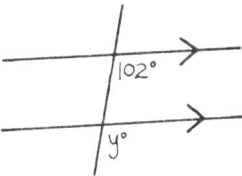

(b)

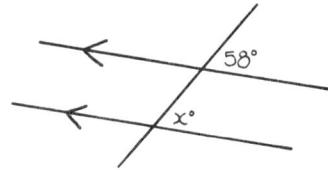

(c)

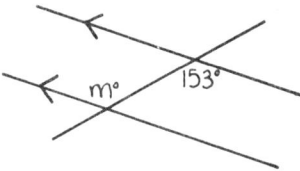

(d)

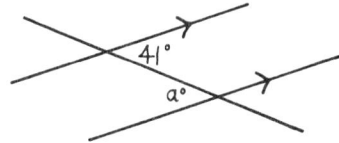

(e)

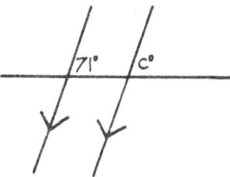

(f)

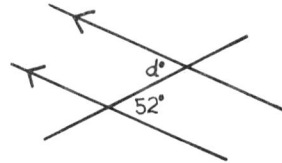

(g)

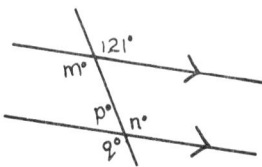

(h)

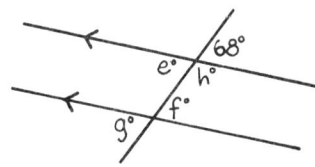

6. Find the value of each of the following pronumerals:

(a)

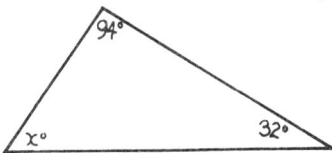

(b)

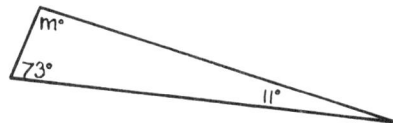

(c)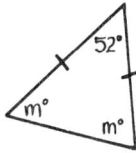

(d)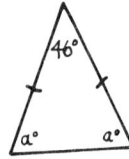

(e)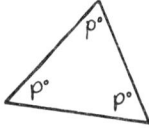

(f)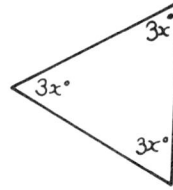

7. Name the triangles below by two names:

(a)

(b)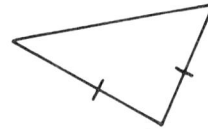

(c)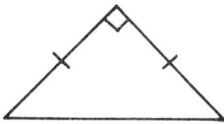

(d)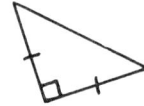

(e)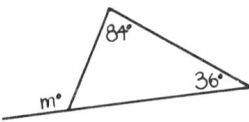

(f)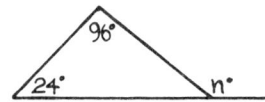

(g)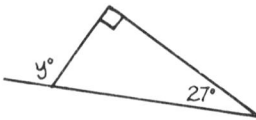

(h)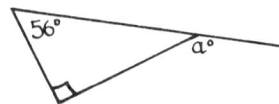

8. Find the value of the pronumeral in each of the following:

(a)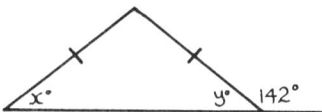

(b)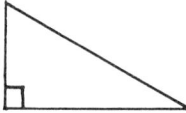

(c)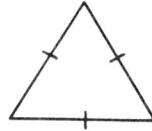

(d)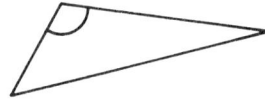

(e)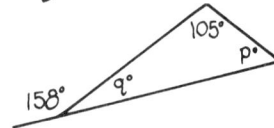

(f)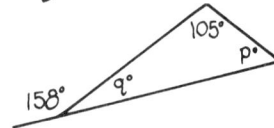

(g)

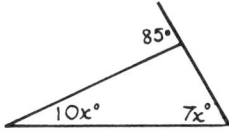

(h)

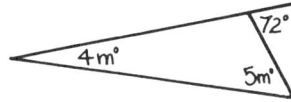

9. Name each of the quadrilaterals below:

(a)

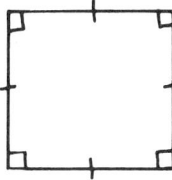

(b)

(c)

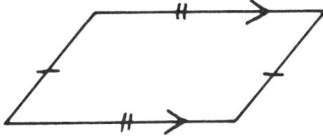

(d)

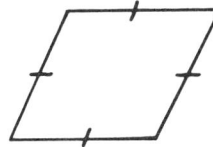

(e)

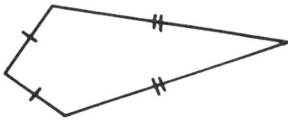

(f)

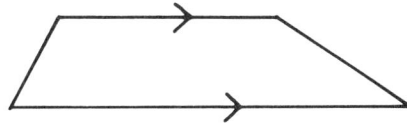

10. Find the value of the pronumerals in each of the following:

(a)

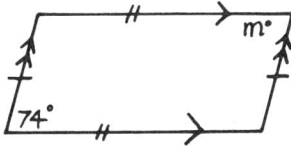

(b)

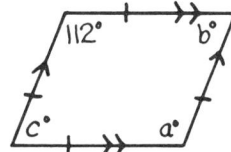

(c)

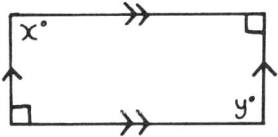

(d)

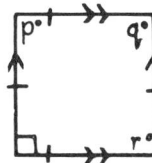

(e)

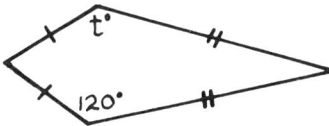

(f)

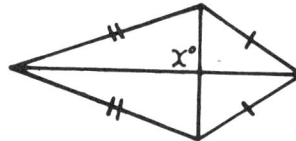

(g)

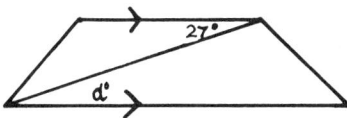

(h)

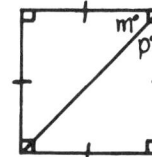

(i)

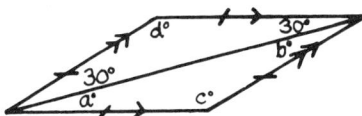

(j)

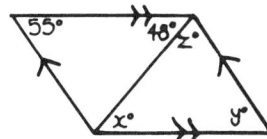

(k)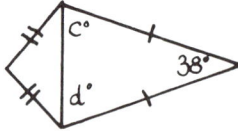

(l)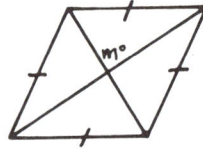

11. Give the (i) x-coordinate (ii) y-coordinate (iii) set of Cartesian coordinates for each of the points *A* to *L* inclusive marked on the Cartesian Plane below:

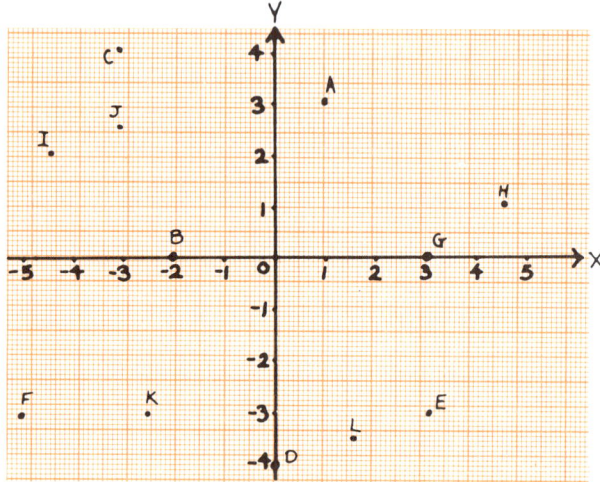

12. On graph paper, draw up a set of Cartesian axes so that the *X* and *Y*-axes can be marked off from −5 to +5.

   On this graph paper mark the following points

   (a) $A(+2, +5)$ (b) $B(−1, +3)$
   (c) $C(4, −2)$ (d) $D(5, 5)$
   (e) $E(0, 3)$ (f) $F(4, 0)$
   (g) $G(−2, 0)$ (h) $H(0, −3)$
   (i) $I(−3, 2)$ (j) $J(−1, −4)$
   (k) $K(1.5, 3)$ (l) $L(1, 2.5)$
   (m) $M(−1, 4.5)$ (n) $N(−3.5, −2)$

13. By plotting each set of points and using a ruler, find the coordinates of two more points in each of the following linear patterns:

   (a) $\{(−3, −9), (−2, −6), (−1, −3), (0, 0), (1, 3)\}$ (b) $\{(−3, 6), (−2, 4), (−1, 2), (0, 0), (1, −2)\}$
   (c) $\{(−3, −4), (−2, −3), (−1, −2), (0, −1), (1, 0)\}$ (d) $\{(−3, 1), (−2, 2), (−1, 3), (0, 4), (1, 5)\}$
   (e) $\{(−3, −5), (−2, −3), (−1, −1), (0, 1), (1, 3)\}$ (f) $\{(−3, 5), (−2, 4), (−1, 3), (0, 2), (1, 1)\}$

14. Find the rule which describes each of the following linear relationships. Give answer in full set notation form.

   (a)

   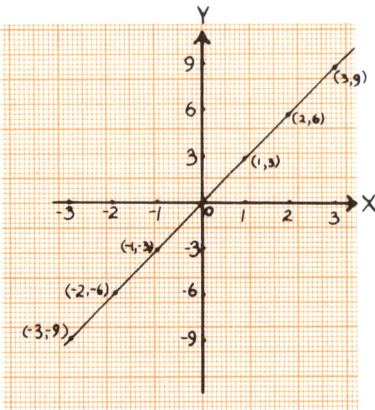

   (b)

   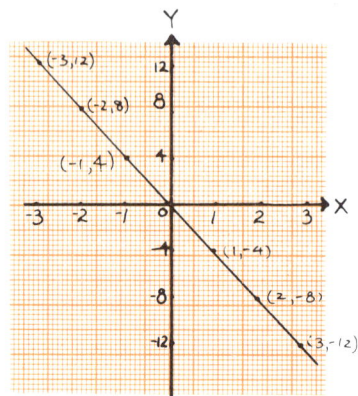

(c)

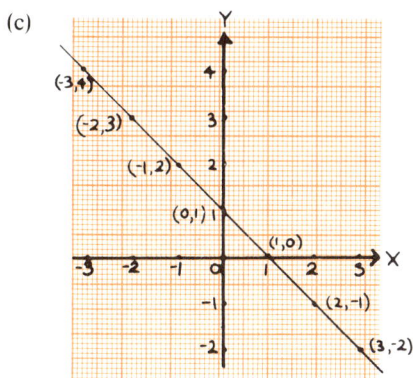

(d)

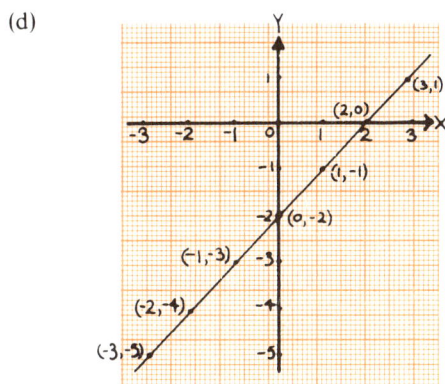

15. Plot each of the following linear relations on a separate sheet of graph paper for $-3 \leq x \leq 3$:
    (a) $\{(x, y): y = x\}$
    (b) $\{(x, y): y = -2x\}$
    (c) $\{(x, y): y = x + 5\}$
    (d) $\{(x, y): y = -x - 2\}$
    (e) $\{(x, y): y = 3x + 1\}$
    (f) $\{(x, y): x + y = -3\}$
    (g) $\{(x, y): y = 6\}$
    (h) $\{(x, y): x = -2\}$

16. (a) On the same set of axes plot the relations
        (i) $\{(x, y): y = x + 3; -3 \leq x \leq 3\}$
        (ii) $\{(x, y): y = 3x - 1; -3 \leq x \leq 3\}$
    (b) Using the graph in (a) find
        $\{(x, y): y = x + 3\} \cap \{(x, y): y = 3x - 1\}$

17. Find, graphically:
    $\{(x, y): y = -2x + 3\} \cap \{(x, y): y = 3x + 8\}$

18. Solve the following pair of equations graphically:
    $y = -x + 4$
    $y = x - 8$

19. (a) Express the following ratios in simplest form:
        (i) $9 : 9$
        (ii) $11 : 11$
        (iii) $5 : 10$
        (iv) $12 : 6$
        (v) $54 : 90$
        (vi) $64 : 104$
        (vii) $3 : 15$
        (viii) $\frac{1}{2} : 5$
    (b) Express the following ratios as fractions in simplest form:
        (i) $2 : 3$
        (ii) $7 : 9$
        (iii) $8 : 4$
        (iv) $3 : 6$
        (v) $12 : 16$
        (vi) $15 : 35$
        (vii) $36 : 48$
        (viii) $56 : 40$
    (c) Express the following fractions as ratios in simplest forms:
        (i) $\frac{1}{4}$
        (ii) $\frac{1}{2}$
        (iii) $\frac{2}{10}$
        (iv) $\frac{6}{9}$
        (v) $\frac{14}{35}$
        (vi) $\frac{16}{96}$
        (vii) $\frac{144}{99}$
        (viii) $\frac{108}{60}$
        (ix) $\frac{117}{39}$
        (x) $\frac{112}{42}$

20. (a) In a class of 26 students, 14 students study a language and the remainder do not. Find the ratios of
        (i)   students studying a language to students not;
        (ii)   students in the class to students studying a language;
        (iii) students not studying a language to students in the class.
    (b) If a car dealer sells a car for £3500, but bought it for only £2800, find the ratios of :
        (i)   the cost price to selling price;
        (ii)   the selling price to the profit;
        (iii) the profit to the cost price.

21. (a) Express the following percentages as fractions in simplest form:

(i)	50%	(ii)	80%
(iii)	10%	(iv)	30%
(v)	75%	(vi)	25%
(vii)	45%	(viii)	15%
(ix)	84%	(x)	36%
(xi)	2.5%	(xii)	7.5%
(xiii)	0.5%	(xiv)	0.8%
(xv)	12.5%	(xvi)	37.5%
(xvii)	$33\frac{1}{3}\%$	(xviii)	$66\frac{2}{3}\%$
(xix)	$87\frac{1}{2}\%$	(xx)	$42\frac{1}{2}\%$

(b) Express the following fractions as percentages:

(i)	$\frac{14}{100}$	(ii)	$\frac{59}{100}$
(iii)	$\frac{50}{200}$	(iv)	$\frac{85}{200}$
(v)	$\frac{140}{500}$	(vi)	$\frac{390}{500}$
(vii)	$\frac{875}{1000}$	(viii)	$\frac{105}{1000}$
(ix)	$\frac{360}{2000}$	(x)	$\frac{1200}{2000}$
(xi)	$\frac{3}{10}$	(xii)	$\frac{7}{20}$
(xiii)	$\frac{39}{50}$	(xiv)	$\frac{18}{20}$
(xv)	$\frac{1}{2}$	(xvi)	$\frac{3}{4}$
(xvii)	$\frac{3}{5}$	(xviii)	$\frac{2}{6}$
(xix)	$\frac{5}{8}$	(xx)	$\frac{7}{12}$

22. (a) Express 12 as a percentage of 48.

(b) Express 18 as a percentage of 108.

(c) Find the percentage 15 is of 40.

(d) Find the percentage 16 is of 48.

(e) If Jenny gained 36 marks out of 42 marks on an exam paper what percentage did she gain? (Answer to 3 decimal places.)

(f) If Tony scored 6 goals out of a total of 10 goals in a football match, what percentage did he score?

(g) Louise paid £14 for a jumper which was originally £21.

Find (i) the percentage she paid of the original price;
     (ii) the percentage she saved.

23. (a) Find 25% of 52.   (b) Find 80% of 65.

(c) Find 87 5% of 96.   (d) Find 0.8% of 350.

(e) If Richard gained 72% for a French exam which was out of 125 marks, find the number of marks he gained out of 125 marks.

24. Find the perimeter of each of the following:

(a)

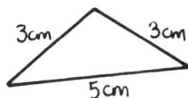

(b)

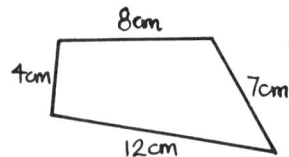

(c)

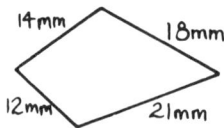

(d)

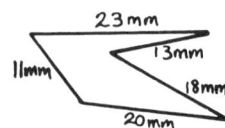

(e)

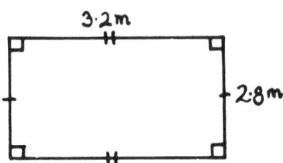

(f)

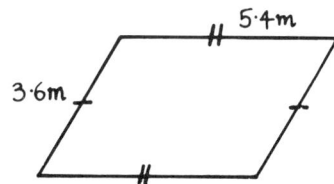

(g)

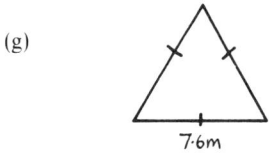

7·6m

(h)

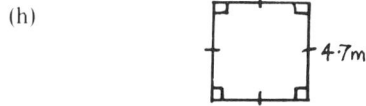

4·7m

25. (a) Convert the following to centimetres:

   (i)   50 mm   (ii)   140 mm

   (iii) 56 mm   (iv) 132 mm

   (v)   2 m   (vi) 3 m

   (vii) 1.6 m   (viii) 2.8 m

   (ix) 3.124 m   (x)   0.671 m

 (b) Convert the following to millimetres:

   (i)   4 cm   (ii)   9 cm

   (iii) 37 cm   (iv) 63 cm

   (v)   0.65 cm   (vi) 0.12 cm

   (vii) 3 m   (viii) 5 m

   (ix) 3.82 m   (x)   4.06 m

 (c) Convert the following to metres:

   (i)   200 cm   (ii)   700 cm

   (iii) 592 cm   (iv)   365 cm

   (v)   43 cm   (vi)   56 cm

   (vii) 3000 mm   (viii) 8600 mm

   (ix) 2059 mm   (x)   1207 mm

   (xi) 145 mm   (xii) 362 mm

   (xiii) 2 km   (xiv) 8 km

   (xv) 5.92 km   (xvi) 3.03 km

   (xvii) 0.014 km   (xviii) 0.302 km

 (d) Convert the following to kilometres:

   (i)   3000 m   (ii)   7600 m

   (iii) 9820 m   (iv)   2651 m

   (v)   234 m   (vi)   68 m

   (vii) 180 000 cm   (viii) 290 000 cm

   (ix) 3 000 000 mm   (x)   5 250 000 mm

26. Find the perimeter of each of the following:

(a)

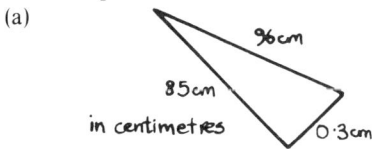

96cm

85cm

in centimetres

0·3cm

(b)

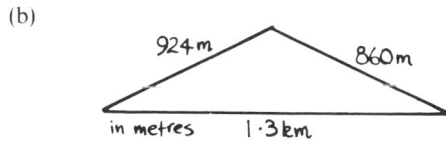

924m   860m

in metres   1·3km

(c)

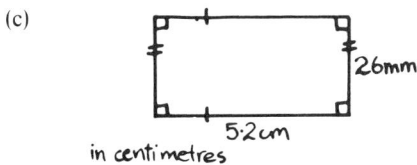

26mm

5·2cm

in centimetres

(d)

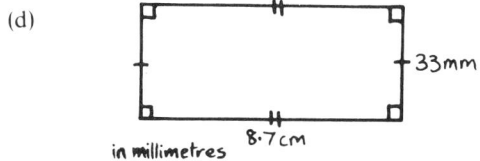

33mm

8·7cm

in millimetres

(e)

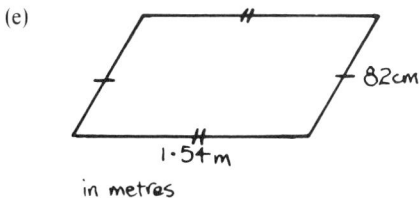

82cm

1·54m

in metres

(f)

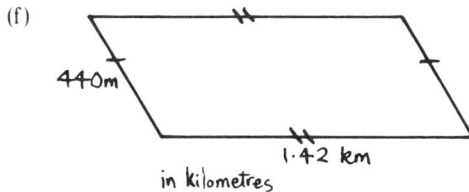

440m

1·42 km

in kilometres

27. Find the area of each of the following:

(a)

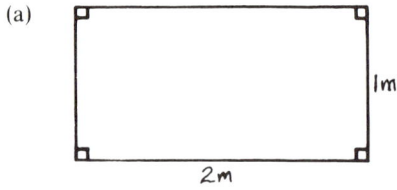

1m
2m

(b)

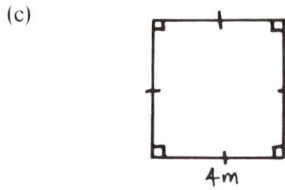

1cm
4cm

(c)

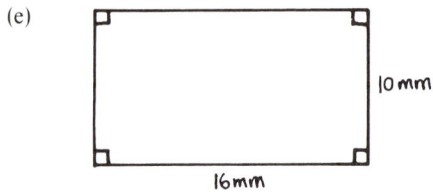

4m

(d)

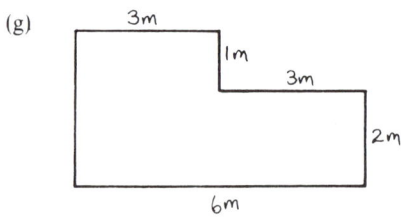

3m

(e)

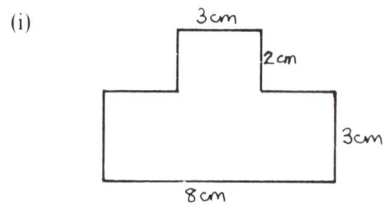

10mm
16mm

(f)

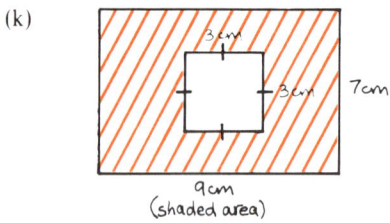

9mm
12mm

(g)

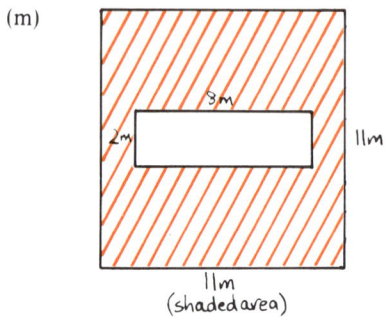

3m
1m
3m
2m
6m

(h)

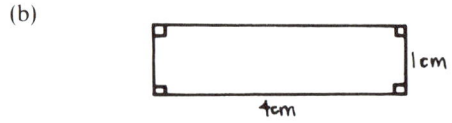

8m
2m
4m
2m

(i)

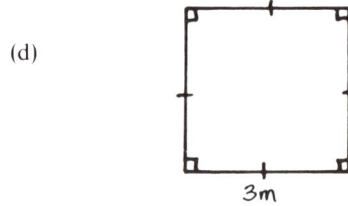

3cm
2cm
3cm
8cm

(j)

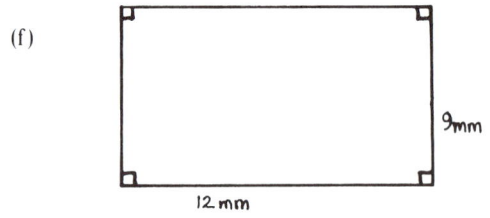

5cm
1cm
3cm
2m
2cm
4cm
3cm
12cm

(k)

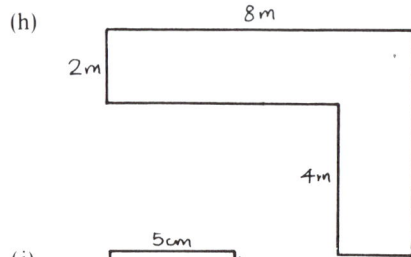

3cm
3cm
7cm
9cm
(shaded area)

(l)

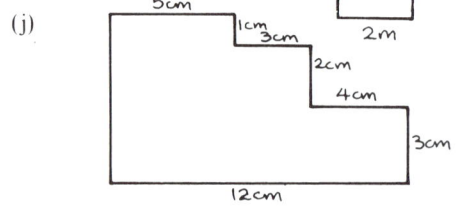

16 mm
7mm
6mm
8mm
(shaded area)

(m)

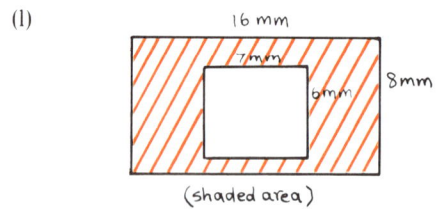

3m
2m
11m
11m
(shaded area)

(n)

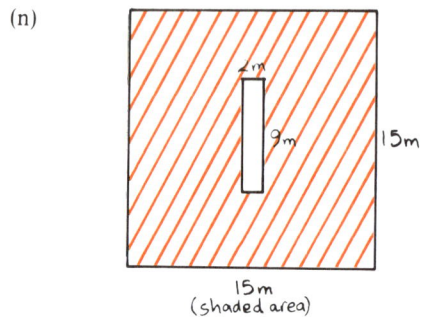

2m
9m
15m
15m
(shaded area)

28. (a) Find the area of a wall, not including the window, if the wall is 8 m long and 3 m high and the window is 2 m long and 1.5 m high.

(b) Find the area of the border around a painting if the painting itself is 45 cm long and 20 cm wide and the backing on which it is mounted is 57 cm long and 32 cm wide.

(c) Find the area of a page not covered by printing if the page is 18 cm long and 15 cm wide and the printed part covers a length of 14 cm and a width of 12 cm.

(d) Find the area of the border round a colour slide if the film part is 22 mm by 20 mm and the total length is 36 mm and the width is 30 mm.

29. Find the area of each of the following:

(a)

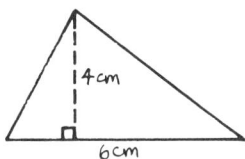

(b)

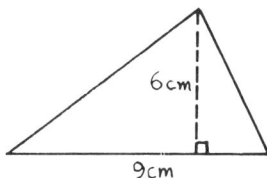

(c)

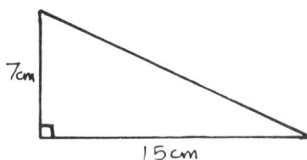

(d)

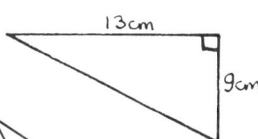

(e)

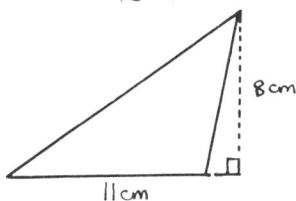

(f)

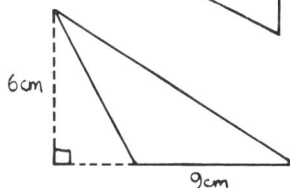

(g)

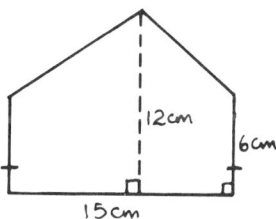

(h)

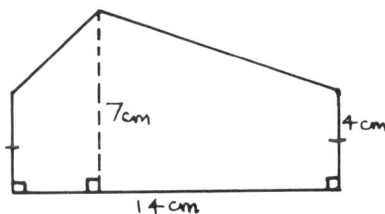

(i)

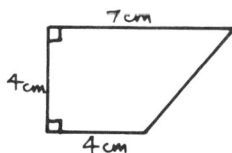

(j)

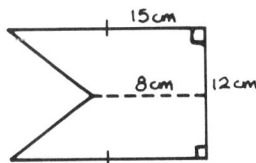

30. Find the volume of each of the following prisms:

(a)

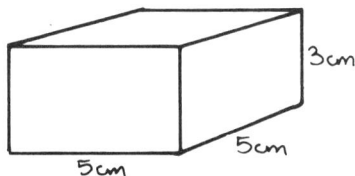

(b)

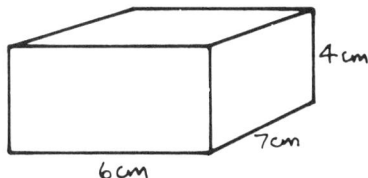

(c)

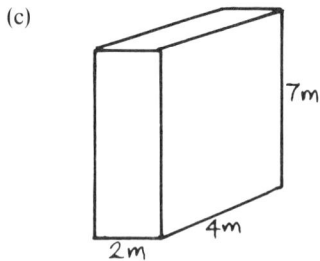

(d)

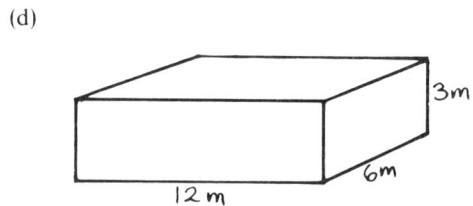

(e)

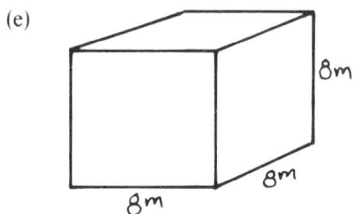

(f)

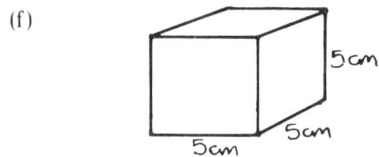

31. (a) Convert the following to millilitres:

(i)	2 ℓ	(ii)	6 ℓ
(iii)	3.5 ℓ	(iv)	7.2 ℓ
(v)	6.43 ℓ	(vi)	2.93 ℓ
(vii)	12.034 ℓ	(viii)	27.061 ℓ
(ix)	0.26 ℓ	(x)	0.314 ℓ
(xi)	0.027 ℓ	(xii)	0.0098 ℓ

   (b) Express the following in litres:

(i)	3000 ml	(ii)	8200 ml
(iii)	7620 ml	(iv)	2146 ml
(v)	532 ml	(vi)	324 ml
(vii)	68 ml	(viii)	95 ml
(ix)	207 ml	(x)	309 ml

32. (a) Express the following measurements in grams:

(i)	2 kg	(ii)	5 kg
(iii)	$1\frac{1}{2}$ kg	(iv)	$2\frac{1}{4}$ kg
(v)	$7\frac{4}{5}$ kg	(vi)	$3\frac{7}{10}$ kg
(vii)	0.25 kg	(viii)	4.5 kg
(ix)	6.032 kg	(x)	0.207 kg

   (b) Express the following measurements in kilograms:

(i)	1000 g	(ii)	2500 g
(iii)	500 g	(iv)	250 g
(v)	380 g	(vi)	928 g
(vii)	42 g	(viii)	91 g
(ix)	26.32 g	(x)	65.41 g
(xi)	237.5 g	(xii)	109.03 g
(xiii)	2 tonne	(xiv)	$4\frac{1}{2}$ tonne
(xv)	$6\frac{1}{4}$ tonne	(xvi)	$8\frac{3}{5}$ tonne
(xvii)	1.72 tonne	(xviii)	0.83 tonne

   (c) Express the following measurements in tonne:

(i)	2000 kg	(ii)	7500 kg
(iii)	3692 kg	(iv)	1820 kg
(v)	453 kg	(vi)	982 kg
(vii)	302 kg	(viii)	701 kg
(ix)	84 kg	(x)	12 kg

33. Find the number of :
    (a) hours in $4\frac{1}{2}$ days;
    (b) minutes in $6\frac{1}{4}$ hours;
    (c) days in $6\frac{3}{7}$ weeks;
    (d) seconds in $28\frac{2}{5}$ minutes;
    (e) weeks in $7\frac{3}{4}$ years;
    (f) months in 4.2 decades;
    (g) minutes in $\frac{1}{2}$ day;
    (h) hours in 2 weeks;
    (i) months in $16\frac{5}{6}$ years;
    (j) years in 9.2 decades.

34. Find the period of time which elapses between the following times:
    (a) 6.00 a.m. and 9.00 p.m.
    (b) 4.00 p.m. and 11.00 a.m.
    (c) 9.15 a.m. and 3.15 a.m.
    (d) 12.20 p.m. and 10.20 a.m.
    (e) 6.45 p.m. and 9.30 a.m.
    (f) 2.16 a.m. and 11.48 p.m.
    (g) 5.37 p.m. and 7.05 a.m.
    (h) 01 00 and 17 00
    (i) 05 15 and 21 15
    (j) 11 18 and 23 26
    (k) 10 03 and 19 51
    (l) 20 16 and 08 08
    (m) 19 43 and 09 21

13 Interpretation of Data

Do you have a weekly lesson timetable which tells you what class you are to attend on a particular day at a particular time? If so then the timetable is usually given in *tabular* (table) form so that you can see this at a glance. When reading that timetable you are required to *interpret the data* presented.

Lesson \ Day	Monday	Tuesday	Wednesday	Thursday	Friday
1	English	Maths	French	Crafts	Maths
2	Maths	Geog.	Science	English	History
3	French	English	History	Geog.	Science
4	Sport	Science	Library	Maths	French
5	History	Music	Maths	Science	English
6	Geog.	Crafts	English	Music	Sport

Table 13.1

Interpreting the lesson timetable given in table 13.1 we find, for example, 3rd lesson on Wednesday is History and 1st lesson on Friday is Maths.

Often, when there is a great deal of information to be given about a particular situation this information is condensed into either tabular or pictorial form and so interpretation of this data is required.

13.1 Tabular Representation

Example 13.1

An interior decorator needs to work out how many mirror wall tiles of a certain type are needed to tile walls or portions of walls of varying sizes. Table 13.2 below gives the number of tiles needed for tiling areas with a specified length (in metres) and specified height (in metres).

Length (m) \ Height (m)	1	2	3	4	5
2	23	45	67	89	112
3	34	67	100	134	167
4	45	89	134	178	223
5	56	112	167	223	278
6	67	134	200	267	334

Table 13.2

Using the information given in table 13.2, answer the following questions.
(a) How many mirror tiles are needed to cover a wall of length 5 metres and height 4 metres?
(b) How many mirror tiles are needed to cover two wall areas: one of length 3 metres and height 1 metre, the other with length 5 metres and height 2 metres?
(c) If a particular wall area needed
     (i)   278 tiles
     (ii) 134 tiles
what are the dimensions of the area tiled?
(d) If the cost of each mirror tile is £4.00, what is the cost of the tiles for a wall area of length 5 metres and height 3 metres?

Solution
(a) Repeating table 13.2:

Length (m) \ Height (m)	1	2	3	4	5
2	23	45	67	89	112
3	34	67	100	134	167
4	45	89	134	178	223
5	56	112	167	223	278
6	67	134	200	267	334

Table 13.3

For a wall of length 5 metres and height 4 metres, 223 tiles would be needed (see table 13.3 above).
(b) Reading from table 13.2, for a wall area of length 3 metres and height 1 metre, 34 tiles are needed. For a wall area of length 5 metres and height 2 metres, 112 tiles are needed.
Thus, total number of tiles needed
$$= 34 + 112$$
$$= 146 \text{ tiles}$$
(c) (i) If 278 tiles are needed, then the area to be covered is of length 5 metres and height 5 metres (see table 13.4 below).

Length (m) \ Height (m)	1	2	3	4	5
2	23	45	67	89	112
3	34	67	100	134	167
4	45	89	134	178	223
5	56	112	167	223	278
6	67	134	200	267	334

Table 13.4

(ii) If 134 tiles are needed then the wall area to be tiled could have three possible dimensions (see table 13.5 below).

Length (m) \ Height (m)	1	2	3	4	5
2	23	45	67	89	112
3	34	67	100	(134)	167
4	45	89	(134)	178	223
5	56	112	167	223	278
6	67	(134)	200	267	334

Table 13.5

The three possible dimensions are:
 length 3 metres and height 4 metres
or length 4 metres and height 3 metres
or length 6 metres and height 2 metres.
(d) To find the cost of the tiles we must first know how many tiles must be bought. Reading from table 13.2, to cover an area of length 5 metres and height 3 metres we need 167 tiles.
Now, each tile costs £4.00.
Thus, total cost of tiles = 167 × £4.00
          = £668.00

Exercise 13A

1. Continental Quilts are frequently used in place of blankets on beds during winter. There are different qualities and sizes of quilts available on the market.
   The prices of the different quilts made by a certain manufacturer are shown in table 13.6 below.

Size \ Quality	Super Down	Down	85% Down 15% Feath.	75% Down 25% Feath.	Polyester
Single Bed	£90	£75	£60	£45	£30
Double Bed	£128	£100	£75	£56	£44
King-Size Bed	£150	£124	£99	£68	£58

Table 13.6

Using table 13.6, answer the following questions.
(a) How much would it cost to buy a king size quilt filled with
  (i) polyester?   (ii) 85% Down, 15% Feather?
(b) How much would it cost to buy a single bed quilt of 'Super Down' quality?
(c) How much would it cost to buy two single bed 'Super Down' quilts and a queen size polyester-filled quilt?
(d) How much would it cost to buy three polyester-filled single bed quilts, a 75% down and 25% feather-filled double bed quilt and a king-size quilt of 'Down' quality?
(e) If I bought a king-size quilt for £124, what quality of quilt did I buy?
(f) If I bought a 85% Down 15% Feather quilt for £75, for what size bed was it bought?

2. A particular retailer owns a shop in each of four different suburbs. In each shop he stocks jeans, T-shirts, jumpers and shoes. At the end of June in one particular year his remaining stock is as given in table 13.7 below.

Clothes Shop	Jeans	T-shirts	Jumpers	Pairs of Shoes
1	50	80	15	10
2	25	20	10	12
3	5	8	3	7
4	30	24	21	15

Table 13.7

(a) How many T-shirts are still in stock at shop 3?

(b) How many jumpers are still in stock at shop 1?

(c) What is the total number of pairs of jeans still in stock?

(d) What is the total number of items still in stock at shop 2?

(e) Because of the large amount of stock remaining at the end of June, the retailer has a sale at all of his shops. He decides to sell each pair of jeans at £10, each T-shirt at £3, each jumper at £8 and each pair of shoes at £9. If all goods are sold during the sale, how much money was taken

   (i) at shop 1   (ii) at shop 2

   (iii) at shop 3   (iv) at shop 4

   (v) altogether   (vi) on the sale of jeans

   (vii) on the sale of jumpers?

3. The following table gives the approximate time it takes for each of the planets to travel around the sun.

Planet	Time Taken
Earth	1 year
Pluto	248 years
Mercury	88 days
Saturn	29 years
Mars	2 years
Venus	125 days
Jupiter	12 years
Uranus	84 years
Neptune	165 years

Table 13.8

(a) Which planet takes the longest time to travel around the sun?

(b) Which planet takes the shortest time to travel around the sun?

(c) Rewrite the table so that it places the planets in order of the time (from shortest to longest) taken to travel around the sun.

(d) Which planets are closer to the sun than the earth?

(e) Which planets are further from the sun than the earth?

(f) How many times does Mercury go around the sun in 440 days?

(g) How many years does it take for Venus to travel around the sun 5 times?

4. The following table shows the number of rolls of wallpaper which need to be bought in order to paper a wall of given length (in metres) and given height (in metres).

Length (m) \ Height (m)	1	2	3	4	5
2	1	1	2	2	2
3	1	2	2	2	3
4	1	2	3	4	4
5	1	2	3	4	5
6	2	3	4	5	6

Table 13.9

Using the information in table 13.9, answer the following questions.

(a) How many rolls of wallpaper are needed to cover a wall of length 4 metres and height 5 metres?

(b) How many rolls of wallpaper are needed to cover three walls of a room 3 metres high if one wall is 3 metres long and the other two are 5 metres long?

(c) A particular wall of height 5 metres needed 3 rolls of wallpaper. How long was the room?

(d) Wallpaper is £11.50 per roll. What would be the cost of papering all four walls of a room measuring 4 m by 6 m and with walls 3 m high?

5. The volumes (in cm^3) of cylinders of various radii (in cm) and heights (in cm) are given in the following table:

Radius (cm) \ Height (cm)	1	2	3	4	5	6	7	8	9	10
1	3.1	6.3	9.4	12.6	15.7	18.8	22	25.1	28.3	31.4
2	12.6	25.1	37.7	50.3	62.8	75.4	88	100.5	113.1	125.7
3	28.3	56.5	84.8	113.1	141.4	169.6	198	226.2	254.5	282.7
4	50.3	100.5	150.8	201.1	251.3	301.6	351.9	402.1	452.4	502.7
5	78.5	157.1	235.6	314.2	392.7	471.2	549.8	628.3	706.9	785.4

Table 13.10

(a) From the table find the volumes of the cylinders with the following radii and heights:
    (i) radius 3 cm, height 2 cm;   (ii) radius 4 cm, height 6 cm;
    (iii) radius 1 cm, height 8 cm;   (iv) radius 5 cm, height 9 cm.

(b) Use the body of the table to find the dimensions of cylinders with the following volumes:
    (i)   251.3 cm^3 (ii) 471.2 cm^3
    (iii) 226.2 cm^3 (iv) 56.5 cm^3

(c) Which cylinder has the same volume as that with dimensions
    (i) radius 1 cm, height 4 cm;   (ii) radius 4 cm, height 1 cm?

(d) A cylinder of radius 3 cm and height 3 cm is doubled in height. What happens to its volume?

6. The following table gives the cost *per person* for tours round the United States. The cost of a tour depends on the number of people in the group, the number of hotel rooms and the size of hire car required.

Number of people	2	3	3	4	4	5	6	6	Reduction for children under 12
Number of hotel rooms	1	1	2	1	2	2	2	3	
Size of hire car	A	B	B	B	C	C	D	D	
Cost (in dollars)									
21 days	1460	1340	1550	1260	1410	1340	1290	1390	470
28 days	1600	1430	1710	1320	1520	1430	1360	1500	420
35 days	1830	1620	1980	1490	1740	1620	1540	1710	480

Table 13.11

(a) From the table find the cost *per person* of the following tours:
    (i)   a family of 4 people in 2 hotel rooms in a type *C* hire car for 35 days;
    (ii) a family of 6 people in 2 hotel rooms in a type *D* hire car for 21 days;
    (iii) a family of 3 people in 1 hotel room in a type *B* hire car for 28 days.

(b) Using the cost per person on the tours described in (a), find the cost of the whole family on each of these tours.

(c) A family of 2 adults and 1 child under 12 wish to have a 35-day holiday in America hiring a type *B* car and sharing a hotel room.
    (i) How much will it cost for each adult?   (ii) How much will it cost for the child?
    (iii) How much will it cost the family?

7. A zookeeper wishes to make a 'monkey-mix' which is a mixture of nuts and fruits. Since the monkey-mix must provide the monkeys with an adequate diet the zookeeper must consult the following table when making up the mixture. The table gives the number of grams of protein, fat and carbohydrate in each kilogram of the food.

Food (1 kg)	Protein (gm)	Carbohydrate (gm)	Fat (gm)
Apples	1	33	0
Grapes	2	44	0
Melons	3	15	0
Pears	1	30	0
Apricots	2	17	0
Bananas	2	55	0
Peanuts	80	139	24
Coconut	11	102	11
Almonds	58	152	12

Table 13.12

A fully grown monkey requires at least 70 gm of protein, 200 gm of carbohydrate and 20 gm of fat.
(a) How much protein, carbohydrate and fat is contained in
   (i)   0.5 kg of peanuts;   (ii) 0.2 kg of apples;
   (iii) 0.1 kg of coconut;   (iv) 0.3 kg of apricots;
   (v) 0.1 kg of almonds;   (vi) 1.2 kg of bananas;
   (vii) 1 kg of grapes?
(b) Would a mixture made up using the quantities of foods in parts (a) give the monkeys sufficient
   (i)   protein;   (ii) carbohydrate;
   (iii) fat?

8.

From \ To	Melbourne	Auckland	Los Angeles	New York	London	Paris	Tokyo
Melbourne	0	+2	−18	−15	−10	−9	−1
Auckland	−2	0	−20	−17	−12	−11	−3
Los Angeles	+18	+20	0	+3	+8	+9	+17
New York	+15	+17	−3	0	+5	+6	+14
London	+10	+12	−8	−5	0	+1	+9
Paris	+9	+11	−9	−6	−1	0	+8
Tokyo	+1	+3	−17	−14	−9	−8	0

Time differences in hours
Table 13.13

Table 13.13 above gives the time differences (in hours) between various pairs of cities in the world. For example, if it is 12 noon in Melbourne on a Monday then the time in New York is 15 hours *behind* (−15); i.e. 12 noon Monday − 15 hours = 9.00 pm Sunday. Thus, it would be 9.00 pm on Sunday in New York.
Using table 13.13 answer the following questions.
(a) If it is 12 noon on a Tuesday in Melbourne what is the time and day in
   (i)   Auckland   (ii) London
   (iii) Tokyo?
(b) If it is 10.00 am on a Wednesday in Los Angeles, what time and day is it in
   (i)   Melbourne   (ii) New York
   (iii) Tokyo?

(c) If it is 6 pm on a Friday in Tokyo, what time and day is it in
  (i)   Melbourne   (ii) Los Angeles
  (iii) London?
(d) If a person in London wants to make a phone call to someone in Melbourne so that it is 7 am Melbourne time, what time should he place the call?
(e) A man takes a 10 hour plane flight from Paris to New York. He left Paris at 3.00 pm on a Wednesday.
  (i)   What time and day would his watch display on arriving at New York? (Assume that his watch is still on Paris time).
  (ii) What time and day would it be in New York on his arrival?

13.2   Bar Charts

A bar chart is a pictorial representation of data. The information it contains is given in the form of bars (rectangles) arranged side by side.

Example 13.2

The bar chart given in fig 13.1 gives the average rainfall (in mm) for each month of the year in a particular city.

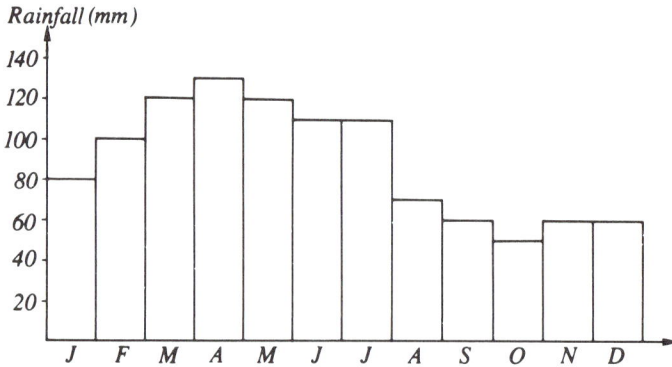

Fig 13.1

Using the information given in fig 13.1 answer the following questions.
(a) In which month would you expect the
  (i)   most rain   (ii) least rain
  to fall?
(b) What is the average rainfall for the month of
  (i)   February   (ii) June
  (iii) October?
(c) For which month is there an average rainfall of
  (i)   70 mm   (ii) 120 mm?
(d) If, for this city, the highest rainfall occurs during the autumn and winter months, is the city to be found in the northern or southern hemisphere?

Solution

(a) Since the amount of rain falling is represented by the heights of the rectangles in the bar chart given in fig 13.1 then the higher the rectangle, the higher the rainfall. Thus,
  (i)   the most rain could be expected to fall in the month of April;
  (ii) the least rain could be expected to fall in the month of October.
(b) (i)   The average rainfall for February is 100 mm.
  (ii) The average rainfall for June is 110 mm.
  (iii) The average rainfall for October is 50 mm.
(c) (i)   August is the month for which there is an average rainfall of 70 mm.
  (ii) Both March and May have average rainfalls of 120 mm.

(d) In the northern hemisphere the autumn and winter months are from September to February. In the southern hemisphere the autumn and winter months are from March to August. Since this city has its heaviest rainfall in the March–July period and this is during their autumn and winter, then the city must be found in the southern hemisphere.

Exercise 13B

1. A survey was taken in a particular street concerning the type and number of pets that are kept by the people who live in that street. The bar chart given in fig 13.2 gives the results of the survey.

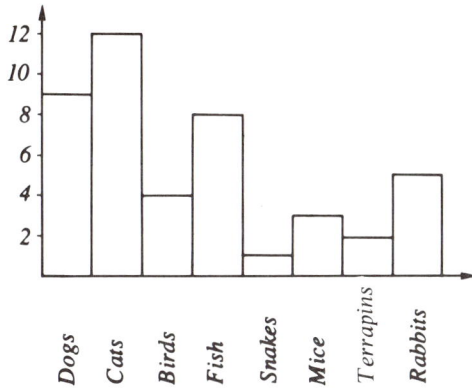

Fig 13.2

Using the information given in fig 13.2, answer the following questions.
(a) What type of pet was most popular?
(b) What type of pet was least popular?
(c) How many dogs were kept in the street?
(d) How many rabbits were kept in the street?
(e) Which type of pet was counted 4 times?
(f) What is the total number of pets in the street?

2. The bar chart given in fig 13.3 below shows the number of Form 4 students in a particular school studying various subjects.

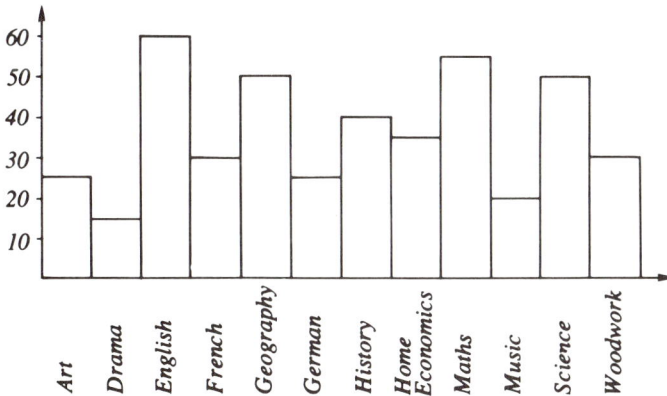

Fig 13.3

(a) If all students must study English, how many Form 4 students are there?
(b) Apart from the compulsory subject of English, which subject is the most popular?
(c) Which subject has the least number of students?
(d) How many students study
   (i) Science,   (ii) Drama?
(e) How many students study a foreign language? (Assume students can study only one foreign language.)
(f) Which subjects contain the same number of students?
(g) If one particular teacher teaches both Geography and Maths, how many end-of-year reports must the teacher write?

3. The bar chart given in fig 13.4 gives a week's sales of different types of books from a particular book shop.

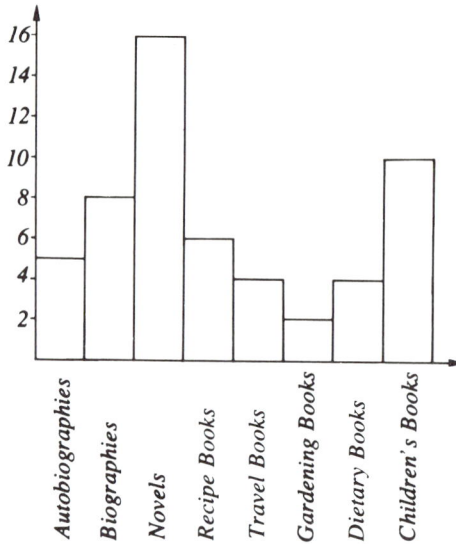

Fig 13.4

Using the information given in fig 13.4 answer the following questions.
(a) What was the most popular type of book? (b) What was the least popular type of book?
(c) How many recipe books were sold?   (d) Which type of book represented 8 sales?
(e) Which two types of book had the same sales?
(f) Which type of book had double the sales of autobiographies.
(g) What were the total sales for the week?

4. A garage had 90 cars on display. The bar chart in fig 13.5 shows how many of each type of car was available at the garage.

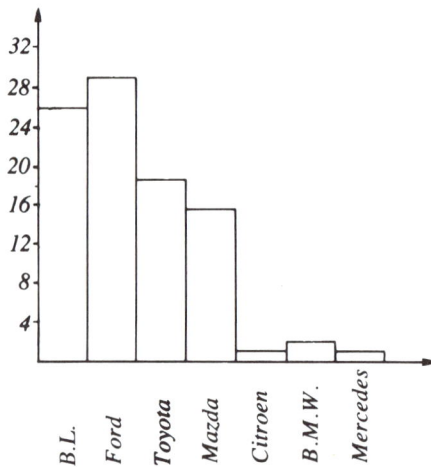

Fig 13.5

Using the information given in fig 13.5, answer the following questions:
(a) What type of car has the greatest representation in the display?
(b) There are 18 cars of one particular type. What type is this?
(c) What percentage of the display are B.L. cars?
(d) If B.M.W.s and Mercedes are both German, what percentage of the display is made up of German cars?
(e) What percentage of the cars are of Japanese origin?

5.

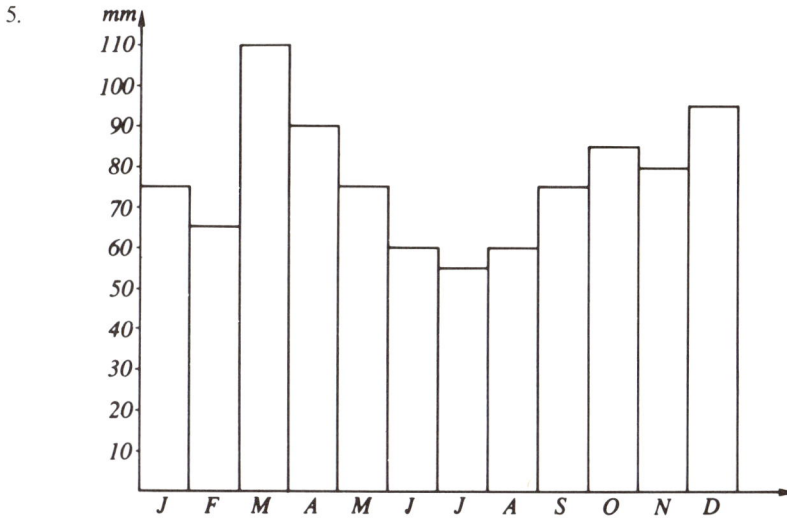

Fig 13.6

The bar chart given in fig 13.6 gives the average monthly rainfall (in mm) for a city in Argentina. Using the information given, answer the following questions.

(a) What month has the (i) greatest (ii) least average rainfall?

(b) In which month can this city expect to have a rainfall of 85 mm?

(c) In which months can the city expect to have the same rainfall?

(d) Knowing that Argentina lies in the southern hemisphere, in what season is the least rain expected to fall?

(e) What is the total expected rainfall per year?

6.

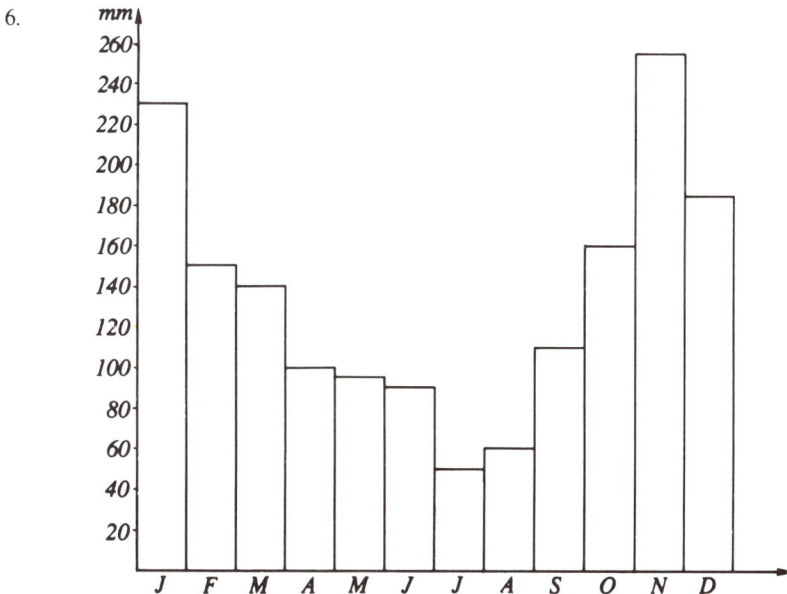

Fig 13.7

The bar chart given in fig 13.7 gives the average monthly rainfall (in mm) for a city in Canada. Using the information given, answer the following questions.

(a) What month has the (i) greatest (ii) least average rainfall?

(b) In which month can this city expect to have a rainfall of 110 mm?

(c) In which two consecutive months does the average rainfall drop by only 5 mm from the month before?

(d) In which two consecutive months is there the greatest difference in average rainfall?

(e) Knowing that Canada is in the northern hemisphere, in what season is the least rain expected to fall?

(f) What is the total expected rainfall per year?

7.

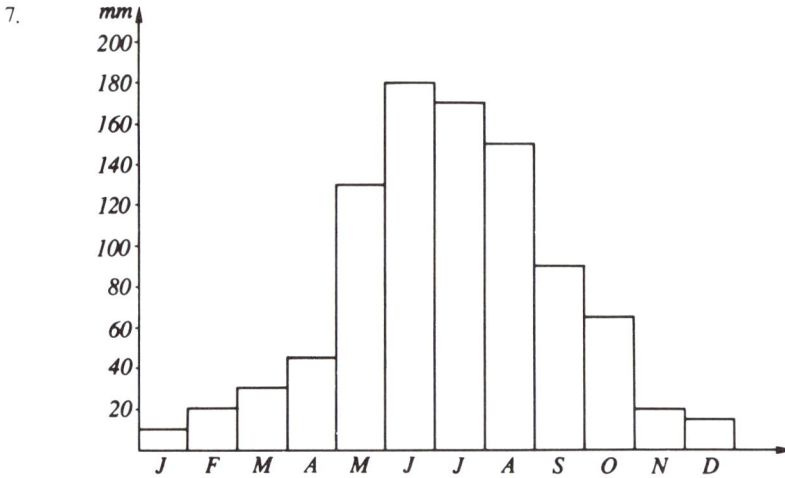

Fig 13.8

The bar chart given in fig 13.8 gives the average monthly rainfall (in mm) for a city in Western Australia. Using the information given, answer the following questions.

(a) What month has the (i) greatest (ii) least average rainfall?

(b) In which months can this city expect to have a rainfall of 20 mm?

(c) Which month can expect half the average rainfall of June?

(d) In which two consecutive months is there the greatest difference in average rainfall?

(e) Knowing that Western Australia is in the southern hemisphere, in what season is the most rain expected to fall?

(f) What is the total expected rainfall per year?

8. The following questions involve a comparison of the bar charts given in questions 5, 6 and 7; that is, comparing the rainfall in the three cities: one in Argentina, one in Canada and one in Western Australia.

(a) Which city has the highest average month's rainfall and in which month does this occur?

(b) Which city has the lowest average month's rainfall and in which month does this occur?

(c) Which city has the (i) greatest (ii) least *total* yearly expected rainfall?

(d) If you were to holiday in one of these cities during the month of (i) March (ii) September and you wanted as little rain as possible, which city would you choose?

(e) If you were to go to the city in Argentina in December, the city in Canada in June and the city in Western Australia in September, where would you expect to have the most rainfall?

13.3   Pictographs

A pictograph is similar to a bar chart but, instead of rectangles arranged side by side, the pictograph consists of columns of identical symbols. Each symbol drawn represents a given number of items. This is given in a 'key' at the side of the pictograph.

Example 13.3

A football team plays 8 matches. The team's goal score for each match is shown in the following pictograph:

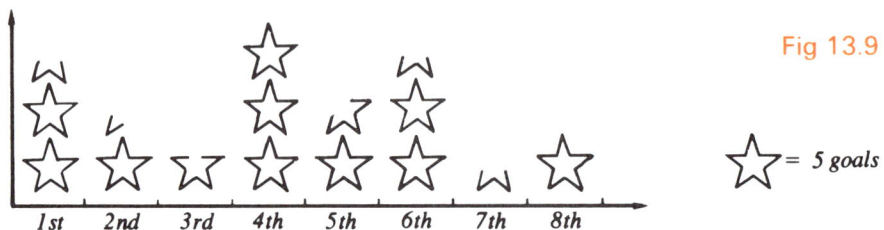

Fig 13.9

☆ = 5 goals

Use the information given in fig 13.9 to answer the following questions:

(a) How many goals were scored in the

(i)   4th match,   (ii) 5th match,

(iii) 7th match?

(b) In which match did the team score the (i) most (ii) least number of goals?

(c) How often and when did the team score the same number of goals?

(d) What was the total number of goals scored for the eight matches?

Solution

The 'key' at the side of the pictograph in fig 13.9 tells us that each star represents 5 goals.

(a) (i) The number of goals scored in the fourth match is represented by 3 stars (see fig 13.10). Each star represents 5 goals.

Thus, number of goals scored = ☆ + ☆ + ☆

$$= 5 + 5 + 5$$
$$= 15$$

Fig 13.10

4th

(ii) The number of goals scored in the fifth match is given by:

☆ + ⌁

i.e.   1 star + $\frac{3}{5}$ of one star (it has 3 of the 5 points of the star)

$$= 5 \text{ goals} + \frac{3}{5} \text{ of 5 goals}$$

$$= \left(5 + \frac{3}{\cancel{5}_1} \times \cancel{5}^1 \right) \text{goals}$$

$$= 5 + 3 \quad \text{goals}$$
$$= 8 \text{ goals}$$

(iii) The number of goals scored in the seventh match is given by

⌁

i.e. $\frac{2}{5}$ of one star (it has 2 of the 5 points of the star)

$$= \frac{2}{5} \text{ of 5 goals}$$

$$= \frac{2}{\cancel{5}_1} \times \cancel{5}^1 \text{ goals}$$

$$= 2 \text{ goals}$$

(b) (i) The most goals were scored in the fourth match because this match is represented by the largest column of stars.

(ii) Since the smallest column of stars corresponds to the seventh match, then it was in this match that the least number of goals were scored.

(c) The team scored the same number of goals in two matches: the first and the sixth. The team scored twelve goals (i.e. ☆ + ☆ + ⌁ = 12) in both matches.

(d) For the total number of goals scored in the eight matches:

1st match	=	☆ ☆ ⌁	= 12
2nd match	=	☆ ⌄	= 6
3rd match	=	⌁	= 4
4th match	=	☆ ☆ ☆	= 15
5th match	=	☆ ⌁	= 8

6th match =  = 12

7th match = 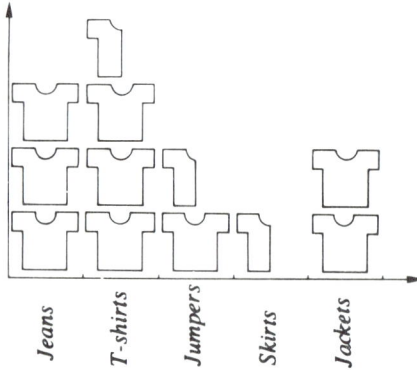 = 2

8th match = = $\frac{5}{}$

Total = 64

∴ the team scored a total of 64 goals in the eight matches.

Exercise 13C

1. Copy and complete the tables adjacent to each of the pictographs given below.

(a)

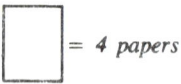

= 2 *pieces of clothing*

Type	Number
Jeans	
T-shirts	
Jumpers	
Skirts	
Jackets	
Total =	

(b)

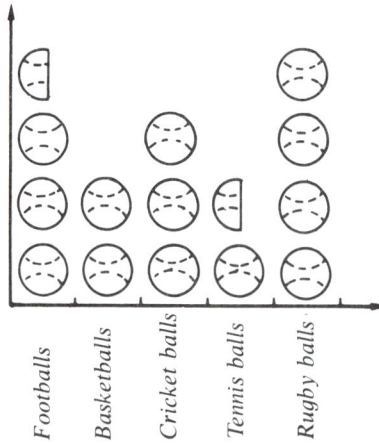 = 2 *balls*

Type	Number
Footballs	
Basketballs	
Cricket balls	
Tennis balls	
Rugby balls	
Total =	

(c)

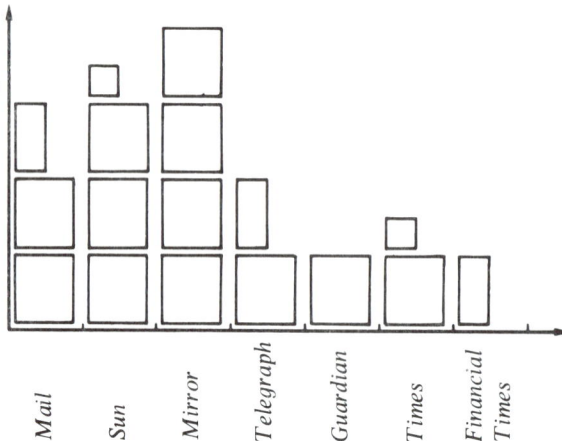 = 4 *papers*

Type	Number
Mail	
Sun	
Mirror	
Telegraph	
Guardian	
Times	
Financial Times	
Total =	

(d)

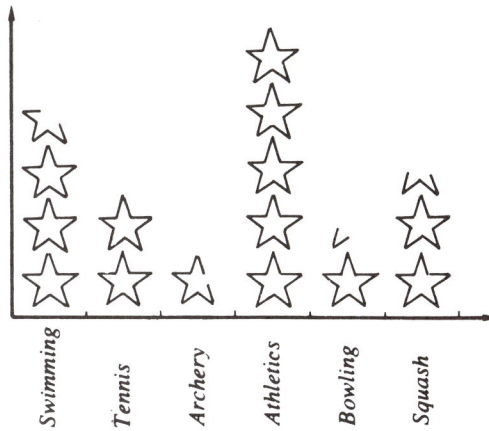

Type	Number
Swimming	
Tennis	
Archery	
Athletics	
Bowling	
Squash	
Total =	

2. Draw a pictograph for each of the following sets of information:

(a) On a school bus trip there are:

12	year 1 students	9	year 2 students
11	year 3 students	16	year 4 students
15	year 5 students	3	year 6 students

Use the symbol ⚲ to represent two students.
(Suggestion: For one student use the symbol ⚲)

(b) The 'Home Made Bakery' shop has the following goods left at the end of a particular day:

8	pastries	22	scones
5	cakes	7	loaves of bread
10	nut loaves	13	sandwiches
9	pies		

Use the symbol � to represent two items of baking.
(Suggestion: For one item use the symbol ⌂)

(c) At an international conference there were:

6	Australians	5	Englishmen
1	Austrian	8	Americans
5	Germans	7	Japanese

Use the symbol ⊠ to represent four people.
(Suggestion: △ = 1 person, ◁ = 2 people, ⊠ = 3 people)

(d) On a vintage car rally there were:

12	Austins	5	Rolls Royces
9	Vauxhalls	4	Daimlers
7	T-model Fords	3	Buicks

Use the symbol ⚮ to represent three cars.
(Suggestion: ○ = 1 car, ∞ = 2 cars)

3. The number of children in a particular street was counted according to their age groups. The results are shown in fig 13.11.

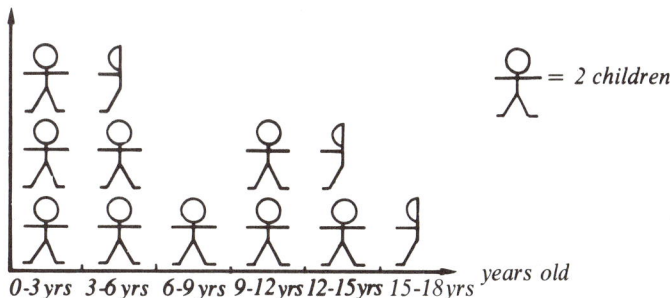

Fig 13.11

(a) Which age group has the (i) greatest (ii) least representation?
(b) How many children are there in the age group 3–6 years?
(c) How many children are there in the age group 12–15 years?
(d) Which age group has 4 members?
(e) What is the total number of children in the street?
(f) If one family with a 5-year old child and an 8-year old child leaves the street and is replaced by a family in which there is only one child (a toddler of 2 years of age), draw a pictograph to represent the new situation.

4. A refrigerated section of a supermarket contains different types of packaged cheeses. The number of packets of each type of cheese is given in fig 13.12.

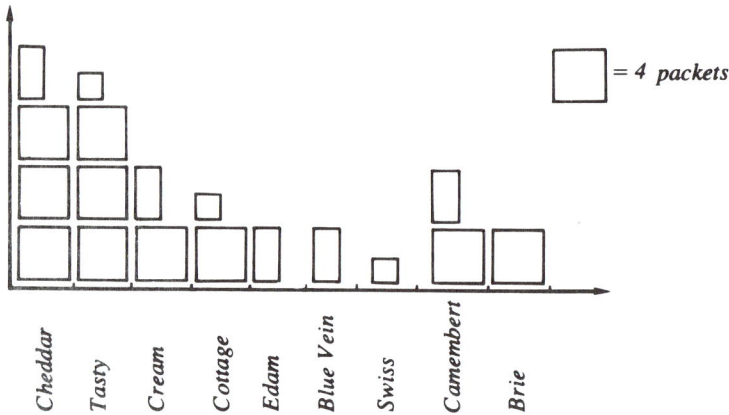

Fig 13.12

(a) Which type of cheese has the (i) greatest (ii) least number of packets on display?
(b) How many packets of tasty cheese are on display?
(c) How many packets of cottage cheese are on display?
(d) What type of cheese has the same number of packets on display as cream cheese?
(e) What is the total number of packets of cheese in the refrigerated section?
(f) A customer buys 3 packets of cheddar cheese, 1 packet of cream cheese, 1 packet of Swiss cheese and 2 packets of Camembert cheese. Draw a pictograph to show how many of each type of cheese are *left* on display after the customer has left the shop.

5. Fig 13.13 shows the number of each type of plant in stock at a certain plant nursery.

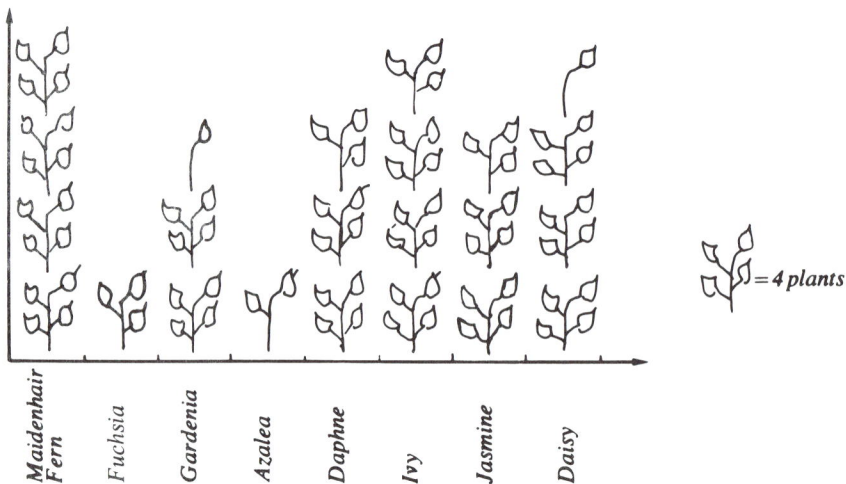

= 4 plants   Fig 13.13

(a) Of which type of plant has the nursery the (i) greatest and (ii) least number in stock?
(b) Which types of plant have the same number of their kind in stock?
(c) How many ivy plants are in stock?
(d) There are 13 of which type of plant in stock?
(e) What is the total number of plants in stock?

(f) What percentage of the stock is made up of gardenia plants?
(g) What percentage of the stock is made up of fuchsias and azaleas?
(h) What type of plant represents 20% of the stock?
(i) Three maiden hair ferns, one fuchsia, two daphne bushes and 5 daisy bushes are sold. Draw a pictograph to show the *remaining* stock.

6. A truck carries 50 crates of soft drink. The number of crates of the different types of soft drink are given in fig 13.14 below.

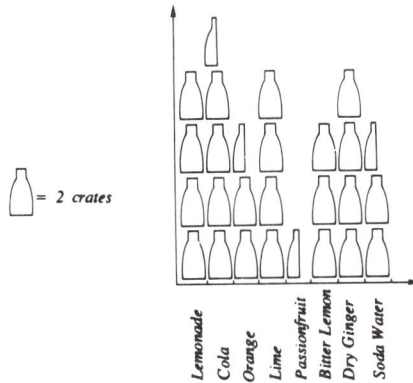

Fig 13.14

(a) What type of soft drink has the (i) greatest and (ii) least number of crates on the truck?
(b) How many crates of (i) lime (ii) soda water are on the truck?
(c) Which types of soft drink each have 5 crates on the truck?
(d) What percentage of the truck's load is made up of crates of lemonade?
(e) What percentage of the truck's load is made up of crates of orange flavoured soft drink?
(f) The truck makes a delivery of 3 crates of lemonade, 5 crates of cola, 4 crates of bitter lemon and 1 crate of dry ginger. Draw a pictograph to show how many crates of each type of soft drink are *remaining* on the truck.

13.4 Graphs

Often information can be given in graphical form.

Example 13.4

The graph given in fig 13.15 below can be used to convert temperatures given in degrees Celsius (°C) to corresponding temperatures in degrees Fahrenheit (°F) and vice versa. Using the graph given, make the following conversions

(a) $15°C = ?°F$ (b) $-10°C = ?°F$
(c) $80°F = ?°C$ (d) $5°F = ?°C$

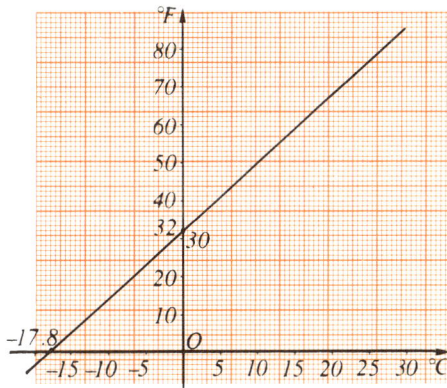

Fig 13.15

Solution

The graph given in fig 13.15 is a *linear* graph (see 10.3) and the coordinates of any point on the graph give corresponding temperatures in degrees Celsius and degrees Fahrenheit. For example, the point (0, 32) can be read as '0°C is equivalent to 32°F'.

(a)

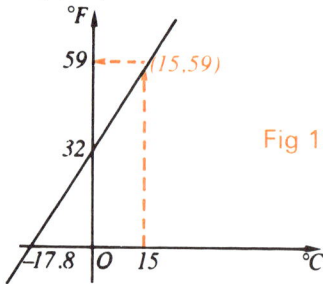

Fig 13.16

To convert 15°C to its equivalent Fahrenheit form, we must:
1. look for the number 15 on the Celsius (horizontal) axis,
2. move upwards until we reach a point on the given straight line,
3. move left from that point until we reach the Fahrenheit (vertical) axis (see fig 13.16).

The number arrived at on the Fahrenheit axis is the equivalent number of degrees Fahrenheit to 15°C. Thus, we find that 59°F is equivalent to 15°C.

∴ 15°C = 59°F

Note that, by following the three steps above, we are in fact finding the coordinates of the relevant point on the straight line; i.e. (15, 59). Remember that the first coordinate is the number of degrees Celsius (horizontal axis) and the second coordinate is the corresponding number of degrees Fahrenheit (vertical axis).

(b)

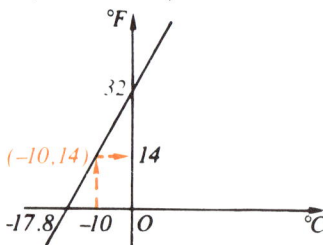

Fig 13.17

Reading from the graph, as in part (a), the point that has a Celsius coordinate of −10, has a Fahrenheit coordinate of 14. Thus,

−10°C = 14°F

(c)

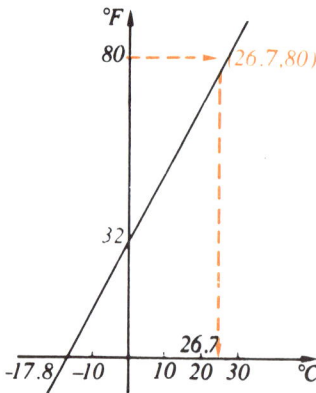

Fig 13.18

This time we want to convert degrees Fahrenheit to degrees Celsius and so we start from the Fahrenheit axis, move across to the straight line and then down to the Celsius axis (see fig 13.18). By doing this we find that the point on the straight line that has a Fahrenheit coordinate of 80 (i.e. 80°F) has a Celsius coordinate of 26.7 (i.e. 26.7°C). Thus,

80°F = 26.7°C

(d)

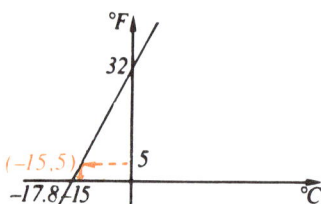

Fig 13.19

Reading from the graph, as in part (c), the point that has a Fahrenheit coordinate of 5 has a Celsius coordinate of −15. Thus,

5°F = −15°C

Example 13.5

A family leaves Manchester by car at 9.00 a.m. one Sunday. The graph given in fig 13.20 below shows how far the car is from Manchester at any time between the hours of 9.00 a.m. and 1.00 p.m. on that day.

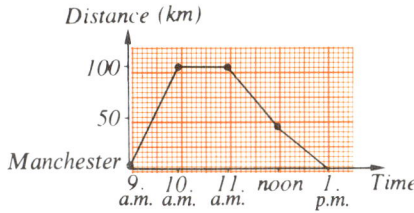

Fig 13.20

Using the information given in fig 13.20, answer the following questions.

(a) (i)   Where is the car at 9.00 a.m.?   (ii) Where is the car at 10.00 a.m.?
   (iii) How far has the car travelled in the first hour?
(b) (i)   Where is the car at 10.00 a.m.?   (ii) Where is the car at 11.00 a.m.?
   (iii) Where is the car at any time between the hours of 10.00 a.m. and 11.00 a.m.?
   (iv) How far has the car travelled in this hour?
(c) (i)   Where is the car at 11.00 a.m.?   (ii) Where is the car at noon?
   (iii) How far has the car travelled in this hour?
(d) (i)   Where is the car at noon?   (ii) Where is the car at 1.00 p.m.?
   (iii) How far has the car travelled in this hour?
(e) During what times is the car
   (i)   travelling *away from* Manchester,   (ii) stationary,
   (iii) travelling *back to* Manchester?

Solution

(a)

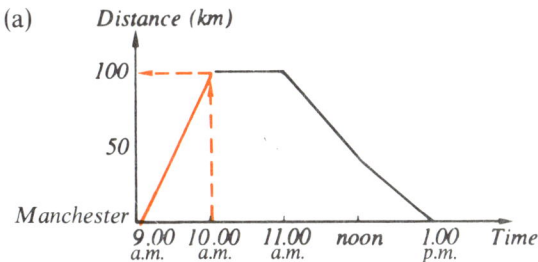

Fig 13.21

(i) We were told that the family leaves Manchester at 9.00 a.m. Hence, at 9.00 a.m. the car is in Manchester.
(ii) To find out where the car is at 10.00 a.m. we must
   1. look for 10.00 a.m. on the time (horizontal) axis,
   2. move upwards until we reach a point on the graph,
   3. move across from this point to the distance (vertical) axis.

The figure arrived at on the distance axis tells us where the car is at 10.00 a.m. In this case we find that the car is 100 km away from Manchester (see fig. 13.21).

Also, the coordinates of the point on the graph found in step 2 above can be described as (10.00 a.m., 100 km); that is, the first coordinate gives the time and the second coordinate gives the position of the car at that time.

(iii) Between 9.00 a.m. and 10.00 a.m. the distance away from Manchester is steadily increasing (see fig 13.21) and at 10.00 a.m. the car is 100 km from Manchester. Therefore, in the first hour the car has travelled 100 km.

(b)

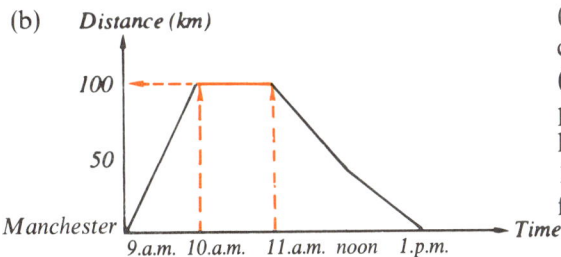

Fig 13.22

(i) At 10.00 a.m. the car is 100 km from Manchester (see (a) (ii)).
(ii) Reading from the graph as in (a) (ii), the point that has a time coordinate of 11.00 a.m. has a distance coordinate of 100 km (see fig 13.22). Therefore, at 11.00 a.m. the car is 100 km from Manchester.

(c)

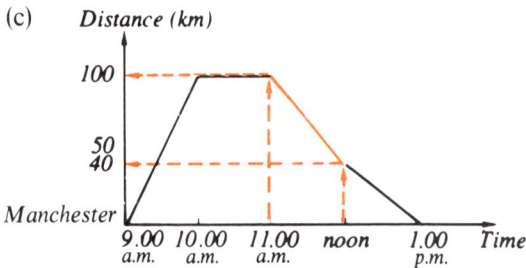

Fig 13.23

(d)

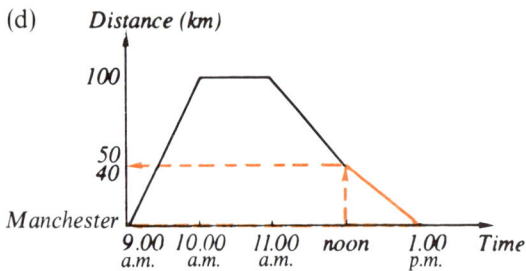

Fig 13.24

(iii) Looking at the graph, at any time between 10.00 a.m. and 11.00 a.m. the car remains 100 km from Manchester.

(iv) The car has not travelled at all in this hour; that is, the car is *stationary*.

(i) At 11.00 a.m. the car is 100 km from Manchester (see (b) (ii)).

(ii) Reading from the graph as in (a) (ii), the point that has a time coordinate of noon has a distance coordinate of 40 km (see fig 13.23). Therefore, at noon the car is 40 km from Manchester.

(iii) At 11.00 a.m. the car is 100 km from Manchester and for the next hour the distance *decreases* steadily (see fig 13.23) until at noon it is only 40 km from Manchester. Hence, the car has travelled 60 km (100 km – 40 km) in this hour.

(i) At noon the car is 40 km from Manchester (see (c) (ii)).

(ii) Reading from the graph as in (a) (ii), the point that has a time coordinate of 1.00 p.m. has a distance coordinate of 0 km (see fig 13.24). Therefore, at 1.00 p.m. the car is 0 km from Manchester; that is, the car is back in Manchester.

(iii) At noon the car is 40 km from Manchester and during the next hour the distance steadily decreases until at 1.00 p.m. the car is back in Manchester. Therefore, the car has travelled 40 km in this hour.

(e) (i)   The car is travelling *away from* Manchester between 9.00 a.m. and 10.00 a.m.

   (ii)   The car is *stationary* from 10.00 a.m. to 11.00 a.m.

   (iii)   The car is travelling *back to* Manchester from 11.00 a.m. to 1.00 p.m.

Exercise 13D

1. The total cost of drinking glasses (in £) depends upon the number bought. This is shown in the following graph.

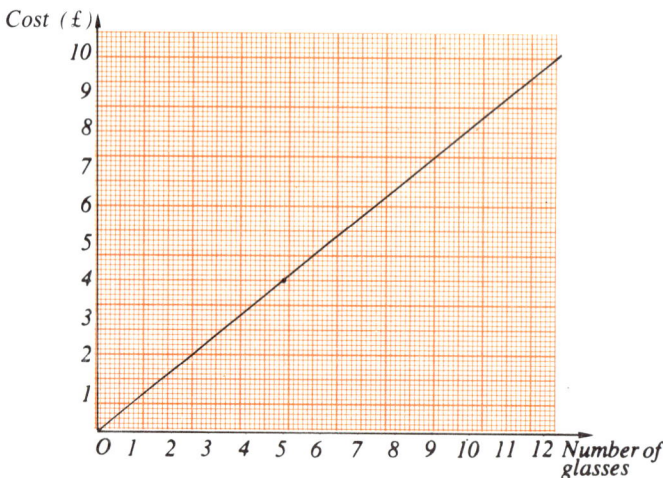

Fig 13.25

From the graph, find:
(a) the cost of
 (i) 6 glasses
 (ii) 8 glasses
 (iii) 10 glasses
 (iv) 12 glasses
(b) The number of glasses which can be bought for
 (i)   £4
 (ii) £8
 (iii) £5.60
 (iv) £8.80
 (v) £2.40
 (vi) £6.40
 (vii) £7
 (viii) £10
(c) If 6 glasses cost £4.80, how much would 12 glasses cost?

2. The following graph shows the comparison between an Australian dollar ($) and an English pound (£).

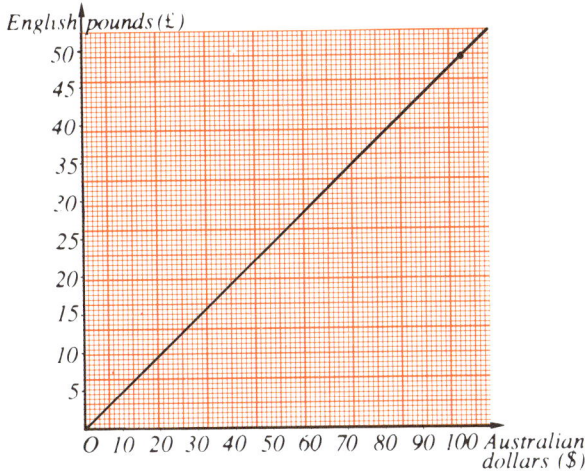

Fig 13.26

(a) Convert each of the following to English pounds
 (i) $50
 (ii) $80
 (iii) $70
 (iv) $30
 (v) $55
 (vi) $95
 (vii) $23
 (viii) $72
(b) Convert each of the following to Australian dollars
 (i)   £10
 (ii) £30
 (iii) £40
 (iv) £20
 (v) £25
 (vi) £45
 (vii) £18
 (viii) £33
(c) An Australian tourist in England bought a souvenir for £7. How much is this in Australian dollars?
(d) A set of cutlery costs $160 in Australia. The same set of cutlery costs £72 in England. Would it be cheaper to buy the set of cutlery in Australia or England?

3. The Brown family leaves home at 9.00 am to go for a Sunday drive. Fig 13.27 shows where the family is at any given time.

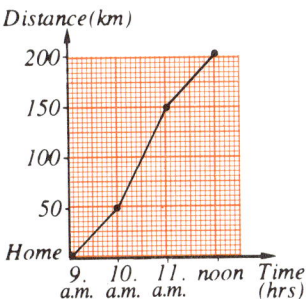

Fig 13.27

(a) Where is the family at (i) 10.00 a.m. (ii) 11.00 a.m. (iii) noon?
(b) How far does the family travel in the (i) first (ii) second (iii) third hour?
(c) At what time is the family 100 km from home?

4. John leaves Keighley at 9.00 a.m. and cycles into the countryside. Fig. 13.28 shows where John is at any given time.

Distance (km)

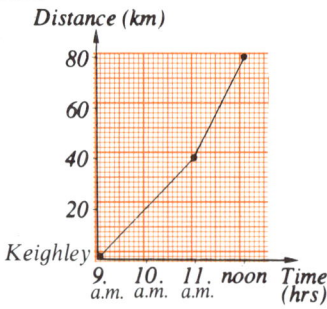

(a) How far does John travel in the (i) first (ii) second (iii) third hour?
(b) What is the total distance travelled by John in the three hours?
(c) Where is John at (i) 11.00 a.m. (ii) 11.30 a.m.?
(d) At what time is John 30 km from Keighley?

Fig 13.28

5. A train takes 4 hours to travel from Town *A* to Town *C*, making a stop at Town *B* along the way. Fig. 13.29 shows the position of the train at any given time.

Distance (km)

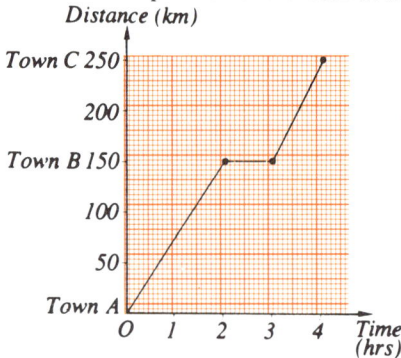

(a) How far is (i) Town *B* (ii) Town *C* from Town *A*?
(b) How long does it take the train to reach Town *B*?
(c) How long does the train stop at Town *B*?
(d) How long does the train take to get from Town *B* to Town *C*?
(e) Where is the train after it has been travelling for 1 hour?

Fig 13.29

6. A man drives from Swansea to Preston, stopping along the way for lunch. His journey takes him a total of four hours. Fig. 13.30 shows where the man is at any given time.

Distance (km)

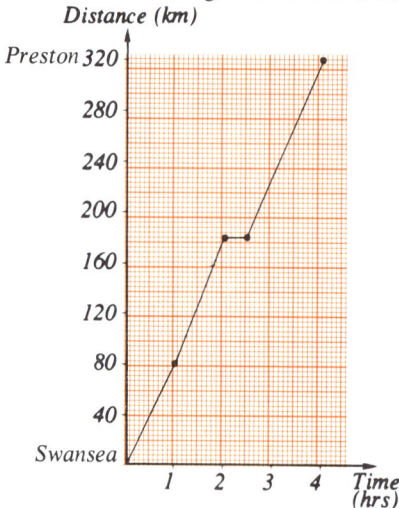

(a) How far does the man travel in the (i) first (ii) second hour?
(b) Where is the man at the end of the second hour?
(c) When does the man stop for lunch? How far is he from Swansea at this time?
(d) How long does he stop for lunch?
(e) After his lunch break how long does it take him to get to Preston and how much farther must he travel?

Fig 13.30

7. A young boy goes for a walk in the hills. He leaves his family's cottage at 9.00 a.m. and returns at 2.00 p.m. Fig. 13.31 shows where the boy is at any given time during his walk.

Distance (km)

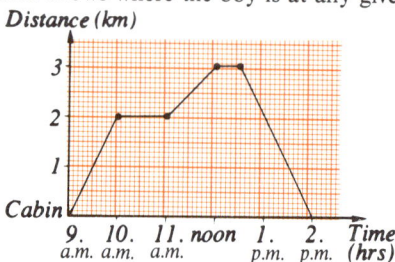

(a) How far does the boy walk in the first hour?
(b) How far does the boy walk between 10.00 a.m. and 11.00 a.m.?
(c) How far does the boy walk between 11.00 a.m. and noon?
(d) For how long does the boy rest when he stops at noon?
(e) How long does it take for the boy to return home?

Fig 13.31

(f) Does the boy stop for any rest periods on his return journey?

(g) How far from the cottage is the boy at
(i) 1.00 p.m. (ii) 2.00 p.m. ?

8. A group of Geography students go on a bus excursion, leaving their school in Town *A* at 9.00 a.m. and returning at 3.00 p.m. During the excursion they make stops at Town *B*, Town *C* and Town *D*. The following graph shows where the bus is at any given time during the excursion.

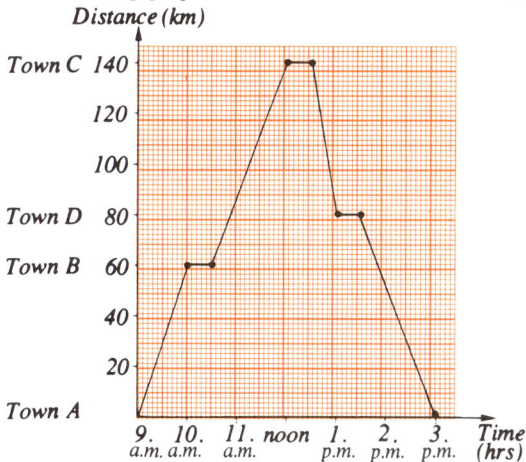

(a) Where is the bus at 10.00 a.m.?

(b) How far is Town *B* from Town *A*?

(c) How long do the students spend in Town *B*?

(d) How far does the bus travel between the time 10.30 a.m. and 11.30 a.m.?

(e) How far is Town *C* from (i) Town *B* (ii) Town *A*?

(f) When does the bus set off on its return journey?

(g) For how long does the bus travel after leaving Town *C* before it arrives at Town *D*?

(h) How far is Town *D* from Town *C*?

(i) For how long does the bus stay in Town *D*?

(j) Where is the bus at 2.30 p.m.?

(k) What is the total time taken for the return journey?

13.5   Pie Charts

A pie chart is a pictorial representation of percentages (or fractions) given in the form of a circle divided into a certain number of sectors of varying sizes. The size of each sector depends upon the percentage or fraction it is representing. Before we draw or analyse a pie chart we should check both our knowledge of angles in a circle and of percentages.

13.5.1   Angles in a Circle

(a) There are 360° in a circle.

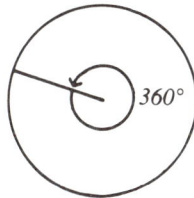

Fig 13.32

(b) A *sector* of a circle is a 'wedge' of a circle.

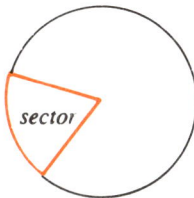

Fig 13.33

(c) Any sector of a circle contains *a part of* the whole centre angle (360°) of a circle.

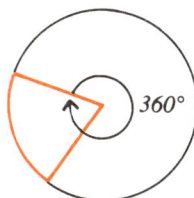

Fig 13.34

(d) If a sector of a circle contains a centre angle of 180° (i.e. one half the total angle of the circle of 360°), then that sector represents one half of the entire circle.

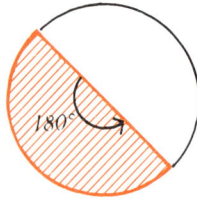

Fig 13.35

$$\text{Sector Angle} = 180°$$
$$= \tfrac{1}{2} \text{ of } 360°$$
$$\therefore \text{Sector} = \tfrac{1}{2} \text{ of circle}$$

(e) If a sector forms one half of a circle, then it can be said that the sector is 50% of the circle. I.e.

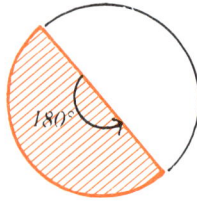

Fig 13.36

$$\text{Sector} = \tfrac{1}{2} \text{ of circle}$$
$$\therefore \text{Sector} = 50\% \text{ of circle}$$

Some other common sector angles and percentages are:

(i)

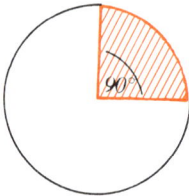

Fig 13.37

$$\text{Sector Angle} = 90°$$
$$= \tfrac{1}{4} \text{ of } 360°$$
$$\therefore \text{Sector} = \tfrac{1}{4} \text{ of circle}$$
$$\therefore \text{Sector} = 25\% \text{ of circle}$$

(ii)

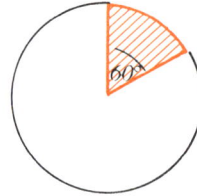

Fig 13.38

$$\text{Sector Angle} = 60°$$
$$= \tfrac{1}{6} \text{ of } 360°$$
$$\therefore \text{Sector} = \tfrac{1}{6} \text{ of circle}$$
$$\therefore \text{Sector} = (\tfrac{1}{6} \times \tfrac{100}{1})\% \text{ of circle}$$
$$= 16\tfrac{2}{3}\% \text{ of circle}$$

(iii)

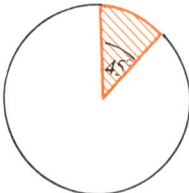

Fig 13.39

$$\text{Sector Angle} = 45°$$
$$= \tfrac{1}{8} \text{ of } 360°$$
$$\therefore \text{Sector} = \tfrac{1}{8} \text{ of circle}$$
$$\therefore \text{Sector} = (\tfrac{1}{8} \times \tfrac{100}{1})\% \text{ of circle}$$
$$= 12\tfrac{1}{2}\% \text{ of circle}$$

(iv)

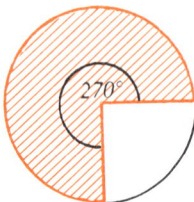

Fig 13.40

$$\text{Sector Angle} = 270°$$
$$= \tfrac{3}{4} \text{ of } 360°$$
$$\therefore \text{Sector} = \tfrac{3}{4} \text{ of circle}$$
$$\therefore \text{Sector} = 75\% \text{ of circle}$$

Example 13.6

(a) Calculate what percentage of the circle is enclosed in the sector given below.

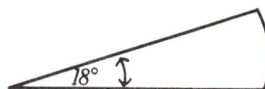

Fig 13.41

(b) A sector of a circle is 20% of the circle (see fig 13.42 below). Find the angle enclosed in the sector.

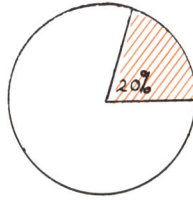

Fig 13.42

Solution

(a)

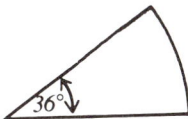

Fig 13.43

The full circle contains an angle of 360°. To find how much of the circle is enclosed in the sector we must find out what fraction of 360° is given by the sector angle of 18°.

I.e. fraction of circle angle = ratio of sector angle to circle angle

$$= \frac{18°}{360°}$$

$$= \frac{18^1}{360_{20}} \quad \text{(remember: a ratio has no units of measurement)}$$

$$= \tfrac{1}{20}$$

Therefore, sector represents $\tfrac{1}{20}$ of the whole circle.

Now, if the sector $= \tfrac{1}{20}$ of circle

then, rewriting $\tfrac{1}{20}$ as a percentage:

the sector $= (\tfrac{1}{20} \times \tfrac{100}{1})\%$ of circle (see 11.2.2)

$= 5\%$ of circle.

Thus, a sector with a centre angle of 18° represents 5% of the circle from which the sector was taken.

(b)

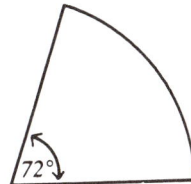

Fig 13.44

If the sector takes up 20% of the whole circle then the angle in the sector will be 20% of the whole circle angle (360°).

Thus, sector angle = 20% of 360°

$$= \frac{20^1}{100_5{}_1} \times {}^{72}360°$$

$$= 72°$$

Hence, the angle in the sector is 72°.

Exercise 13E

1. Calculate what (i) fraction (ii) percentage of a circle is represented by the following sectors.

(a)

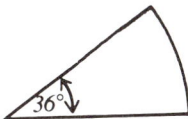

(b)

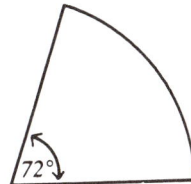

(c)

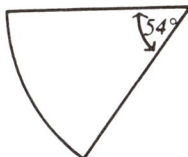

(d)

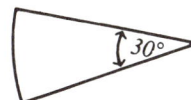

(e)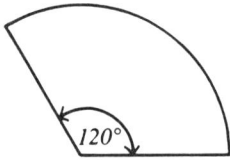

(f)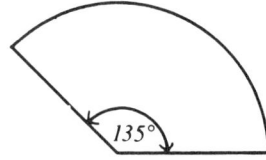

(g)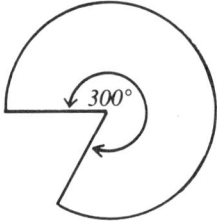

(h)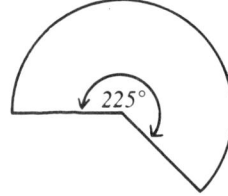

(i)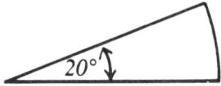

(j)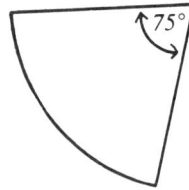

2. Find the centre angle enclosed in each of the followed shaded sectors of circles.

(a)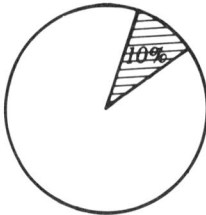

(b)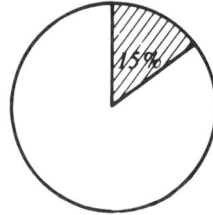

(c)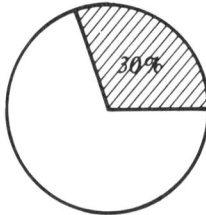

(d)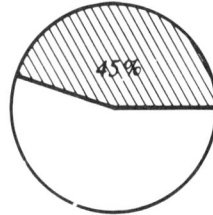

(e)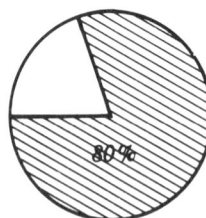

(f)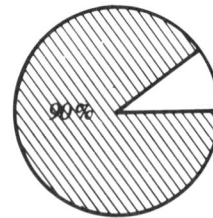

(g)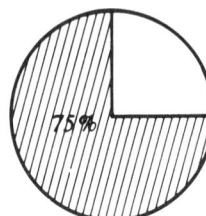

(h)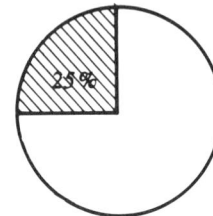

(i) 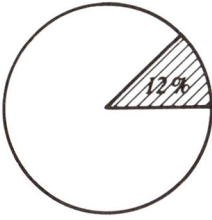   (j)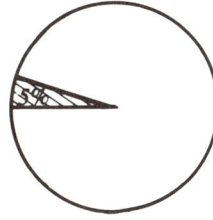

13.5.2   Interpretation and Formation of Pie Charts

Example 13.7

In a group of 200 students, some were 12 years old, some were 13 years old and the rest were 14 years old. The percentages of 12, 13, and 14 year olds are represented in the following pie chart.

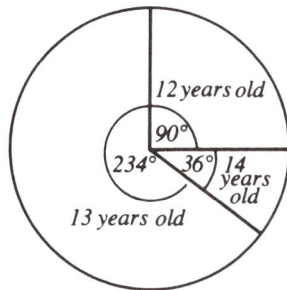

Fig 13.45

Using the pie chart given, answer the following questions.

(a) What percentage of the students were
    (i)   12 years of age?   (ii) 13 years of age?
    (iii) 14 years of age?

(b) How many students (out of the 200) were
    (i)   12 years of age?   (ii) 13 years of age?
    (iii) 14 years of age?

Solution

(a) (i)   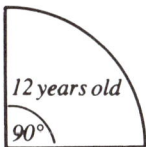   The sector representing the 12 year olds contains an angle of $90°$.

Fig 13.46

Therefore, sector represents $\frac{1}{4}\left(\frac{90°}{360°}\right)$ of whole circle. Hence, 25% $(\frac{1}{4} \times \frac{100}{1}\%)$ of the students are 12 years of age.

(ii)   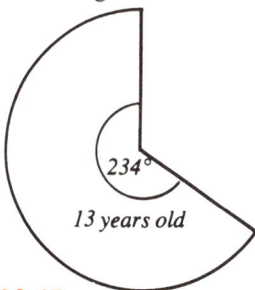   The sector representing the 13 year olds contains an angle of $234°$.

$$\text{Therefore, sector} = \frac{{}^{13}\cancel{234}°}{{}_{20}\cancel{360}°} \quad \text{of circle}$$

$$= \frac{13}{20} \quad \text{of circle}$$

$$\text{Thus, sector} = \frac{13}{\cancel{20}_{1}} \times \frac{{}^{5}\cancel{100}}{1}\% \text{ of circle}$$

$$= 65\% \text{ of circle.}$$

Fig 13.47   Hence, 65% of the students are 13 year olds.

(iii) Since 25% of the students are 12 years old and 65% are 13 years old (that is, a total of 90%) then the remaining 14 year olds must represent 10%; i.e. $100\% - (25 + 65)\%$. Alternatively, the percentage of students who are 14 years old could have been calculated using the same method given in parts (i) and (ii).

(b) (i) If 25% of the students are 12 years old (see answer for (a) (i)) and there are a total of 200 students, then the *number* of 12 year olds is calculated by:

$$25\% \text{ of } 200 \text{ are } 12 \text{ years old}$$

$$\therefore \frac{25}{\cancel{100}_1} \times \frac{\cancel{200}^2}{1} \text{ are } 12 \text{ years old}$$

$$\therefore 50 \text{ students are } 12 \text{ years old.}$$

(ii) From (a) (ii), 65% of the 200 students are 13 years old.

$$\therefore \text{Number of 13 year olds} = \frac{65}{\cancel{100}_1} \times \frac{\cancel{200}^2}{1}$$

$$= 130$$

(iii) Number of 14 year olds

= Total number of students − students aged 12 or 13.

$$= \qquad 200 \qquad - (50 \qquad + \qquad 130)$$
$$= 200 - 180$$
$$= 20$$

Alternatively, number of 14 year olds = 10% of 200

$$= 20$$

Exercise 13F

1. For each of the following pie charts copy and complete the adjacent table:

   (a)

   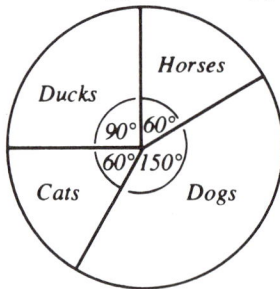

Animals	Fraction of Animals	Percentage of Animals
Horses		
Dogs		
Cats		
Ducks		

   (b)

   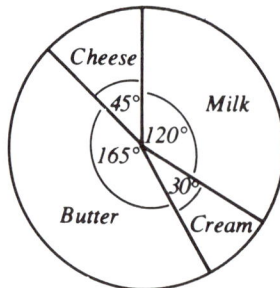

Dairy Product	Fraction of Dairy Prod.	Percentage of Dairy Prods.
Milk		
Cream		
Butter		
Cheese		

   (c)

   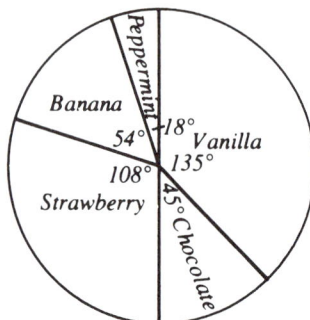

Icecream flavours	Fraction of flavours	Percentage of flavours
Vanilla		
Chocolate		
Strawberry		
Banana		
Peppermint		

(d)

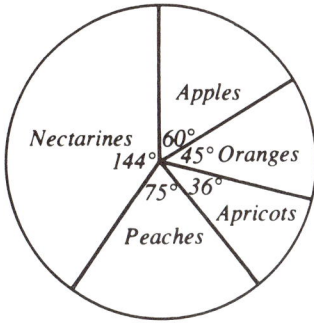

Types of fruit	Fraction of fruit	Percentage of fruit
Apples Oranges Apricots Peaches Nectarines		

2. A particular country exports wheat, wool, beef and minerals. The percentages of these quantities of the total export programme are represented in the pie chart given in fig 13.48.

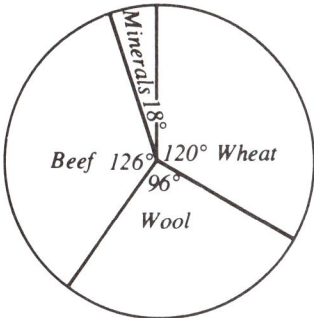

(a) What percentage of exported commodities is taken up with
  (i) wheat;   (ii) wool;
  (iii) beef;   (iv) minerals?
(b) What percentage of the exports directly involve animals?

Fig 13.48

3. A car salesman sells B.L.s, Mazdas, B.M.W.s and Lancias. The pie chart given in fig 13.49 shows an analysis of his year's sales.

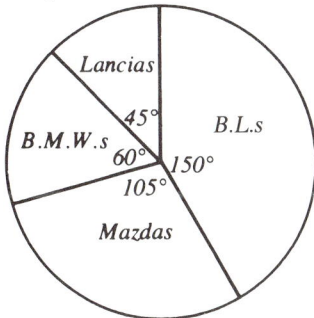

(a) What percentage of his sales were made up of
  (i) B.L.s;   (ii) Mazdas;
  (iii) B.M.W.s;   (iv) Lancias?
(b) What percentage of his sales were made up of cars imported from Europe?

Fig 13.49

4. In a sports programme at school the students must choose to do one of the following sports—tennis, basketball, hockey, or squash. There is a total of 600 students involved in the sports programme. The pie chart given in fig 13.50 depicts the percentage of students choosing to do each sport.

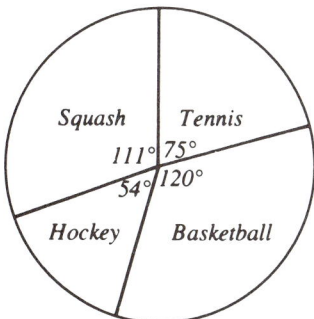

(a) What (i) fraction (ii) percentage of the students chooses tennis?
(b) What (i) fraction (ii) percentage of the students chooses basketball?
(c) What (i) fraction (ii) percentage of the students chooses hockey?
(d) What (i) fraction (ii) percentage of the students chooses squash?
(e) Out of the 600 students how many play (i) tennis, (ii) basketball, (iii) hockey, (iv) squash?

Fig 13.50

5. A teacher arrives at school at 8.00 am on a particular day and leaves at 4.00 pm. During the day his time is spent teaching, correcting students' homework, having discussions with other teachers and taking refreshment breaks. The pie chart in fig 13.51 shows the break-down of the time spent on each of these activities.

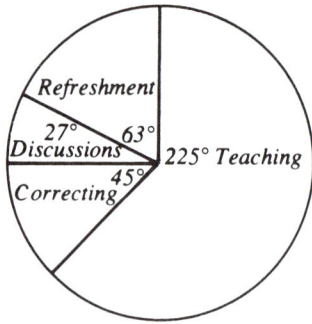

Fig 13.51

(a) What (i) fraction (ii) percentage of his time is spent teaching?
(b) What (i) fraction (ii) percentage of his time is spent correcting homework?
(c) What (i) fraction (ii) percentage of his time is spent having discussions with other teachers?
(d) What (i) fraction (ii) percentage of his time is spent on refreshment breaks?
(e) Of the 8 hours spent at school, how much time is spent (i) teaching; (ii) correcting homework; (iii) having discussions with other teachers; (iv) taking refreshment?

Example 13.8

During one particular day, a student spends 8 hours sleeping, 6 hours in class, 2 hours eating, 3 hours playing sport and the remaining hours relaxing in some way. Depict his day's activities on a pie chart.

Solution

A day consists of 24 hours.

	Sleeping	Class	Eating	Sport	Relaxing
Hours	8	6	2	3	5
Fraction of day	$\frac{8}{24} = \frac{1}{3}$	$\frac{6}{24} = \frac{1}{4}$	$\frac{2}{24} = \frac{1}{12}$	$\frac{3}{24} = \frac{1}{8}$	$\frac{5}{24}$

(add up to 24)

To depict the above information on a pie chart we need a compass and a protractor.

Step 1: Draw a large circle (representing a full day) and mark in a starting radius.

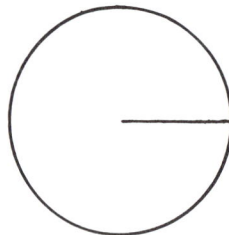

Fig 13.52

Step 2: Calculate the angle needed in the sector which is to represent the time spent sleeping.
I.e. Spends $\frac{1}{3}$ of day sleeping
∴ $\frac{1}{3}$ of circle represents sleeping
∴ $\frac{1}{3}$ of 360° in sleeping sector

$$= \frac{1}{3_1} \times 360°^{120°}$$

$= 120°$ in sleeping sector

Step 3: Place bottom line of protractor on starting radius of circle and mark off the sleeping sector angle of 120°.

Fig 13.53

Step 4: Calculate the sector angle needed for time spent in class.

I.e. spends $\frac{1}{4}$ of day in class

$\therefore$ $\frac{1}{4}$ of circle represents class time.

$\therefore$ $\frac{1}{4}$ of 360° in class time sector

$$= \frac{1}{\cancel{4}_1} \times \cancel{360°}^{\,90°}$$

$= 90°$ in class time sector

Step 5: Now place bottom line of protractor along the radius which *finishes* the sleeping sector (and thus *starts* the class time sector). Mark off class time sector angle of 90°.

Fig 13.54

Continue in this manner until all activities are represented in an appropiate sector of this circle. Calculating the sector angles for the remaining activities:

Eating: $\frac{1}{12}$ of day

$\qquad$ $\therefore$ sector angle $= \frac{1}{12}$ of 360°

$$= \frac{1}{\cancel{12}} \times \cancel{360°}^{\,30°}$$

$\qquad$ $= 30°$

Sport: $\frac{1}{8}$ of day

$\qquad$ $\therefore$ sector angle $= \frac{1}{8}$ of 360°

$$= \frac{1}{\cancel{8}_1} \times \cancel{360°}^{\,45°}$$

$$= 45°$$

Relaxing: $\frac{5}{24}$ of day

$\qquad$ $\therefore$ sector angle $= \frac{5}{24}$ of 360°

$$= \frac{5}{\cancel{24}_1} \times \cancel{360°}^{\,15°}$$

$$= 75°$$

Note: If the first four sector angles have been measured accurately, then the angle in the last sector should automatically be 75°. I.e. $75° = 360° - (120° + 90° + 30° + 45°)$.

The complete pie chart depicting the student's activities on that day should appear as:

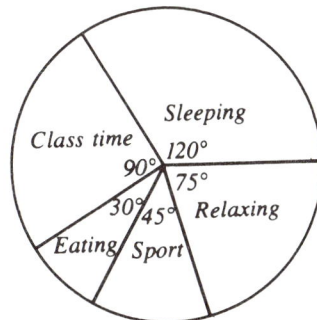

Fig 13.55

Exercise 13G

Draw a pie chart to represent each of the following sets of information.

1. A salesgirl selling cosmetics finds that, for one day's sales:
   10% were bottles of perfume   35% were lipsticks
   25% were mascaras   30% were eyeshadows

2. At 6.00 p.m., the choice of T.V. programmes is to be made from the news, a game show, a situation comedy and a soap opera serial. A survey of viewers at this time showed that:
   30% watched the news   5% watched the game show
   10% watched the situation comedy   55% watched the soap opera serial

3. The pattern for a multicoloured jumper states that 30 balls of wool are needed. Of these 30 balls:
   6 balls are white   12 balls are navy blue
   2 balls are pink   5 balls are red
   5 balls are green

4. A man has 150 books in his library. Of these 150 books:
   80 books are novels   15 books are autobiographies
   5 books are biographies   40 books are educational texts
   10 books are science fiction

5. In a small country town there are 80 pets kept by the various families that live there. Of these 80 pets:
   36 are cats   20 are dogs
   4 are horses   14 are birds
   6 are goats

6. At an athletics meeting one team had 40 competitors. Of these competitors:
   28 were runners   2 were high jumpers
   5 were long jumpers   3 were discus throwers
   2 were shot putters

Applications and Diversions 14

14.1 Applications

14.1.1 Arithmetic

Exercise 14A

1. Sally Smith bought 2 kg of corned beef which costs £2.99 per kg and 3 kg of chuck steak which costs £3.99 per kg.
   (a) How much did she pay for the corned beef?
   (b) How much did she pay for the chuck steak?
   (c) What was her total meat bill?
   (d) How much change did she receive from £20?

2. Bertha Baloney bought 1.5 kg of cheese at £1.98 per kg, 2.5 kg of sausages at £1.76 per kg and 0.2 kg of ham at £6.75 per kg.
   (a) How much did she pay for the cheese?
   (b) How much did she pay for the sausages?
   (c) How much did she pay for the ham?
   (d) What was her total bill?
   (e) How much change did she receive from £10?

3. At a fete, pony-rides are 25p, a ride on the merry-go-round is 20p, coconut shy is 15p and lucky-dip is 10p.
   (a) What are the takings if 50 people have a pony-ride, 100 have rides on the merry-go-round, 30 try the coconut shy and 80 have a lucky dip?
   (b) If the organisers have costs of £20, how much will they make on these side-shows?

4. The menu at Pat and Geoff's Hamburger Shop is as follows:

Plain hamburger	£1
Hamburger with the lot	£1.50
Chicken roll	£0.80
Fish 'n chips	£1.10

   On one particular day, 16 people bought plain hamburgers, 31 people bought hamburgers with the lot, 18 people bought chicken rolls and 21 people bought a portion of fish 'n chips. If each customer bought only one item, find:
   (a) how many customers Geoff served that day;
   (b) how much money Geoff took that day;
   (c) if the food cost Geoff £40 and the rent was £30, how much profit Geoff made;
   (d) the profit as a percentage of his takings for the day.

5. The following table shows the distance in kilometres between some major British cities. Use this table to answer the questions below it.

	London	Birmingham	Bristol	Derby	Exeter	Glasgow	Liverpool	Manchester	Oxford	Sheffield	Southampton
Birmingham	179										
Bristol	183	132									
Derby	198	63	195								
Exeter	275	259	121	315							
Glasgow	634	459	589	454	721						
Liverpool	330	151	264	142	383	340					
Manchester	309	130	269	93	385	338	55				
Oxford	90	100	106	158	224	547	251	230			
Sheffield	261	121	262	58	399	386	119	64	217		
Southampton	127	206	121	264	179	668	357	325	106	323	
York	315	216	346	143	469	340	158	103	291	90	398

(a) What is the distance from London to Southampton?

(b) What is the distance from Exeter to York?

(c) Rodney Rep has to drive from Birmingham to Manchester and back in a day. How far does he drive?

(d) A lorry driver is in Glasgow on Monday. In the week he drives to London, from there to Oxford and on to Birmingham. He then drives to Liverpool and back to Glasgow. How far does he travel that week?

(e) I took five hours to drive from Derby to Exeter. What was my average speed?

(f) Which two towns are 93 km apart?

(g) Find two pairs of towns which are 121 km apart.

(h) I travel from Oxford to Birmingham at an average speed of 50 km per hour. How long does my journey take?

14.1.2   Symbolic Expression

It has already been seen that Algebra uses pronumerals and symbols in the same way as Arithmetic uses numbers. One major difference between Algebra and Arithmetic is the fact that in Algebra it is only possible to *simplify* an expression whereas arithmetic enables us to *evaluate* an answer. The value of a particular pronumeral in an *equation* can be found.

Example 14.1

If $x + y = 180$
which of the following statements is true:

$A. \; y = 180 - x$ $B. \; x = y - 180$ $C. \; x = y$

Solution

$x + y = 180$
$-x$ both sides $x - x + y = 180 - x$
$y = 180 - x$ $\therefore A$ is true.

Similarly,
if $x + y = 180$
$x + y - y = 180 - y$
$\therefore x = 180 - y$ $\therefore B$ is false.
As we do not know the exact value of x and y, we do not know that $x = y$.
$\therefore C$ is false.

Example 14.2

If the sum of two angles is 90°, and one angle is $y°$, what does the other angle measure? Draw a diagram for your answer.

Solution

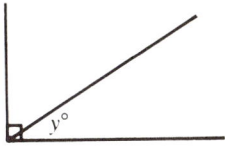

If one angle is $y°$ then the other angle is $y°$ less than 90°.
i.e. $90° - y°$

Fig 14.1

Example 14.3

If a kilogram of tomatoes costs £P how much will I pay for
(a) 2 kilogram;   (b) g kilogram?

Solution

(a) 1 kg costs £P   (b) 1 kg costs £P
$\therefore$ 2 kg costs $2 \times £P = £2P$ $\therefore$ g kg costs $£g \times P$ or $£gP$

Exercise 14B

1. If $a + b = 90$, which of the following is necessarily true?
   A. $b = 90 - a$ B. $a = b$ C. $a = b - 90$

2. If $x + y = 360$, which of the following is necessarily true?
   A. $x = y$ B. $y = x - 360$ C. $x = 360 - y$

3.

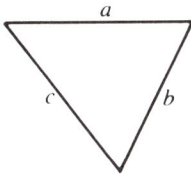

   What is the perimeter of the triangle in terms of a, b, c.?

4. What is the perimeter of the rectangle whose length is L cm and width is W cm?

5. If the angles of a triangle are $a°$, $b°$ and $c°$, express the sum of the angles in terms of a, b and c.

6.

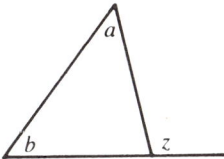

   State a relationship between a, b and z.

7.

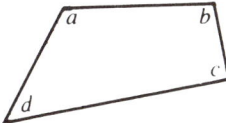

   State a relationship between a, b, c and d.

8.

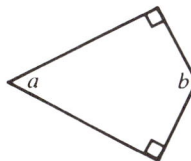

   State a relationship between a and b.

9.

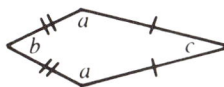

   State a relationship between a, b, and c.

10.

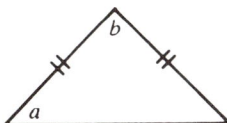

   State a relationship between a and b.

11. If the sum of two angles is 180° and one angle measures $y°$, what does the other angle measure?

12. If the sum of two angles is 270° and one angle measures $2z°$, what does the other angle measure?

13. If two angles of a triangle measure $a°$ and $b°$ respectively, what does the third angle measure in terms of a and b?

14. If three angles of a rhombus measure $m°$, $n°$ and $p°$ respectively, what does the fourth angle measure in terms of m, n and p?

15. If I weigh x kg now, find my weight in terms of x if:
    (a) I put on 2 kg;   (b) I lose 3 kg;
    (c) I lose d kg;   (d) I put on p kg.

16. If a litre of icecream costs £L and a bottle of chocolate sauce costs £S, find the cost of:
    (a) 2 litres of icecream   (b) 3 bottles of chocolate sauce
    (c) a litre of icecream and a bottle of chocolate sauce.

17. If a packet of Pup dog biscuits cost £D and a tin of Mew cat food costs £C how much would I pay for:
    (a) 2 packets of Pup biscuits;   (b) 5 tins of Mew;
    (c) 3 packets of Pup biscuits and 2 tins of Mew;
    (d) z packets of Pup biscuits and y tins of Mew?

14.1.2.1   Substitution in Formulae

In Chapter 12 it was shown that the area of a rectangle was found by multiplying its length by its width. If we use the pronumerals A for area, L for length and W for width, we can then write

Area = length × width

$$A = L \times W$$

This is called the *formula* for finding the area of the rectangle.

Consider a rectangle 6 cm long and 4 cm wide.

We wish to find its area. We may find the area by replacing L by 6 and W by 4 in the formula $A = L \times W$

$$A = L \times W$$
$$= 6 \times 4$$
$$= 24 \text{ cm}^2$$

Thus, by substituting (replacing the pronumerals by numerals) in the formula, we may find the area of a rectangle whose length and width is known.

Consider now a rectangle whose area is 42 cm². If it is 7 cm long, can we find its width?

Using the formula for area,

$$A = L \times W$$

we know $A = 42$ cm² and $L = 7$ cm.

Substituting these values into the formula, we have

$$A = L \times W$$
$$\therefore 42 = 7 \times W$$

This equation may be solved using the methods shown in Chapter 8.

$$42 = 7W$$
$$\therefore \frac{42}{7} = \frac{7W}{7}$$
$$6 = W$$

$\therefore$ the width of the rectangle is 6 cm.

Formulae are used frequently in various areas of science. For example, the formula $d = st$ gives the distance (d km) travelled in time (t hrs) by a car travelling at a speed of s km per hour. Similarly, electrical current in a wire can be calculated using the formula $I = \dfrac{E}{R}$.

Care must be taken when substituting, and the correct order of operations (see 1.1) must be used when calculating values.

Example 14.4
Given that $a = bc - d$, find
(a) a if $b = 3, c = 4, d = 5$ (b) a if $b = 2, c = 5, d = -6$
(c) b if $a = 20, c = 2, d = 6$

Solution
(a) $a = bc - d$ where $b = 3, c = 4, d = 5$
$\quad\quad = b \times c - d$
$\quad\quad = 3 \times 4 - 5$
$\quad\quad = 12 - 5$
$\quad\quad = 7$
$\therefore a = 7$

(b) $a = bc - d$ where $b = 2, c = 5, d = -6$
$\quad a = b \times c - d$
$\quad\quad = 2 \times 5 - {}^-6$
$\quad\quad = 10 - {}^-6$
$\quad\quad = 10 + 6$
$\quad\quad = 16$
$\therefore a = 16$

(c) $\quad a = bc - d$ where $b = 20, c = 2, d = 6$
$\quad\quad a = b \times c - d$
$\quad 20 = b \times 2 - 6$
$\therefore 20 = 2b - 6$
$20 + 6 = 2b - 6 + 6$
$\quad 26 = 2b$
$\quad \dfrac{26}{2} = \dfrac{2b}{2}$
$\quad 13 = b$
$\therefore b = 13$

Exercise 14C
1. Given that $s = vt$ find the value of s if
   (a) $v = 100$ and $t = 2$ (b) $v = 75$ and $t = 3$
   (c) $v = 50$ and $t = 4$ (d) $v = 25$ and $t = 9$
   (e) $v = 80$ and $t = 5$ (f) $v = 60$ and $t = 5$
   (g) $v = 100$ and $t = 4$ (h) $v = 100$ and $t = 3$

2. Given that $s = vt$ find the value of
   (a) v if $s = 200$ and $t = 2$ (b) v if $s = 400$ and $t = 8$
   (c) v if $s = 800$ and $t = 5$ (d) v if $s = 500$ and $t = 5$
   (e) t if $s = 600$ and $v = 100$ (f) t if $s = 300$ and $v = 60$

3. Given that $y = 2x + 3$, find the value of y if
   (a) $x = 3$ (b) $x = 5$
   (c) $x = -2$ (d) $x = -4$
   (e) $x = 0$ (f) $x = -1$

4. Given that $y = x^2$, find the value of y if
   (a) $x = 3$ (b) $x = 7$
   (c) $x = -5$ (d) $x = -3$
   (e) $x = -\frac{1}{2}$ (f) $x = \frac{1}{4}$

5. Given that $P = 2(l + w)$ find P if
   (a) $l = 8$ and $w = 5$ (b) $l = 10$ and $w = 3$
   (c) $l = 9$ and $w = 4$ (d) $l = 7$ and $w = 6$
   (e) $l = 14.3$ and $w = 6.5$ (f) $l = 8.7$ and $w = 4.5$

6. Given that $y = mx + c$ find
   (a) y if $m = 3, x = 5, c = 4$
   (b) y if $m = 4, x = 2, c = 5$
   (c) y if $m = -1, x = 3, c = 7$
   (d) y if $m = -2, x = 3, c = -4$
   (e) c if $y = 14, m = 2, x = 5$
   (f) c if $y = 4, m = 3, x = 2$
   (g) x if $y = 3, m = 2, c = -1$
   (h) x if $y = 9, m = 4, c = -3$

7. Given that $a = bc - d$, find
   (a) a if $b = 2, c = 4, d = 3$
   (b) a if $b = 7, c = 5, d = 1$
   (c) a if $b = -2, c = 3, d = -2$
   (d) a if $b = -3, c = -1, d = 3$
   (e) b if $a = 4, c = 3, d = 2$
   (f) b if $a = 16, c = -2, d = -4$

8. If the length of a paddock is 92 metres and the width is 30 metres, find the area, A, in square metres, using $A = l \times w$.

9. Find the circumference of a circle (c) if the radius (r) is 14 cm and π is $\frac{22}{7}$ using $c = 2\pi r$.

10. Find the area of a trapezium (A) when $a = 4, b = 6, h = 6$, using $A = \frac{1}{2}(a + b)h$.

11. In electrical engineering the formula $I = \dfrac{E}{R}$ is used. Find I if $E = 24$ and $R = 3$.

12. The expression $n = \dfrac{100(T - t)}{T}$, relates to the efficiency of a steam engine. If $T = 50$ and $t = 30$ find n.

13. The formula $v = u + at$ is used in Physics to find the velocity (motion or speed). Find v if $u = 20$, $a = 8$ and $t = 10$.

14. Find the value of P in the formula $P = A + \dfrac{B}{r^2}$ when $A = 3.06, B = 2.16$ and $r = 0.2$.

15. In the following problems use the formulae given to calculate the value of the unknown pronumeral.
    (a) The formula for finding the distance travelled (d km), in the number of hours (t) at a speed of r km per hour is $d = rt$. Find the average speed (r km/hour) of the QE2 if she sails 3000 km in 90 hours.
    (b) The formula $F = \frac{9}{5}C + 32$ is used for changing temperature in degrees Celsius to degrees Fahrenheit. Express $59°F$ in degrees Celsius.
    (c) The normal weight for an adult can be found from $W = 5.5(h - 40)$ where W is the weight in pounds and h is the height in inches. For a normal person whose weight is 165 pounds, what is his height?
    (d) What is the value of P in $a = \dfrac{P}{Wm}$ when $a = 42, m = 7$ and $W = 10$.
    (e) Find d in the formula $W = 12t(d - t)$ if $W = 180, t = 30$.

14.1.3   Geometry

14.1.3.1   Polygons
A polygon is a many (straight) sided, closed plane figure.

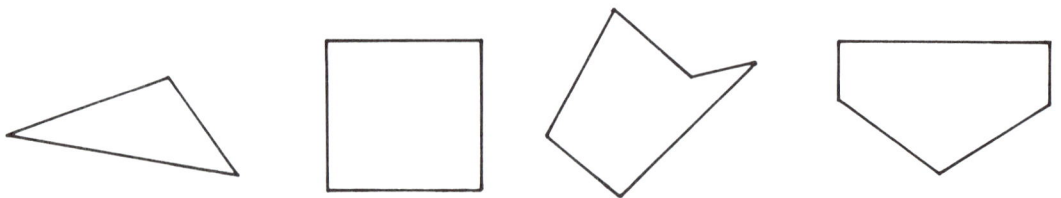

Fig 14.2

are all polygons.
But

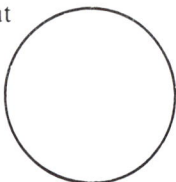 and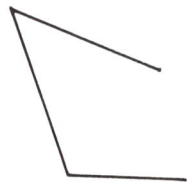

are *not* polygons because, in the first example, a circle does not have straight sides and in the second example, the figure is not closed.

The *least* number of sides which a polygon can have is 3.
Polygons can be grouped according to their numbers of sides.
The most common polygons are as follows:

Figure	*Number of Sides*	*Name*
△	3	triangle
□	4	quadrilateral
⬠	5	pentagon
⬡	6	hexagon
⬡	7	heptagon
⬡	8	octagon
⬡	9	nonagon
⬡	10	decagon
⬡	12	dodecagon

Table 14.1

In Chapter 9 we saw that triangles can be further grouped according to their side or angle properties (see 9.4). Also quadrilaterals can be further grouped according to their properties (see 9.5).
Polygons which have a reflex angle are called concave polygons.
E.g.

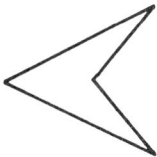 and  are concave polygons.

Polygons which do not have a reflex angle are said to be convex polygons.
E.g.

 and  are convex polygons.

Exercise 14D

1. Which of the following are polygons?

(a)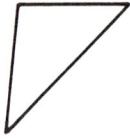

(b)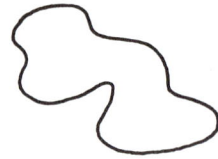

(c)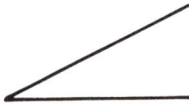

(d)

(e)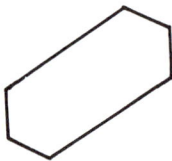

(f)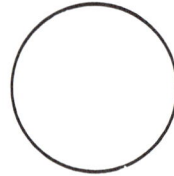

(g)

(h)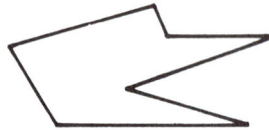

(i)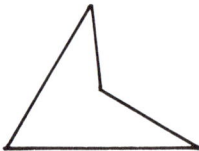

(j)

2. Which of the figures in question 1 are convex polygons?

3. Which of the figures in question 1 are concave polygons?

4. Measure (i)   the length of each side
   and   (ii) each angle
   in each of the following polygons:
   (a)

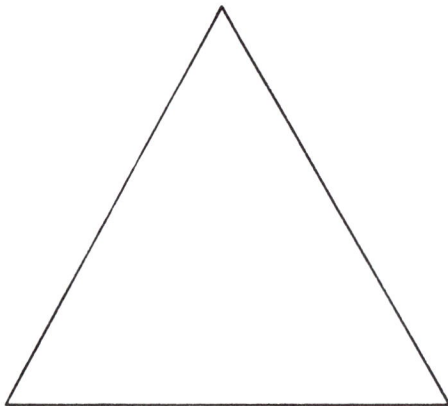

(b)

(c)

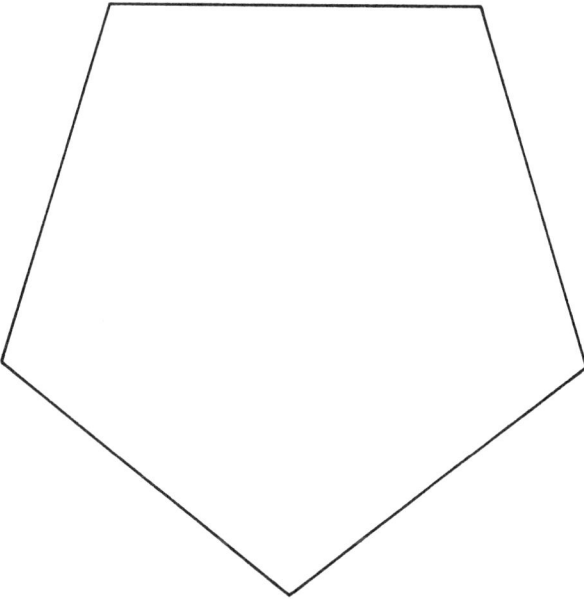

(d)

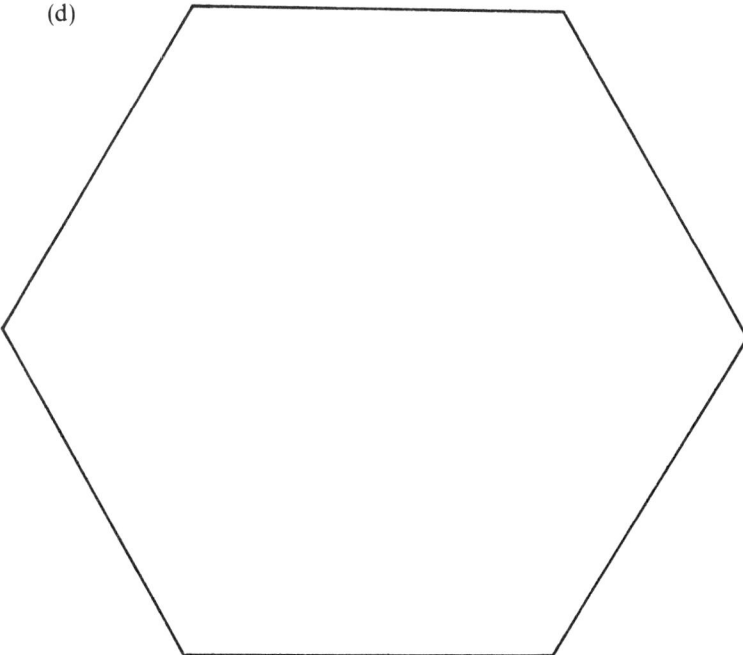

What are the common features of the polygons in question 4, above? You should have found that in each polygon all the sides were the same length and all the angles were the same—in the triangle each angle was 60°, in the pentagon 108°, etc. Such polygons are called *regular polygons*.

> A regular polygon has all sides the same length *and* all angles of equal measure.

All other polygons are said to be *irregular*.

Exercise 14E

1. Name each of the following polygons according to the number of sides:

(a)

(b)

(c)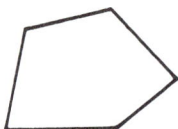

(d)

(e)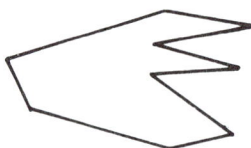

(f)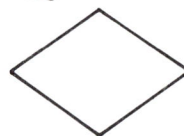

(g)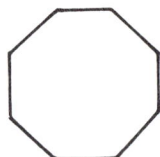

(h)

(i)

(j)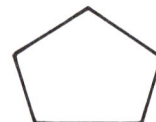

2. State whether each figure in question 1 is a regular or irregular polygon.

In 9.4 it was found that the sum of the angles of a triangle is 180°. Can we find the angle sum of other types of polygons?

Consider the quadrilateral $ABCD$ in fig 14.3.

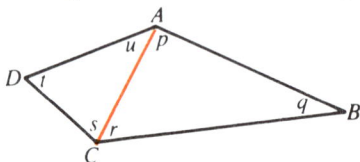

If the diagonal $\overline{AC}$ is drawn, the quadrilateral is divided into 2 triangles.

Fig 14.3

In $\triangle ABC$ $p + q + r = 180°$
In $\triangle ACD$ $u + s + t = 180°$
Thus $p + q + r$ + $u + s + t$

$=$ $180°$ + $180°$
$= 2 \times 180° = 360°.$

Thus the angle sum of a quadrilateral is 360°.
Now consider the pentagon *ABCDE* in fig 14.4.

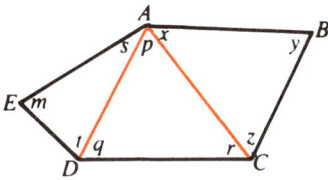

Diagonals $\overline{AD}$ and $\overline{AC}$ divide the pentagon into 3 separate triangles.

Fig 14.4

In $\triangle ABC$ $x + y + z = 180°$
In $\triangle ACD$ $p + r + q = 180°$
In $\triangle ADE$ $s + t + m = 180°$
Thus $\underbrace{x + y + z}_{180°}$ $+$ $\underbrace{p + q + r}_{180°}$ $+$ $\underbrace{s + t + m}_{180°}$
     $= 3 \times 180° = 540°$

Thus the angle sum of a pentagon is 540°.
Similarly we may find the angle sum of any polygon by dividing the figure into triangles:

No. of Sides	Figure	No. of Triangles	Angle Sum
3		1	180°
4		2	$2 \times 180° = 360°$
5		3	$3 \times 180° = 540°$
6		4	$4 \times 180° = 720°$
7		5	$5 \times 180° = 900°$
$\vdots$		$\vdots$	$\vdots$
n		$n - 2$	$(n - 2) \times 180°$

Table 14.2

As $180° = 2 \times 90° = 2$ right angles, $(n - 2) \times 180° = (n - 2) \times 2$ right angles
                                     $= 2n - 4$ right angles.

Exercise 14F

1. Copy Table 14.2 and extend it to include polygons with:
   (a) 8 sides   (b) 9 sides
   (c) 10 sides   (d) 12 sides

2. Find the value of the pronumeral in each of the following polygons:
   (a)   (b)

(c)

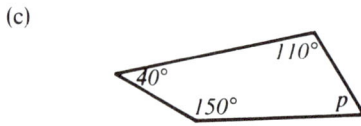

(d)

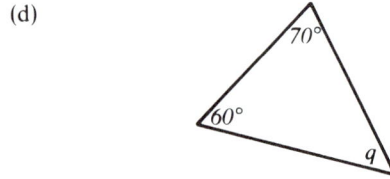

(e)

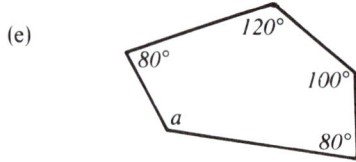

(f)

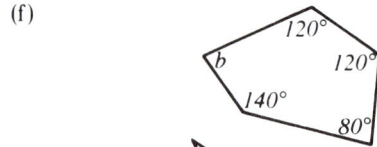

(g)

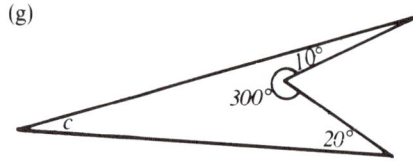

(h)

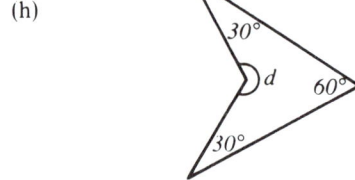

(i)

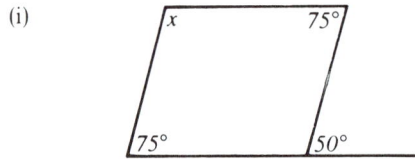

(j)

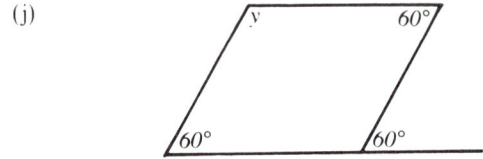

(k)

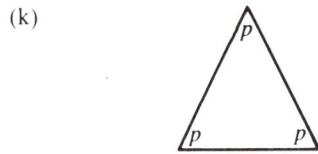

(l)

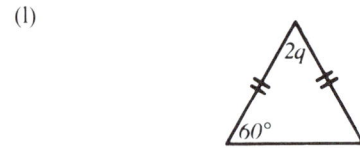

(m)

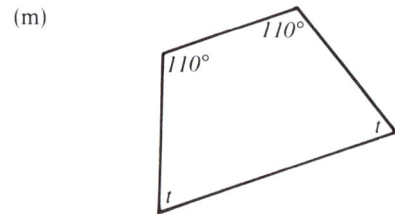

(n)

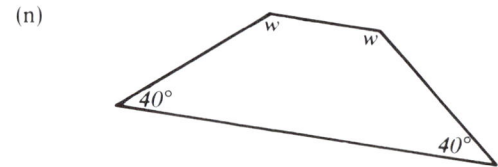

Regular polygons may be constructed within a circle.

Exercise 14G

1. Equilateral triangle

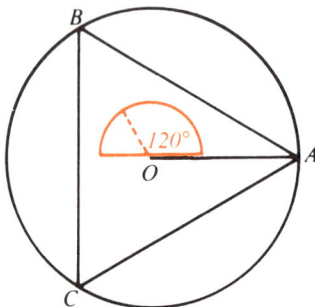

(a) Draw a circle radius 6 cm.
(b) Draw radius OA.
(c) Using a protractor, draw
   $\angle BOA = 120°$ ($\frac{1}{3}$ of $360° = 120°$)
(d) Using a protractor, draw
   $\angle COB = 120°$
(e) Using pencil and ruler, draw lines $\overline{AB}$, $\overline{BC}$ and $\overline{CA}$
(f) Measure the lengths of $\overline{AB}$, $\overline{BC}$ and $\overline{CA}$
(g) Measure $\angle CAB$, $\angle ABC$ and $\angle BCA$.

2. Square

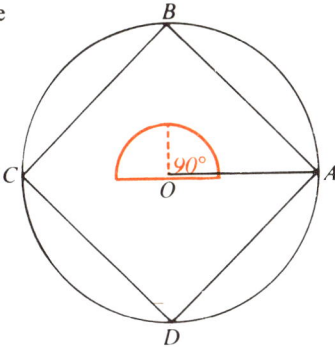

(a) Draw a circle radius 6 cm.
(b) Draw radius OA.
(c) Using a protractor draw
$\angle BOA = 90°$ ($\frac{1}{4}$ of $360° = 90°$)
(d) Using a protractor draw
$\angle BOC = 90°$
(e) Using a protractor draw
$\angle COD = 90°$
(f) Draw lines $\overline{AB}$, $\overline{BC}$, $\overline{CD}$ and $\overline{DA}$.
(g) Measure the lengths of $\overline{AB}$, $\overline{BC}$, $\overline{CD}$ and $\overline{DA}$.
(h) Measure $\angle DAB$, $\angle ABC$, $\angle BCD$ and $\angle CDA$.

3. Pentagon

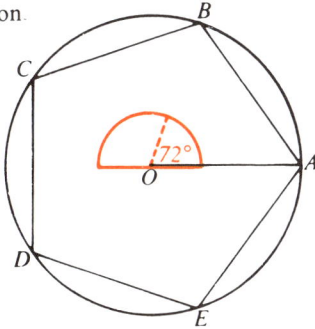

(a) Draw a circle, radius 6 cm.
(b) Draw radius OA.
(c) Using a protractor, draw
$\angle BOA = 72°$ ($\frac{1}{5}$ of $360° = 72°$)
(d) Using a protractor, draw
$\angle COB = \angle DOC = \angle EOD = 72°$
(e) Draw lines $\overline{AB}$, $\overline{BC}$, $\overline{CD}$, $\overline{DE}$, $\overline{EA}$.
(f) Measure the lengths of each side of the pentagon.
(g) Measure the angles of the pentagon.

4. Hexagon

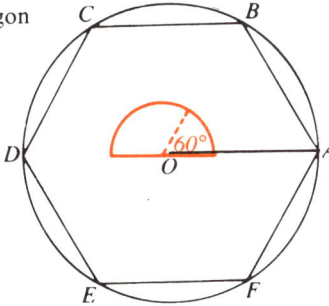

Repeat the construction above with the angle being measured, $\angle BOA$, etc, equal to 60° ($\frac{1}{6}$ of 360° = 60°).

5. Octagon

Repeat the construction above with the angle being measured, $\angle BOA$, etc, equal to 45° ($\frac{1}{8}$ of 360° = 45°).

6. Decagon

Repeat the construction above with the angle being measured, $\angle BOA$, etc, equal to 36° ($\frac{1}{10}$ of 360° = 36°).

14.1.3.2 Construction of Triangles

If we are given the length of each side of a triangle, we can accurately construct the triangle using compasses and a ruler.

For example, to construct a triangle ABC with sides 3 cm, 4 cm and 5 cm respectively:

(a) draw line $\overline{AB}$ 3 cm in length;
(leave room above line for rest of construction)

Fig 14.5 (a)

(b) open compasses 4 cm and, with compass-point on B, mark an arc above line $\overline{AB}$;

Fig 14.5 (b)

$A \underline{\quad 3cm \quad} B$

(c) open compasses 5 cm and, with compass-point on *A*, mark an arc to intersect the first arc at *C*;

Fig 14.5 (c)

A———3cm———B

(d) draw lines $\overline{BC}$ and $\overline{AC}$;

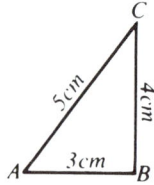

Fig 14.5 (d)

(e) measure the length of each side of the triangle to check the accuracy of your construction. Similarly, if we are given the length of two sides and the magnitude of the angle between the sides, we can accurately construct the triangle using a ruler, protractor and compasses.

Example 14.5
Construct a triangle *ABC* with $\overline{AB}$ = 5 cm, $\overline{BC}$ = 7 cm and $\angle B$ = 50°.

Solution
(a) draw line $\overline{AB}$ 5 cm long. (Make sure you leave plenty of space above the line for the rest of the construction.)   A———5cm———B   Fig 14.6 (a)

(b) using a protractor, measure an angle of 50° at *B*;

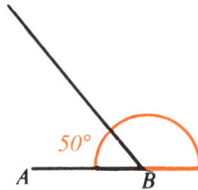

Fig 14.6 (b)

(c) open your compasses 7 cm and with the point on *B*, mark an arc to intersect the ray at *C*;

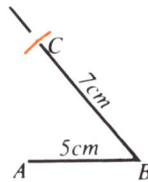

Fig 14.6 (c)

(d) Join $\overline{AC}$.

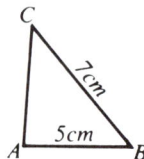

Fig 14.6 (d)

Similarly, if we are given the length of one side of the triangle and the size of the angle at each end of that side, we can use a protractor and ruler to construct the triangle.

Example 14.6

Use a ruler and protractor to construct $\triangle ABC$ with

d$\overline{AB}$ = 5 cm
$\angle A = 40°$
$\angle B = 50°$

Solution

(a) draw line $\overline{AB}$ 5 cm long. (Leave room above the line for the rest of the construction.)

Fig 14.7 (a)

(b) with protractor on A draw an angle of 40°;

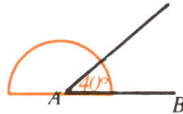

Fig 14.7 (b)

(c) with protractor on B draw an angle of 50°; this ray should intersect the ray drawn in (b) at C.

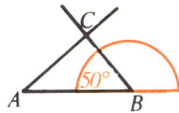

Fig 14.7 (c)

Exercise 14H

1. Use compasses and ruler to construct triangle ABC with all sides 5 cm long.

2. Use compasses and ruler to construct triangle ABC with all sides 7 cm long.

3. Use compasses and ruler to construct triangle ABC with side $\overline{AB}$ 4 cm long and sides $\overline{BC}$ and $\overline{AC}$ each 6 cm long.

4. Use compasses and ruler to construct triangle ABC with side $\overline{AB}$ 5 cm long and sides $\overline{BC}$ and $\overline{AC}$ each 8 cm long.

5. Use compasses and ruler to construct triangle ABC with side $\overline{AB}$ 4 cm long, side $\overline{BC}$ 6 cm long and side $\overline{AC}$ 8 cm long.

6. Use compasses and ruler to construct triangle ABC with side $\overline{AB}$ 10 cm long, side $\overline{BC}$ 7 cm long and side $\overline{AC}$ 5 cm long.

7. Use compasses and ruler to construct triangle ABC with side $\overline{AB}$ 12 cm long, side $\overline{BC}$ 9 cm long and side $\overline{AC}$ 6 cm long.

8. Use compasses and ruler to construct triangle ABC with side $\overline{AB}$ 12 cm long, side $\overline{BC}$ 7 cm long and side $\overline{AC}$ 10 cm long.

9. Use compasses, protractor and ruler to construct triangle ABC with side $\overline{AB}$ 4 cm long, $\angle B = 60°$ and side $\overline{BC}$ 4 cm long.

10. Use compasses, protractor and ruler to construct triangle ABC with side $\overline{AB}$ 6 cm long, $\angle B = 80°$ and side $\overline{BC}$ 5 cm long.

11. Use compasses, protractor and ruler to construct triangle ABC with side $\overline{AB}$ 6 cm long, $\angle B = 120°$ and side $\overline{BC}$ 4 cm long.

12. Use compasses, protractor and ruler to construct triangle ABC with side $\overline{AB}$ 8 cm long, $\angle B = 130°$ and side $BC = 5$ cm long.

13. Use a protractor and ruler to construct triangle ABC with side $\overline{AB}$ 6 cm long, $\angle A = 40°$ and $\angle B = 60°$.

14. Use a protractor and ruler to construct triangle ABC with side $\overline{AB}$ 10 cm long, $\angle A = 60°$ and $\angle B = 60°$.

15. Use a protractor and ruler to construct triangle ABC with side $\overline{AB}$ 8 cm long, $\angle A = 70°$ and $\angle B = 70°$.

16. Use a protractor and ruler to construct triangle ABC with side $\overline{AB} = 8$ cm long, $\angle A = 70°$ and $\angle B = 60°$.

17. Use a protractor and ruler to construct triangle *ABC* with side $\overline{AB}$ 6 cm long, $\angle A = 90°$ and $\angle B = 50°$.

18. Use a protractor and ruler to construct triangle *ABC* with side $\overline{AB}$ 7 cm long, $\angle A = 30°$ and $\angle B = 90°$.

14.1.4   Cartesian Plane

Exercise 14I

1. Plot the following points and join them up in order:
   (a) (4, 1), (11, 1), (11, 5), (14, 5), (11, 10), (10, 10), (10, 12), (9, 12), (9, 10), (4, 10), (1, 5), (4, 5).
   (b) (0, 0), (0, 1), (1, 2). (3, 2), (6, 0), (8, 2), (8, 0), (8, −2), (6, 0), (3, −2), (1, −2), (0, −1), (0, 0).
   (c) (−2, 0), (2, 0), (3, 1), (0, 1), (0, 4), (2, 2), (−2, 2), (0, 4); start again at, (0, 1), (−2, 1), (−2, 0).
   (d) (0, 0), (−2, −1), (−2, −3), (−1, −4), (2, −4), (3, −3), (3, −1), (1, 0), (2, 1), (2, 4), (1, 3), (0, 3), (−1, 4), (−1, 1), (0, 0).
   (e) (−1, 0), (1, 0), (2, 1), (0, 1), (0, 3), (4, 3), (1, 5), (3, 5), (1, 7), (2, 7), (0, 9), (−2, 7), (−1, 7), (−3, 5), (−1, 5), (−4, 2), (0, 2), (0, 1), (−2, 1), (−1, 0).

2. *Treasure Island*
   You have sailed very far and have arrived at Treasure Island where you are looking for the Pirates' Treasure! You consult your directions and begin!
   Start at $(−19, −12\frac{1}{2})$ leaving two men at first base.
   Stop for night at $(−5, −12\frac{1}{2})$.
   Go on to Rope Bridge at Deadman's River at $(10, −12\frac{1}{2})$.
   Stop for night at $(19, −12\frac{1}{2})$.
   Struggle through Burning Desert to (19, 0).
   Then travel onwards and set up second base at (19, 7).
   Move to $(18\frac{1}{2}, 9)$ and then to $(17\frac{1}{2}, 11)$.
   Travel on to $(16\frac{1}{2}, 12\frac{1}{2})$, $(15\frac{1}{2}, 14)$, (13, 16), $(11\frac{1}{2}, 17)$, $(10, 17\frac{1}{2})$, $(8, 18\frac{1}{2})$.
   Come to Bottomless Pit at (6, 19)!
   Stop for night at (1, 19).

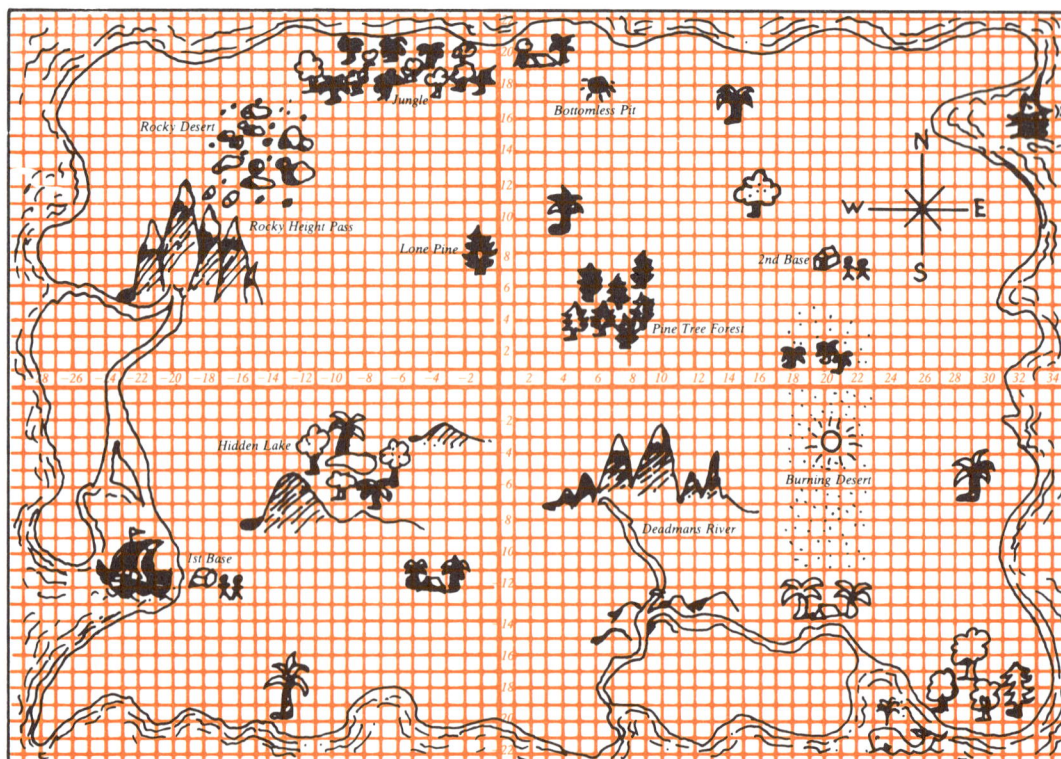

Move through thick jungle to $(-9, 19)$.
Come out at $(-11\frac{1}{2}, 17\frac{1}{2})$.
Travel through rocky desert to $(-13, 16)$; $(-14, 15)$; $(-15, 14)$; $(-16, 12\frac{1}{2})$; $(-18, 9\frac{1}{2})$.
Go through Rocky Height Pass to $(-19, 7)$.
Return to first base for more supplies. Go on to $(-19, 0)$.
Then to $(-9, -12\frac{1}{2})$. Collect supplies and return to $(-19, 7)$.
Climb through mountains to $(-12\frac{1}{2}, 7)$, then on to $(-5, 7)$.
Arrive at Lone Pine at $(0, 7)$. Turn to $(-1, 6)$ and again to $(-\frac{1}{2}, 5\frac{1}{2})$. Go South to $(-\frac{1}{2}, 4)$ and then East 1 unit on the map then North $1\frac{1}{2}$ units and turn to the North West. The pirates' treasure is found just where you cross your path again. You return to second base to collect the two men and a helicopter returns you to first base.
You go home with the treasure!

14.1.5   Ratio and Percentage

14.1.5.1   Scale Drawing

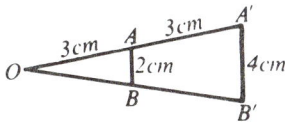

Fig 14.8

If we measure the lengths of lines $\overline{OA}$ and $\overline{AA'}$ in fig 14.8, we find that $d\overline{OA} = d\overline{AA'} = 3$ cm.
$\therefore d\overline{OA'} : d\overline{OA} = 2 : 1$.
Similarly $d\overline{OB} = d\overline{BB'} = 3$ cm
$\therefore d\overline{OB'} : d\overline{OB} = 2 : 1$.
Thus A' and B' are twice as far from O as are A and B.

Now, if we measure the length of $\overline{AB}$ and $\overline{A'B'}$, we find that $d\overline{AB} = 2$ cm and $d\overline{A'B'} = 4$ cm. I.e. $d\overline{A'B'} : d\overline{AB} = 2 : 1$. Thus $\overline{A'B'}$ is twice the length of $\overline{AB}$.
Similarly, measuring the lengths of lines in fig 14.9,

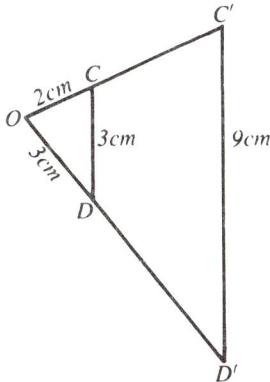

$d\overline{OC} = 2$ cm and $d\overline{OC'} = 6$ cm.
$\therefore d\overline{OC'} : d\overline{OC} = 3 : 1$.
also $d\overline{OD} = 3$ cm and $d\overline{OD'} = 9$ cm.
$\therefore d\overline{OD'} : d\overline{OD} = 3 : 1$.
also $d\overline{CD} = 3$ cm and $d\overline{C'D'} = 9$ cm.
$\therefore d\overline{C'D'} : d\overline{CD} = 3 : 1$.
Notice that, although in fig 14.9 the length of $\overline{OC}$ is *not* the same as the length of $\overline{OD}$, it does not alter the outcome of the construction.

Fig 14.9

This method can be used to enlarge figures to twice or three times their size.
Consider the square $ABCD$ in fig 14.10.

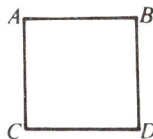

Each side is 2 cm long.

Fig 14.10

We wish to enlarge *ABCD* to twice its size; i.e. construct a square *A'B'C'D'* with each side twice as long as those in *ABCD*. Re-drawing *ABCD* and choosing a point *O* outside the figure, we draw lines from *O* passing through each of *A*, *B*, *C* and *D*:

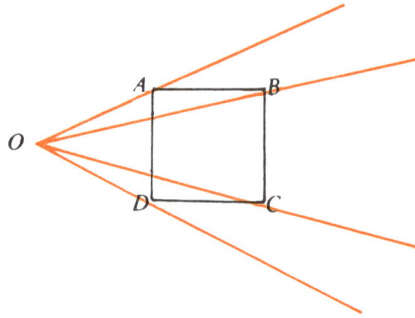

Fig 14.11

We now wish to find, first the point *A'*, such that $d\overline{OA'} : d\overline{OA} = 2:1$. Thus $d\overline{OA} = d\overline{AA'}$. With compass-point on *A*, open the compasses distance $\overline{OA}$. With compass-point still on *A*, mark *A'* on the line $\overline{OA}$ produced. Now put the compass-point on *B* and with compasses open distance $\overline{OB}$, mark *B'* on $\overline{OB}$ produced. Repeat this with compass-point on *C* and then on *D*.

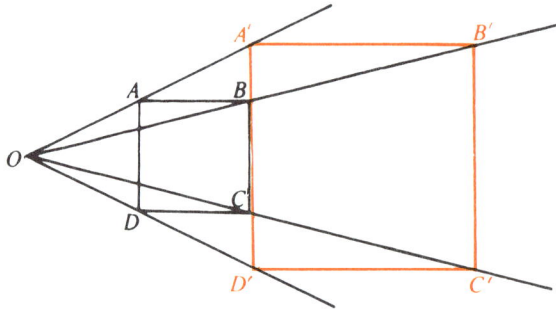

Fig 14.12

Join up *A'B'C'D'*.
Measuring $\overline{A'B'}$, $\overline{B'C'}$, $\overline{C'D'}$ and $\overline{D'A'}$ we find that each line is 4 cm. Thus we have constructed a square which is twice the size of the original square.

The point *O* may be inside the figure.

Example 14.7
Construct a figure twice the size of fig 14.13.

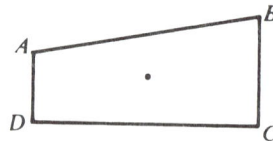

Fig 14.13

Solution

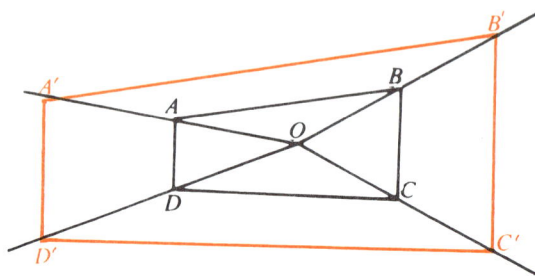

Fig 14.14

Notice that in this construction the enlarged figure surrounds the original figure whereas when *O* was outside the figure the enlarged figure was beside the original figure.

Exercise 14J

1. For each of the following, copy the diagram into your book, and construct a figure twice the size:

(a)
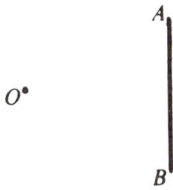

(b)
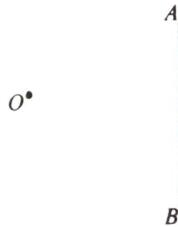

(c)
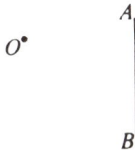

(d)

(e)
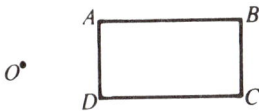

(f)
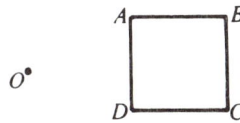

(g)
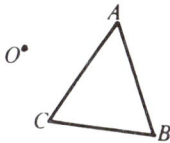

(h)
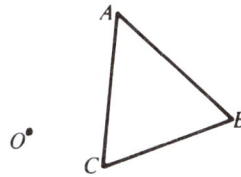

(i)
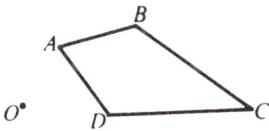

(j)
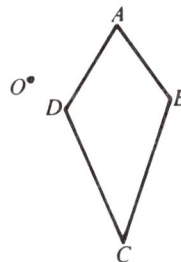

(k)
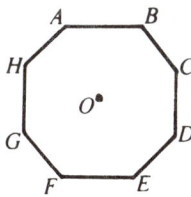

(l)
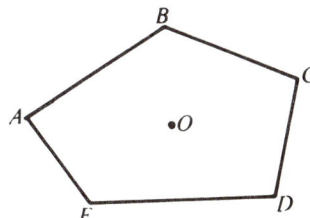

(m)
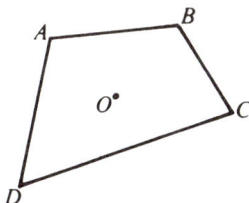

(n)

2. Draw a line $\overline{AB}$ 1 cm long. Mark a point O beside it. Construct a line $\overline{A'B'}$ 3 cm long.

3. Draw a square $ABCD$ of side 1 cm. Mark a point O beside it. Enlarge $ABCD$ to three times its size.

14.2   Diversions

In the following sections are given some interesting mathematical exercises which require little or no prior knowledge.

14.2.1   Three-Dimensional Shapes

Many objects which you see around you are three-dimensional. Boxes, houses, pipes, balls and so on all have length, width and height and so we say they are three-dimensional. Many three-dimensional shapes can be formed by joining up some plane shapes (two-dimensional). This is called the *net* of the figure.

For example, a cube is made up of six squares, and these squares are called the *faces* of the three-dimensional shape.

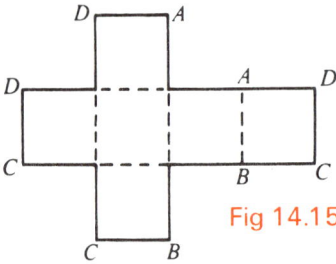

Fig 14.15

This diagram is the net of a cube. It contains the six faces needed. If we were to cut right round the shape and fold along the broken lines, matching points with the same letter, we could make a cube.

Exercise 14K

Below, you will find the nets of some other three-dimensional shapes. Trace them, cut them out, fold in such a way that points which have matching letters meet. Fasten the edges together with glue or adhesive tape.

1. Triangular prism:

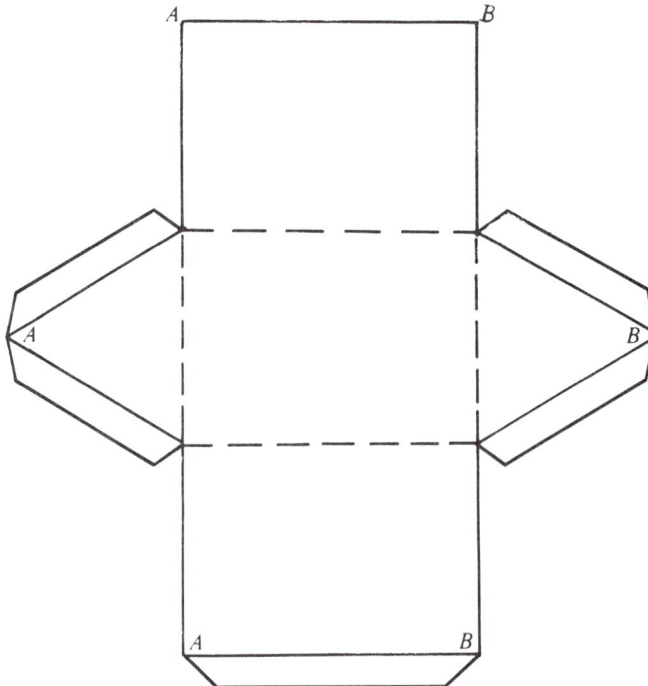

2. Pentangular prism:

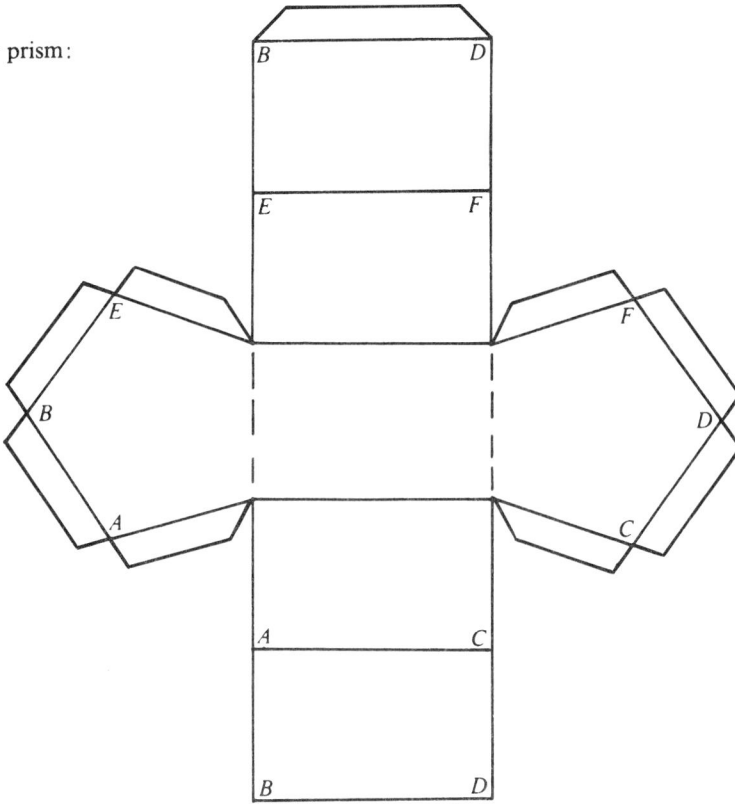

3. Hexagonal prism:

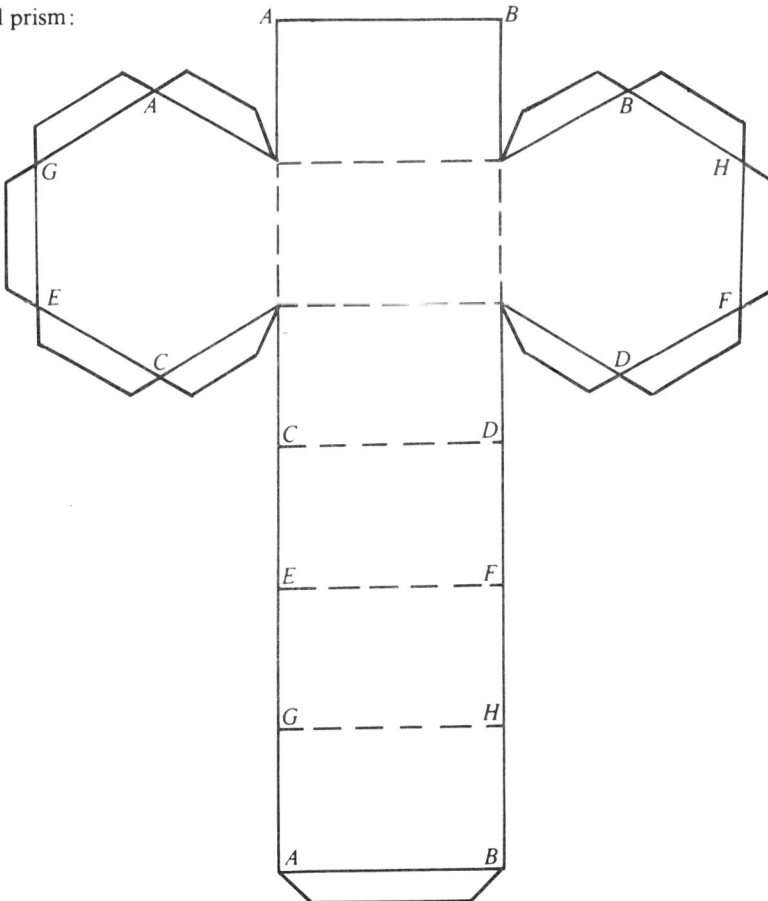

4. Heptangular prism:

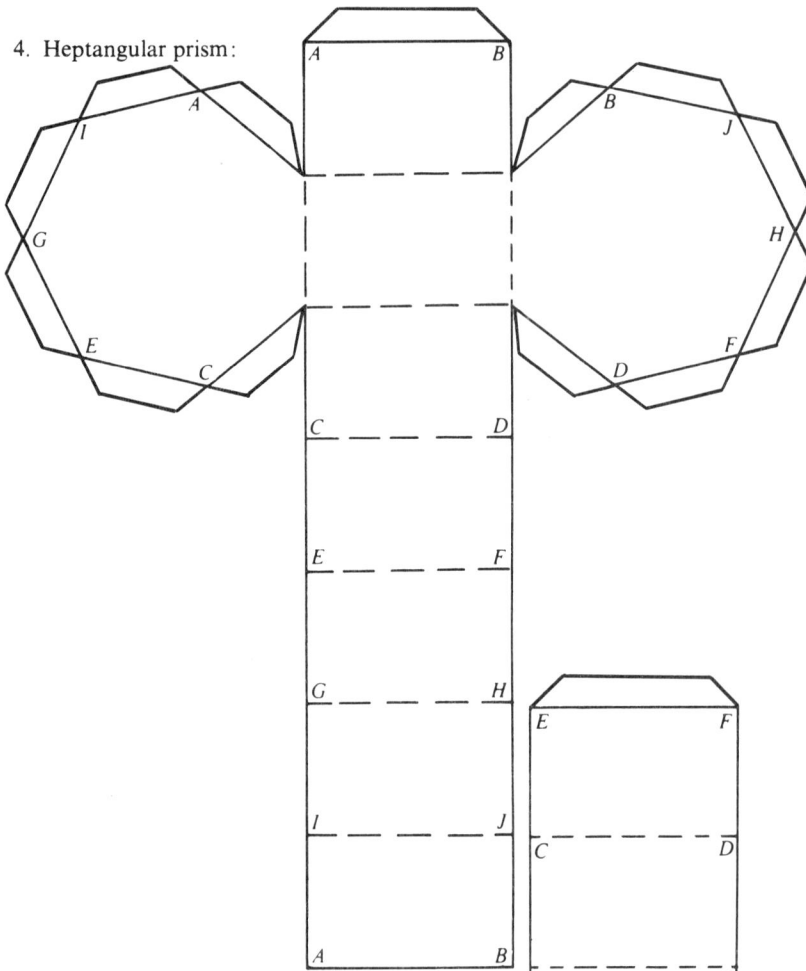

5. Octangular prism:

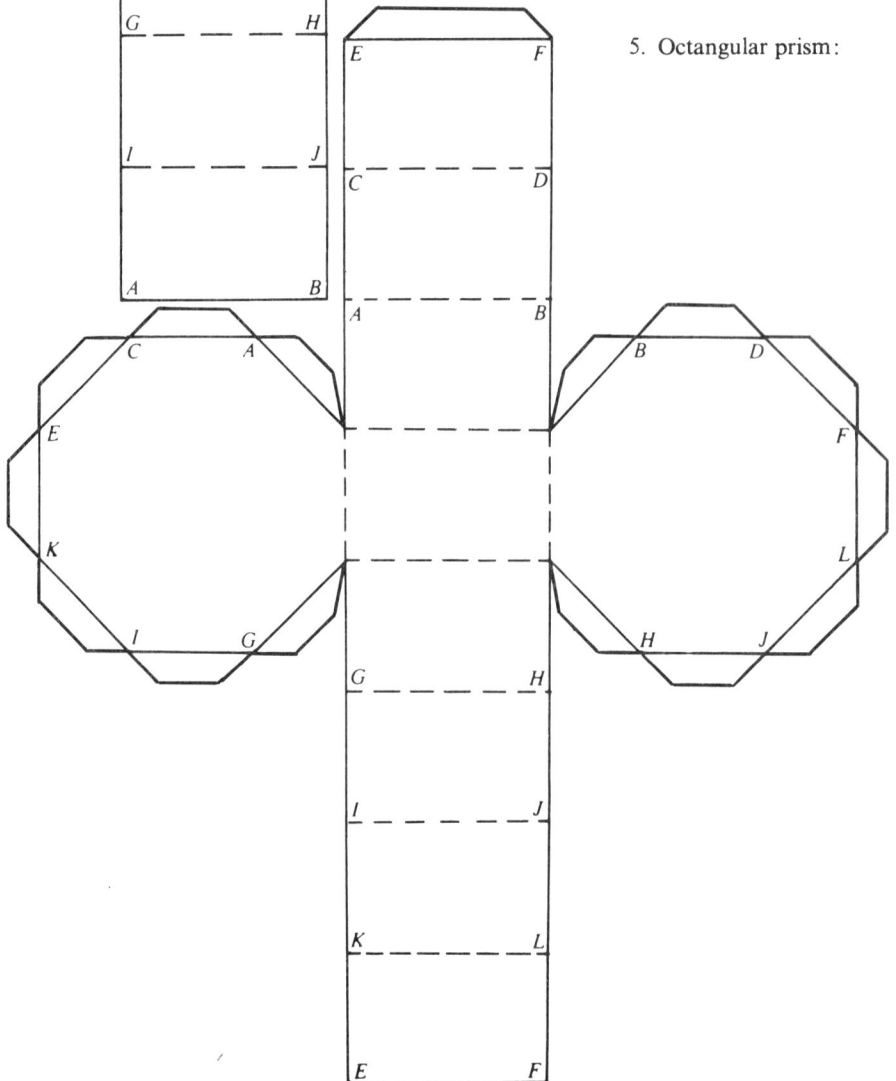

6. Draw a net for each of the following figures, marking the dimensions as shown:

(a)

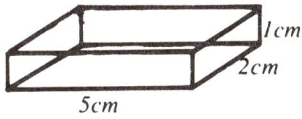

1cm
2cm
5cm

(b)

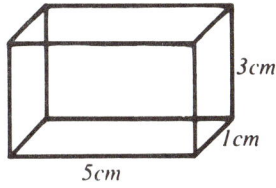

3cm
1cm
5cm

(c)

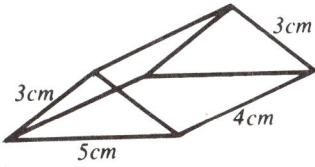

3cm
3cm
4cm
5cm

(d)

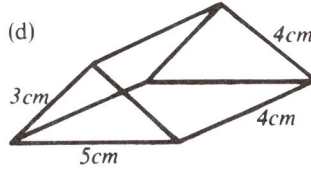

4cm
3cm
4cm
5cm

(e)

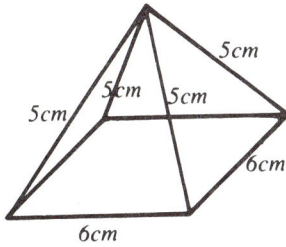

5cm
5cm 5cm
5cm
6cm
6cm

(f)

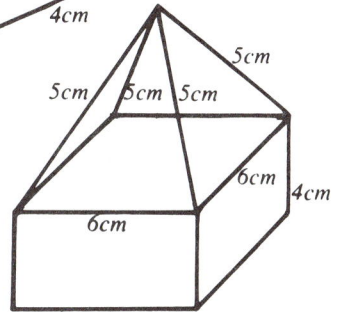

5cm
5cm 5cm 5cm
6cm
6cm
4cm

7. Construct the nets obtained in question 6 from coloured cardboard and fold and glue them to form the given figure.

8. In each of the following, the whole of the outside of the figure has been painted yellow.
   (i) How many cube faces are yellow?
   (ii) How many cubes have 0, 1, 2, 3, 4, 5, 6 yellow faces?

(a)

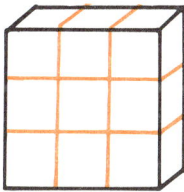

(b)

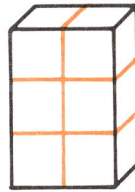

(c)

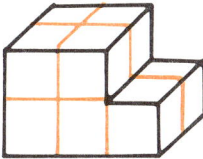

(d)

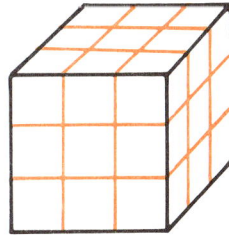

(e)

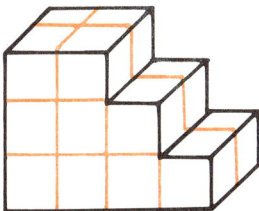

(f)

14.2.2   Circles

We saw in 9.6 that the distance round the border of a plane figure is called the perimeter, and that the perimeter of a circle has a special name—the *circumference*.

How would you find the circumference of a bicycle wheel or the lid of a marmalade jar? (HINT: Use a piece of string.)

Exercise 14L

Measure the circumference of a bicycle wheel and its diameter.

Using a calculator, find the ratio of the circumference to the diameter.

Now do the same with some more circular items and find the ratio in each case.

Set out your results in a table like this:

Item	*Circumference* (C)	*Diameter* (d)	$\dfrac{C}{d}$
Wheel Lid 10p coin			

In exercise 14L, for most of your results you should get a value for $\dfrac{C}{d}$ of just over 3. In fact, if you took the average of all the results obtained by your class it should be near 3.14. This ratio has a special symbol, π (pi). It is a non-terminating decimal. π correct to 2 decimal places is 3.14 or as a fraction $3\frac{1}{7}$. Many calculators have a π button which gives the number correct to six or more decimal places.

The sentence 'May I have a large container of coffee' gives the value of π to seven places of decimals by counting the letters in each word thus 3.1415926. π to fourteen decimal places is given by 'Now I want a drink, alcoholic of course, after the heavy chapters involving quantum mechanics.' Using computers, the value of π has been calculated to many thousands of decimal places.

Consider the circle below.

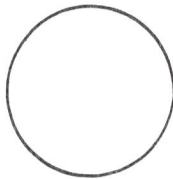

Fig 14.16

To find its area, we could try to count the number of square units it contains, as we did before. However, in the case of the circle, this cannot be done exactly.

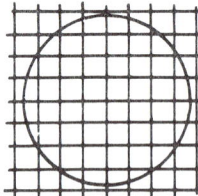

Fig 14.17

The next step in trying to find its area is to relate it to the area of a rectangle. The Greeks decided many thousands of years ago that the way to find the area of a circle was to 'square the circle': that is, find a square of the same area. This cannot, in fact, be done but we can divide the circle into many equal sectors as shown in fig 14.18. Each sector then closely resembles a triangle since the curve of the circle is so small that it appears almost straight.

Fig 14.18

The next step is to arrange them 'top to tail' as shown and cut the final section in half to form almost a right angle at each end.

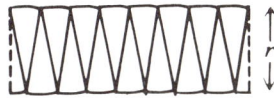

Fig 14.19

The completed shape is a rectangle with a width equal to the radius (r) of the circle.

Exercise 14M

For each of the given radii, draw a circle and cut it up as shown above to form a 'rectangle'. Measure the lengths of the sides and use your results to complete the table below.

Radius	Length (L)	Width (W)	$\frac{L}{W}$
3 cm			
4 cm			
5 cm			
6 cm			
8 cm			
10 cm			

14.2.3 Compass Bearings

Directions on the ground are given with reference to the 'four points of the compass': north (N), south (S) east (E) and west (W). A compass needle points a little to the west of the North Pole of the earth.

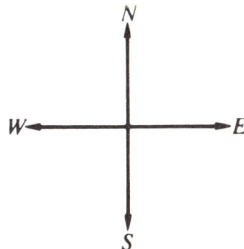

Fig 14.20

These directions are at 90° to one another. If a direction other than N, S, E or W needs to be specified then these basic directions can be divided up in units of 45°.

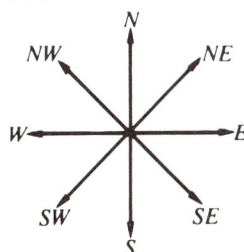

Fig 14.21

Thus NE is 45° east of north
SW is 45° south of west.

Similarly, these directions can be divided even further:

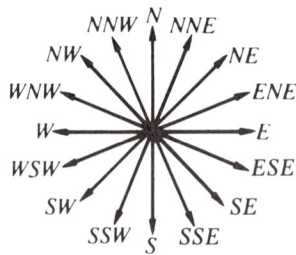

Fig 14.22

Thus *NNE* is $22\frac{1}{2}°$ north of north east

   WNW is $22\frac{1}{2}°$ west of north west.

Compass bearings can be used to indicate the direction in which you are travelling.

Example 14.8

Lucy Lost rode her bicycle from home a distance of 3 km towards the east and then 4 km towards the north. Represent her ride on a Cartesian Plane using a scale of 1 cm = 1 km and so find how far she would have to ride directly back to her home?

Her friend Freda lives 1 km north of Lucy's home. If Lucy decides to ride to Freda's house instead of her own, in what direction should she travel?

Solution

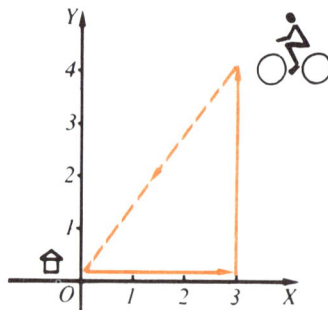

Fig 14.23

Using a ruler to measure the distance back to her home, on the graph we find that this distance is 5 cm.

Since 1 cm = 1 km on the scale

then 5 cm = 5 km

and so Lucy is 5 km from her home.

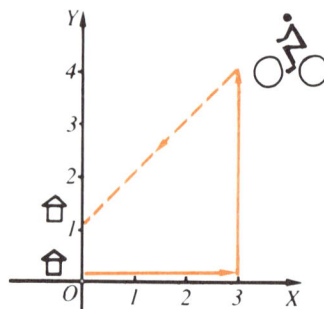

Fig 14.24

To find the direction of the dotted line in fig 14.24 we will need to give it using North, South, East or West. Referring to the compass directions, the direction required seems to be in a south-westerly direction. Measuring the angle between the southerly direction and the dotted line, we find that it is 45°. Thus Lucy must travel in a south-westerly direction to reach Freda's house.

Exercise 14N

1. What is the angle between each of the following directions?
   (a) *N* and *S* (b) *N* and *W*
   (c) *N* and *NW* (d) *W* and *SE*
   (e) *NE* and *SE* (f) *NW* and *SE*
   (g) *NNW* and *NW* (h) *NNE* and *ESE*
   (i) *SSW* and *NNW* (j) *ENE* and *WSW*

2. A jogger goes for a daily run, each time heading for a different destination (*D*). His various routes and destinations are shown below on the Cartesian Plane. All distances are in kilometres. In each case, use a protractor to find in which direction he must run to return home again.

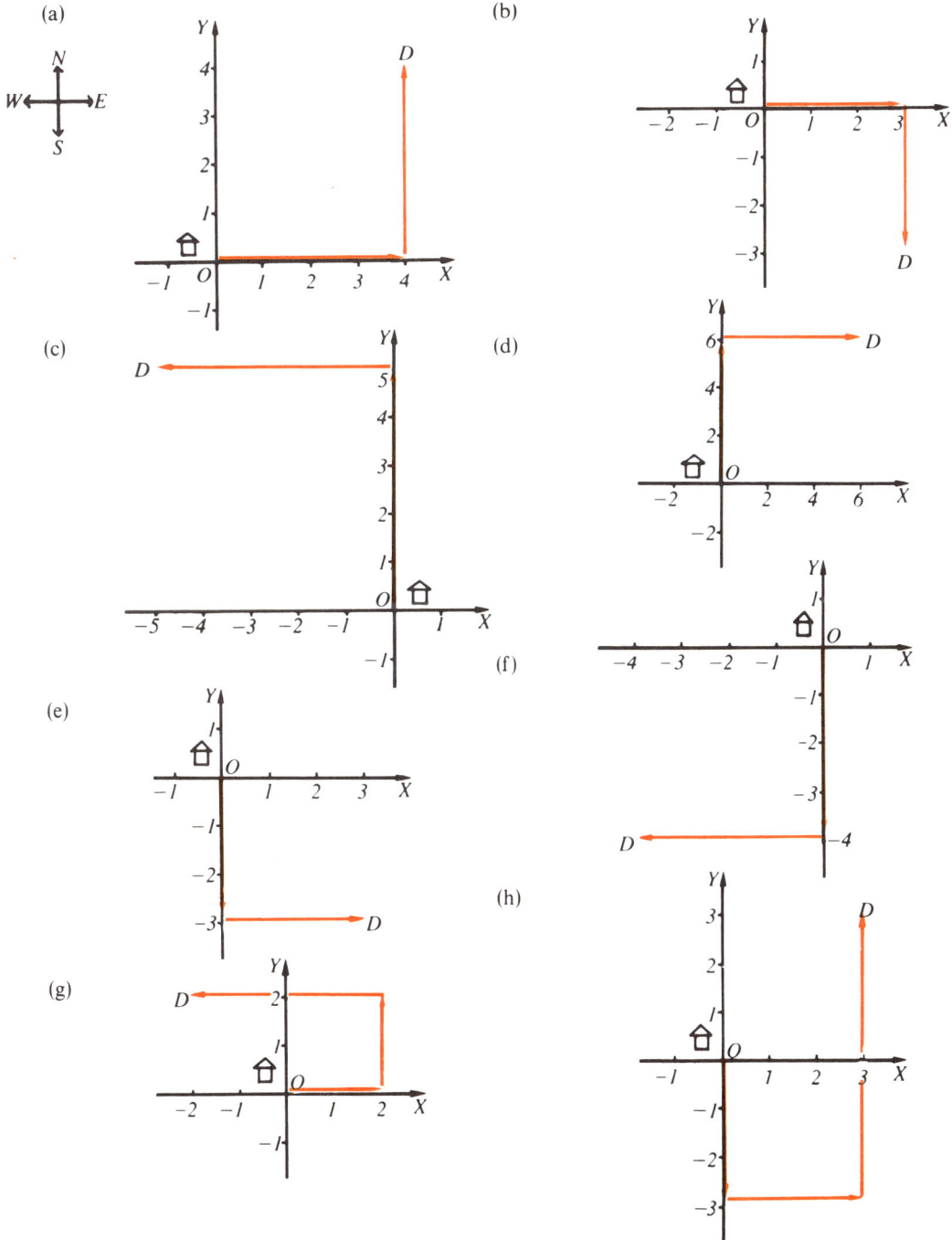

(a)

(b)

(c)

(d)

(e)

(f)

(g)

(h)

(i)

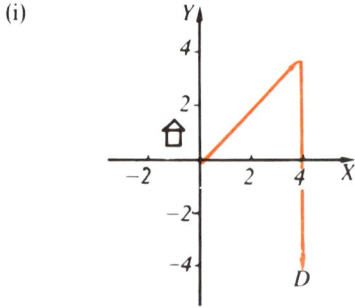

(j)

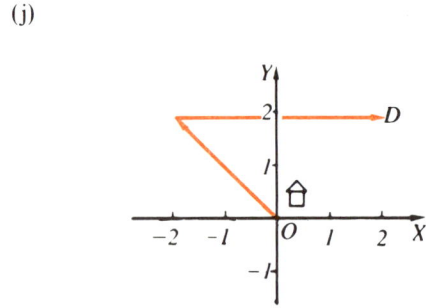

3. Gertrude Golots leaves home and drives to a supermarket. She then goes on to visit a friend. To get to the supermarket she must travel 3 km east and then 2 km north. From the supermarket she travels 1 km west to her friend's house. Depict Gertrude's travels on the Cartesian Plane, using a scale of 1 cm = 1 km. Find the direction in which she must drive in order to get home again, and the distance to her home.

4. Milton Mover is a salesman. In one day he visits two customers. He leaves his office and arrives at his first customer's place after driving 12 km west and 14 km south. He then leaves and travels 2 km east and 4 km north to arrive at the second customer's place of business. Show Milton's route on the Cartesian Plane, using a scale of 1 cm = 4 km. Find the direction in which he must travel to get back to his office, and measure the distance back to his office.

Compass bearings can also be used to indicate position. Thus, your school may be 3 km *NW* of your home. The shopping centre may be 1 km *SSE* of your home.

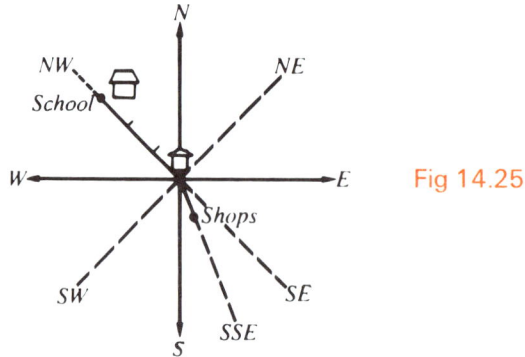

Fig 14.25

Of course, your school may not be exactly in the direction *NW*;
The angle between the line from home to school and the north line may not be exactly 45°.
Let us say that this angle is actually 30°.

Fig 14.26

Standing at home facing north we would have to turn 30° towards the west in order to face the school. We say the bearing of the school from home is

$$N\ 30°\ W$$

i.e.   Face *N* turn 30° towards *W*

Similarly,   *S* 20° *E*

means   Face *S* turn 20° towards *E*.

Exercise 140

1. Use a protractor to measure the bearings from the point O of the points $A - J$ in the following diagram.

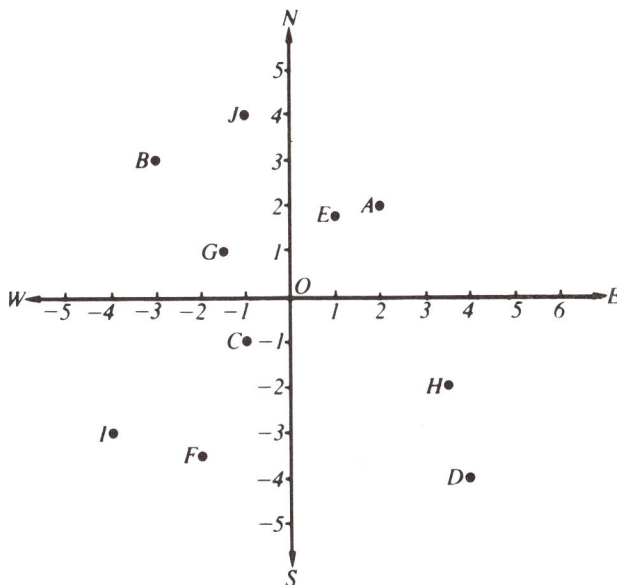

2. Represent the points with the following bearings from O on a circle as below:

(a) $N\ 30°\ E$ (b) $N\ 20°\ W$
(c) $N\ 40°\ W$ (d) $N\ 35°\ E$
(e) $S\ 10°\ W$ (f) $S\ 15°\ E$
(g) $S\ 70°\ E$ (h) $S\ 60°\ W$
(i) $N\ 18°\ E$ (j) $N\ 26°\ W$

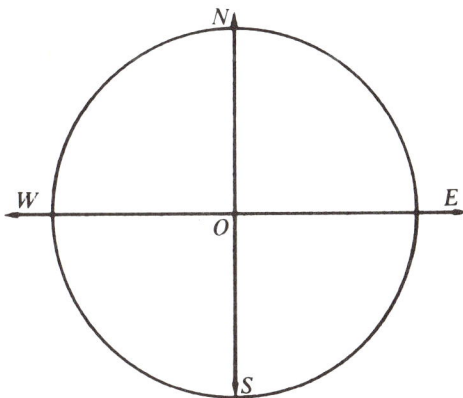

3. Find the angle between each of the following pairs of bearings:

(a) $N\ 30°\ E$ and $N\ 40°\ E$ (b) $N\ 5°\ E$ and $N\ 32°\ E$
(c) $N\ 10°\ E$ and $N\ 20°\ W$ (d) $N\ 30°\ W$ and $N\ 15°\ E$
(e) $S\ 18°\ E$ and $S\ 6°\ W$ (f) $S\ 24°\ W$ and $S\ 30°\ W$
(g) $N\ 25°\ E$ and $S\ 18°\ E$ (h) $N\ 32°\ E$ and $N\ 20°\ W$
(i) $N\ 10°\ W$ and $N\ 18°\ E$ (j) $S\ 18°\ E$ and $S\ 26°\ W$

4.

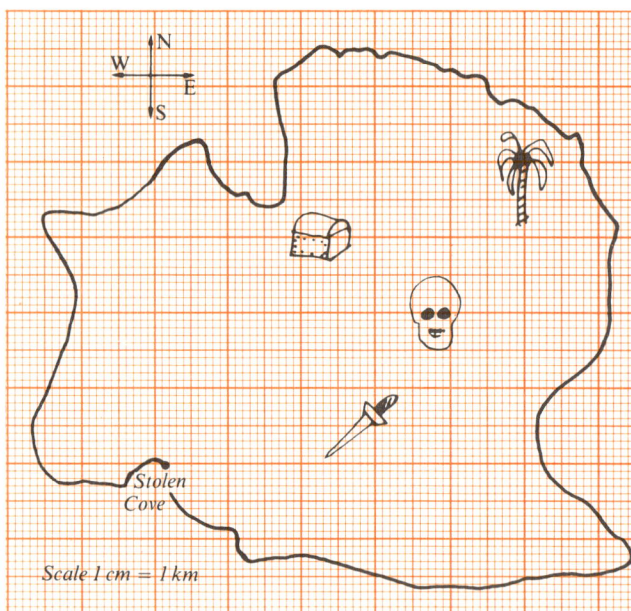

Peg Leg Bill, a smuggler, lands at Stolen Cove and, using the map shown, tries to find the treasure. He walks 3 km in a direction of $N\ 45°\ E$, then 1 km due East. He then walks 2 km in a direction of $S\ 60°\ E$ then 3 km due North. In desperation he walks 4 km $S\ 60°$ W. Success at last! Where was the treasure?

5.

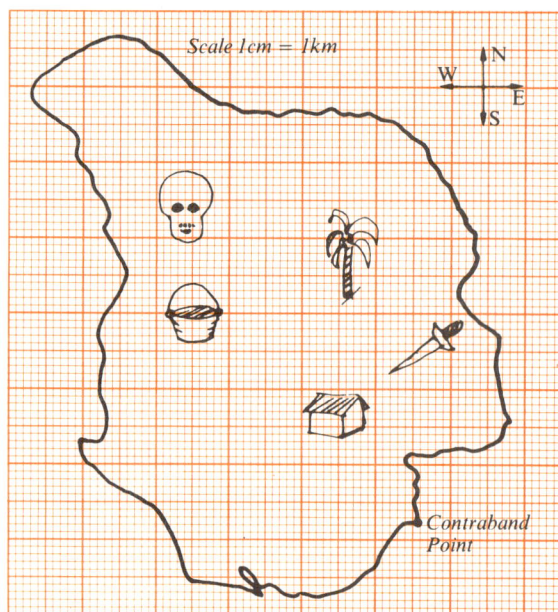

Fearless Fred landed on Forbidden Island at Contraband Point. In search of golden treasure he walked 3 km in a direction $N\ 20°\ W$ then $1\frac{1}{2}$ km $S\ 50°\ W$. Still searching, he walked $1\frac{1}{2}$ km due west, then 3 km $N\ 30°\ E$. Finally 3 km $S\ 20°\ E$ brought him to the treasure. Where was it?

6. Make up your own treasure map and ask your friend to try to find the hidden treasure.

14.2.4   Estimation and Measurement

It is useful to be able to estimate the length, area and volume of differently shaped figures.
Consider

Fig 14.27

and

We wish to estimate which is longer.

We can divide the two lines roughly into centimetre lengths.

is approximately 8 cm

and

is approximately 6 cm.

Fig 14.28

Similarly, to compare areas of different shapes, we can divide the shape into squares, each of side 1 cm.

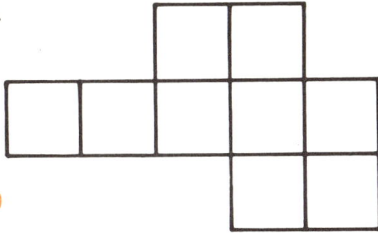

is approximately 9 cm²

Fig 14.29

and

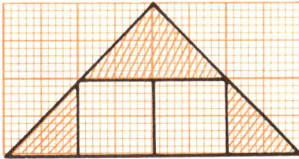

is also approximately 9 cm².

Fig 14.30

It is interesting that different shapes can have the same area.

Also, the shapes in fig 14.29 and fig 14.30 have different perimeters, even though their areas are the same. In fig 14.29 the perimeter is approximately 16 cm but in fig 14.30 it is approximately 20 cm.

If the figure has sides which are not at right angles its approximate area can still be found by counting squares.

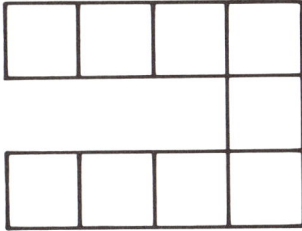

has an area of approximately 4 cm² made up of 2 whole squares and 4 half squares.

Fig 14.31

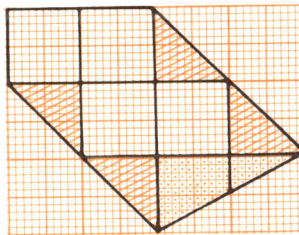

has an area of approximately 7 cm² made up of 4 whole squares, 4 half squares and the remaining area (dotted) which can be cut to form approximately another whole square.

Fig 14.32

We have seen in Chapter 12 that the volume of a prism may be found by counting cubes.

Thus

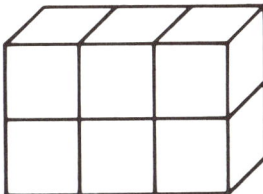

has a volume of 6 cm³ as it is made up of 6 cubes each 1 cm by 1 cm by 1 cm.

Fig 14.33

We can find the area of the surface of this figure by counting the number of square centimetres which can be drawn on its surface.

From fig 14.33 it can be seen that

Area of front face = 6 cm²

∴ area of back face = 6 cm² also.

Area of top face = 3 cm²

∴ area of bottom face = 3 cm² also.

Area of end face $= 2$ cm^2

$\therefore$ area of other end face $= 2$ cm^2 also.

Thus the area of the surface of the prism

$$= 6 + 6 + 3 + 3 + 2 + 2$$
$$= 22 \text{ cm}^2.$$

Exercise 14P

1. Which of the following pieces of string is the longest?

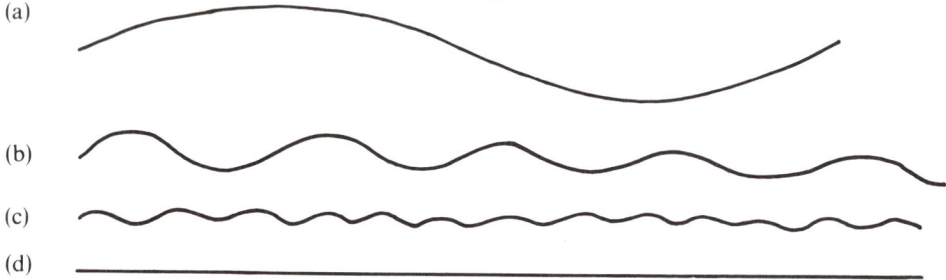

   (a)

   (b)

   (c)

   (d)

2. Which of the following lines is the longest?

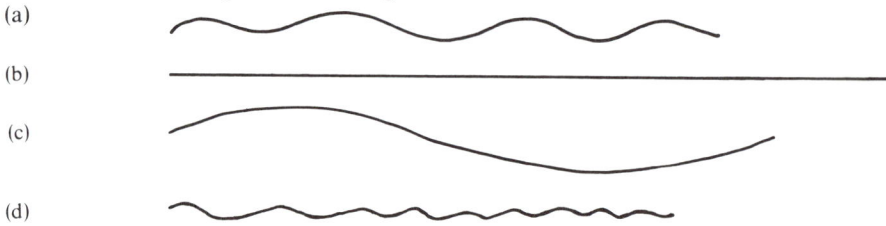

   (a)

   (b)

   (c)

   (d)

3. Copy and complete the following table:

Figure	Area (cm^2)	Perimeter (cm)
(a)		
(b)		
(c)		
(d)		

(e)

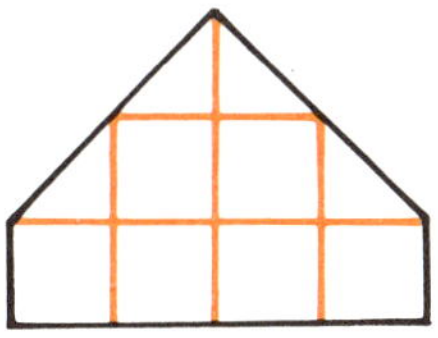

(f)

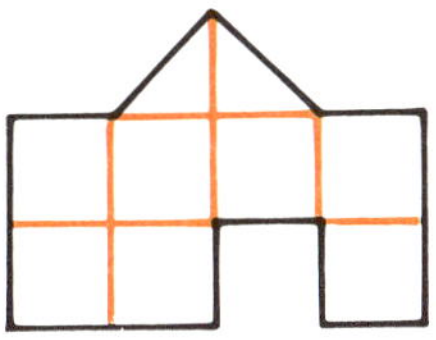

(g)

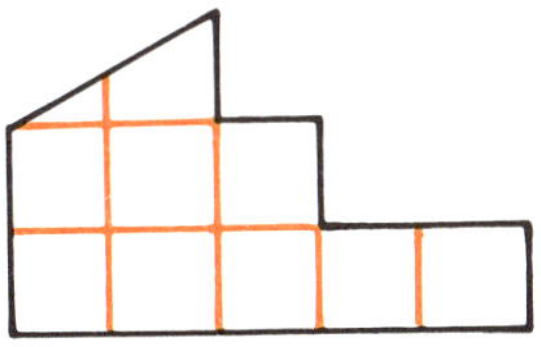

(h)

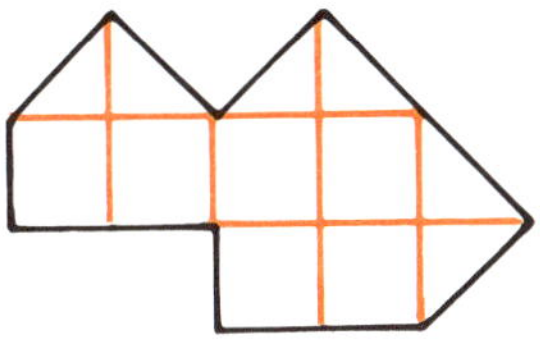

4. Copy and complete the following table:

Figure	Volume (cm³)	Area of Surface (cm²)
(a) 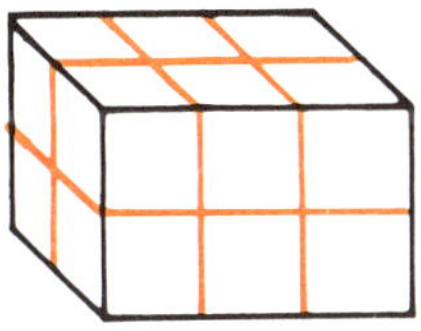		
(b) 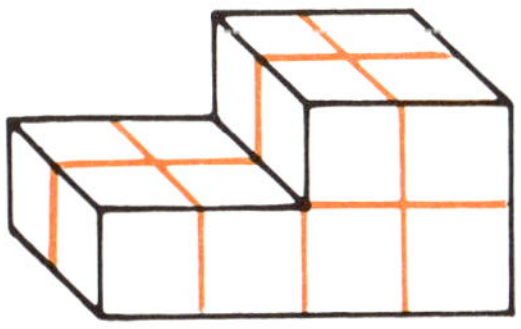		
(c) 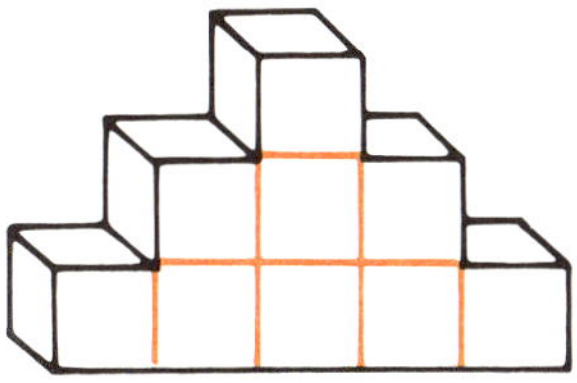		
(d) 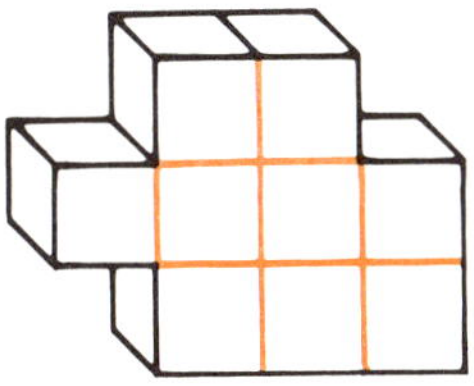		

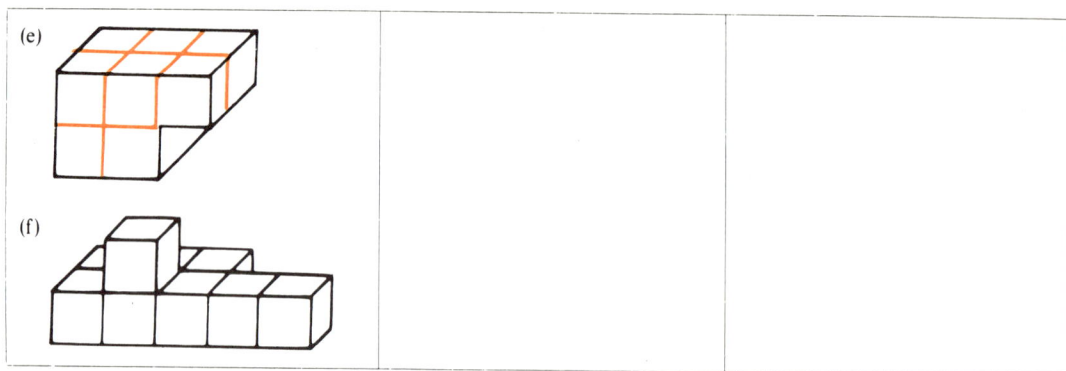

(e)

(f)

14.2.5   Numbers

14.2.5.1   Estimation

It is always a good idea to have an estimate of the answer to an arithmetic problem. For example, if we multiplied 5.4 × 2.6 and the answer we found was 140.4, we should realise that this could not be correct. How would we know?

5.4 is approximately 5 (rounding the decimal down because it is below 0.5)

2.6 is approximately 3 (round the decimal up because it is above 0.5)

so 5.4 × 2.6 is approximately 5 × 3 = 15

Thus, an answer of approximately 140 cannot be correct.

> **Round down decimals if they are below 0.5.**
> **Round up decimals if they are 0.5 or above.**

Exercise 14Q

Find the approximate answer to the following, by rounding the decimals up or down:

1. 6.2 + 5.3
2. 72.5 + 34.1
3. 17.3 + 15.4
4. 9.63 + 6.95
5. 18.2 − 15.5
6. 16.76 -- 13.59
7. 42.61 − 13.39
8. 16.49 − 3.06
9. 4.51 × 4.49
10. 23.16 × 6.92
11. 2.13 × 3.08
12. 5.72 × 1.39
13. 9.39 ÷ 3.42
14. 18.46 ÷ 5.92
15. 32.68 ÷ 10.53
16. 24.85 ÷ 4.96

14.2.5.2   Number Puzzles

Exercise 14R

1. A number square can be formed so that the sum of each of the rows, each of the columns and each of the diagonals is the same.

E.g.

2	7	6
9	5	1
4	3	8

Sum of each row
= Sum of each column
= Sum of each diagonal
= 15

Copy and complete the following magic squares:

(a)

2	8	
	6	
		10

(b)

		1
	4	7
		4

(c)

	3	
4	3	2

(d)

3		2
	4	
6		

(e)

4	8	9
		10

(f)

	6	
8	8	
	10	

(g)

6		
	4	5
3		

(h)

2		4
6		8

(i)

1		2
3		
		3

(j)

		2
	4	
10	0	

2. In each of the following, each letter takes the place of a number and different letters stand for different numbers. Use trial and error to translate the words into arithmetic.

(a)
```
  MAIL
+ MINE
 LATER
```

(b)
```
  PLAY
+ YOUR
 HARP
```

(c)
```
  PASS
+ THE
 SOAP
```

(d)
```
  SLOTS
+ MORE
 MARKS
```

(e)
```
  MAST
×    I
 TOOL
```

(f)
```
   WAIT
×    IN
 SNUTON
```

(g)
```
  MESS
×   TA
 VACME
 MESSY
 CSTAE
```

(h)
```
  TURN
I)CROSS
```

(i)
```
   ART
ON)RATS
```

(j)
```
     AT
AT)FAR
    SO
    OR
    OR
    XX
```

3. Complete the following cross-number puzzle, using the clues given below.

Across:
1. 28 760 + 9832
4. 10 467 − 271
7. 75 × 25
8. 21 425 + 24 101 + 22 412
9. 5093 − 3061
10. 14 958 − 7323

Down:
1. 527 947 − 211 615
2. 2783 + 3215
3. 43 170 − 21 293
4. 2844 − 1319
5. 23 679 + 100 019
6. 8 × 8

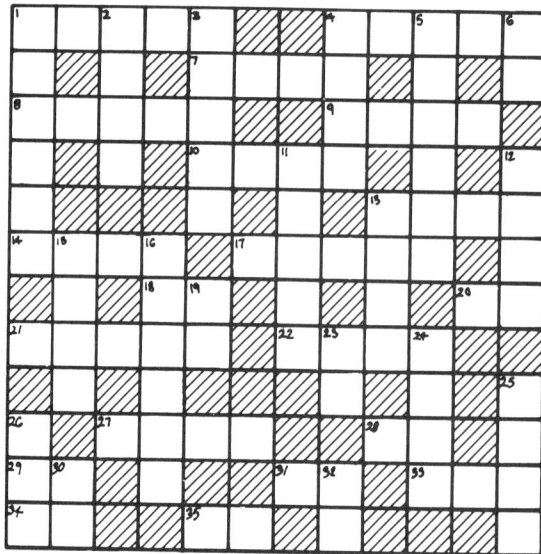

13. $7026 - 5111$
14. $2927 - 564$
17. $76\,291 - 43\,443$
18. $4 \div 10$
20. 7×12
21. $37\,807 + 56\,102$
22. $8163 - 2194$
27. $3742 - 943$
28. 7×7
29. 4×8
31. 8×8
33. $436 + 522$
34. 11×2
35. 9×8

11. $17\,343 + 14\,952$
12. $4343 + 2171$
13. $1247 + 189$
15. $1792 + 1439$
16. $1.5470 + 1.5$
19. 7×7
23. 12×8
24. $5387 + 4412$
25. $3428 + 3361$
26. 11×12
30. 11×2
32. 6×7

4. Complete the following cross-number puzzle using the clues given below.

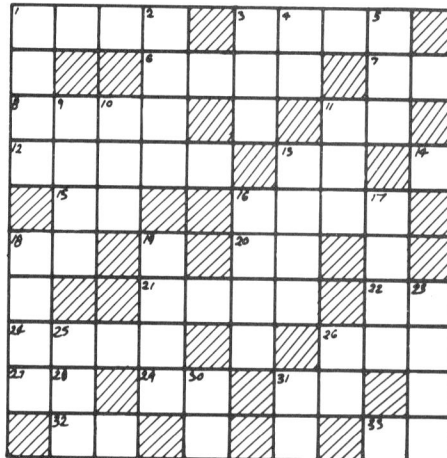

Across:
1. $456 + 159 + 2489 + 363$
3. $9512 - 2165$
6. $7932 \div 2 + 623$
7. Find x if $3x = 102$
8. $0.9385 \times 10\,000$

Down:
1. $4631 - 940$
2. $7000 + 452$
3. $700 + 80 + 3$
4. 3×13
5. $2 \times 350 + 11 \times 3 + 2$

11. The degrees of a right angle − 25
12. 1 million − 983 876
13. The degrees of a line − 127
14. 456 ÷ 8 − 55
15. 9 × 6
16. 7530 − 83
18. $\frac{1}{4}$ of 100 − 1
20. Same as 15 across
21. 347 300 cm = . . . m
22. A prime number between 0 and 20
24. 849 × 9 + 4
26. 8 × 38
27. 12 × 12 − 61
29. 1000 − 906
31. 2 × 52 − 70
32. The perimeter of a triangle with sides 25, 32, 14 units.
33. Find x if $7x = 679$

9. 3 × 1218
10. The area of a rectangle with length 203.5 units and width 4 units
11. 600 + 5 × 6 + 4
13. 2500 × 2 + 400 + 43
16. 7 × 1082
17. 9598 − 1988
18. 45 + 86 + 75 + 1478 + 426 + 256 + 12
19. 0.2359 m^2 = . . . cm^2
23. 396 × 10 − 543
25. 91 × 7
26. Same as 7 across
30. 3870 ÷ 90
31. 7 × 5

5.   1 × 1 =
   11 × 11 =
  111 × 111 =
 1111 × 1111 =

     .

     .

     .

111111111 × 111111111 =

6. 142857 × 1 =
142857 × 2 =
142857 × 3 =
142857 × 4 =
142857 × 5 =
142857 × 6 =
What do you notice?

7. Use exactly five '2's and the signs ÷, ×, +, − to make every number from 0 to 9. Remember BODMAS.
e.g. 2 − 2 ÷ 2 + 2 − 2 = 1

8. Find as many ways as possible to make 30, using three identical digits.
e.g. 6 × 6 − 6 = 30

14.2.6   Scale Drawing

If you look at yourself in a curved mirror, your reflection will be distorted—you may be long and thin or short and fat depending upon the curve of the mirror. Similarly, plotting a pattern on different types of graph paper will create distortions.

Exercise 14S

1. Trace the grids below into your book. For each question, plot the points given and join them up in order on each type of grid.

    (a) $O(0,0)$ $A(2,0)$ $B(2,2)$ $C(0,2)$

    (b) $P(2,0)$ $Q(6,0)$ $R(6,2)$ $S(2,2)$

    (c) $A(0,0)$ $B(1,0)$ $C(1,1)$ $D(2,1)$ $E(2,2)$ $F(0,2)$

    (d) $A(0,0)$ $B(4,0)$ $C(4,1)$ $D(1,1)$ $E(1,2)$ $F(1,4)$ $G(0,4)$

    (e) $A(0,0)$ $B(0,1)$ $C(2,1)$ $D(2,4)$ $E(4,4)$ $F(4,1)$ $G(6,1)$ $H(6,0)$

    (f) $L(2,0)$ $M(4,0)$ $N(4,2)$ $P(6,2)$ $Q(6,4)$ $R(4,4)$ $S(4,6)$ $T(2,6)$ $U(2,4)$ $V(0,4)$ $W(0,2)$
       $X(2,2)$

2. What is the effect of plotting a pattern on grids other than square grids?

3. Make up your own pattern and plot on each type of grid.

    Square:

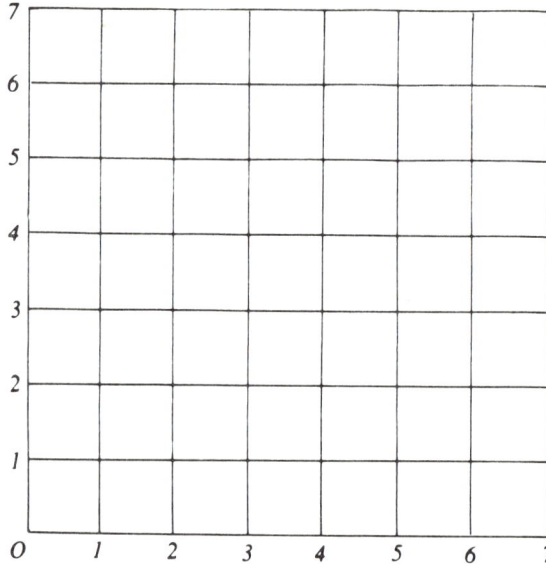

    Rectangle:

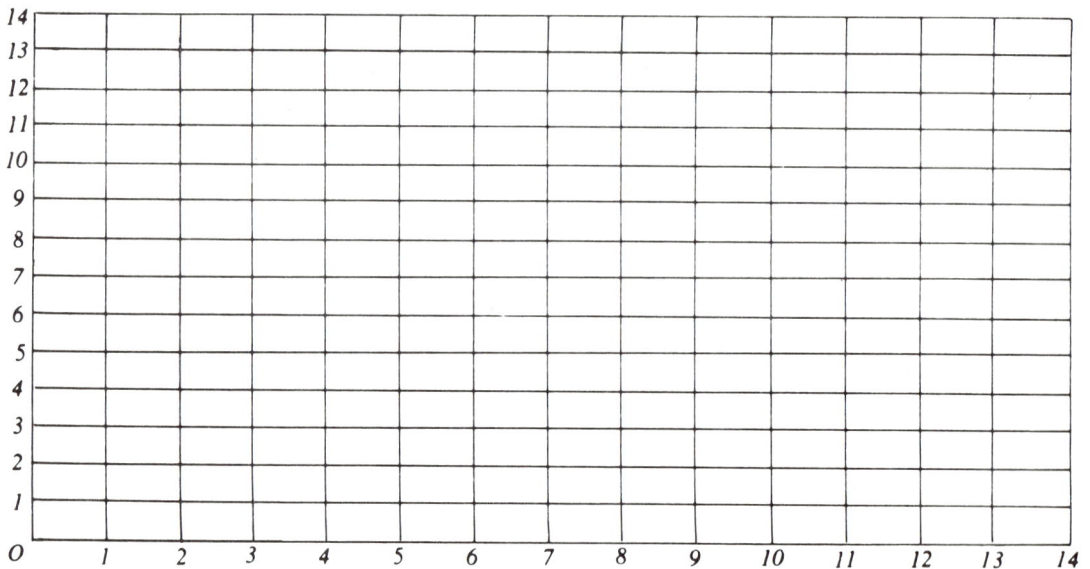

Rhombus:

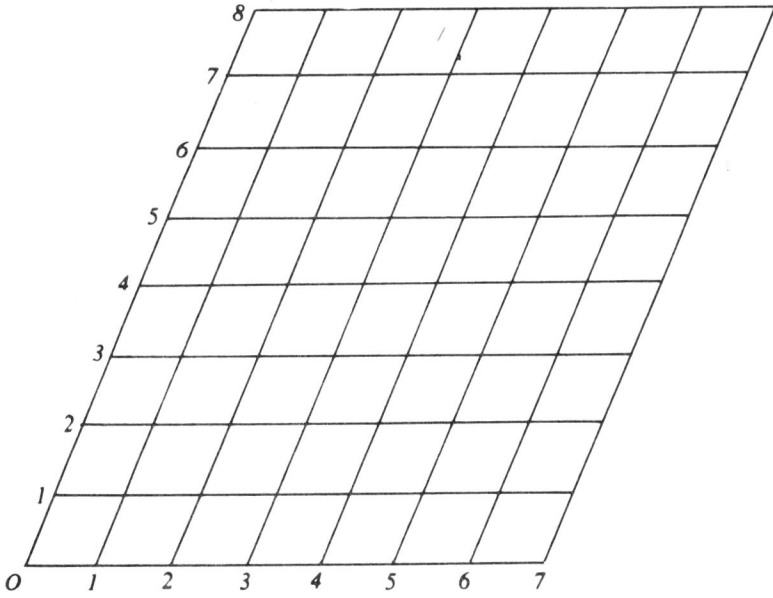

Curve:

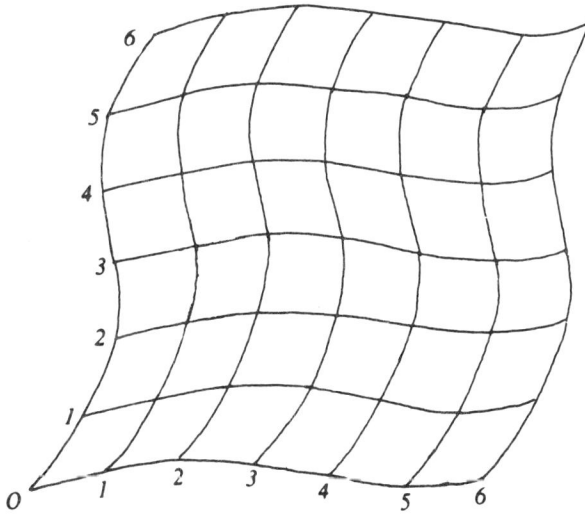

Polar:

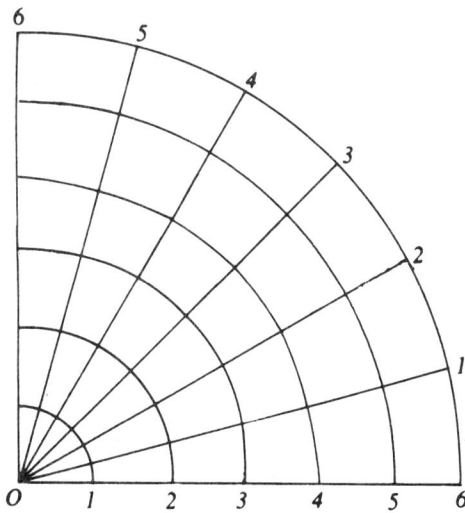

14.2.6.1   Enlargement by 'Squares'

Sometimes it is necessary to enlarge (or reduce) the size of a figure without distorting its shape.
One method, using centres of similarity, has been seen in 14.1.5.1.
Consider the figures in fig 14.34 (a) and fig 14.34 (b):

(a)   (b)

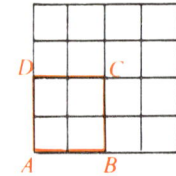

Fig 14.34 (a)

Fig 14.34 (b)

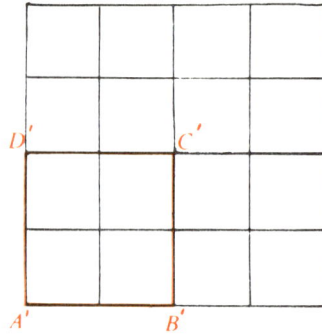

In fig 14.34 (a) the grid is made up of squares $\frac{1}{2}$ cm $\times$ $\frac{1}{2}$ cm and in fig 14.34 (b) the grid is made up of squares 1 cm $\times$ 1 cm. Measuring each side on $ABCD$ in fig 14.34 (a) and $A'B'C'D'$ in fig 14.34 (b) we find that

$d\overline{AB} = 1$ cm	and	$d\overline{A'B'} = 2$ cm
$d\overline{BC} = 1$ cm		$d\overline{B'C'} = 2$ cm
$d\overline{CD} = 1$ cm		$d\overline{C'D'} = 2$ cm
$d\overline{DA} = 1$ cm		$d\overline{D'A'} = 2$ cm

It can be seen that each side in fig 14.34 (b) is twice the length of the corresponding side in fig 14.34 (a). $A'B'C'D'$ is exactly the same shape as $ABCD$ but twice the size. Thus, to double the size of a figure it should be plotted on a grid which has squares twice the size. This method is called the method of squares.

Example 14.9

Enlarge the following figure to twice its present size

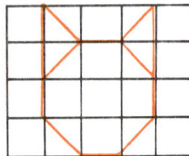

Solution

The grid is made up of squares $\frac{1}{2}$ cm by $\frac{1}{2}$ cm. To double the size of the figure, we must use a grid with squares 1 cm by 1 cm.

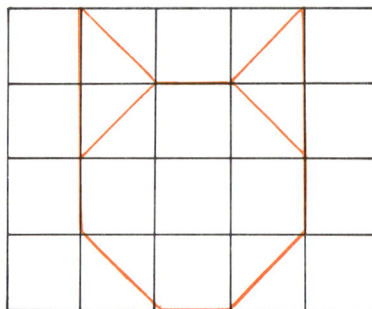

Similarly, a figure may be reduced by plotting it on a grid with smaller squares.

Example 14.10

Reduce the following figure to half its present size:

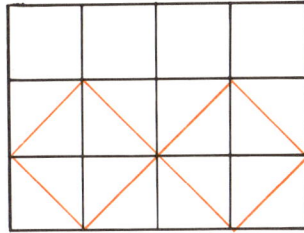

Solution

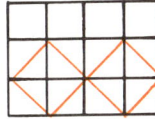

Exercise 14T

Using the method of squares, enlarge each of the following figures so that they are twice the size.

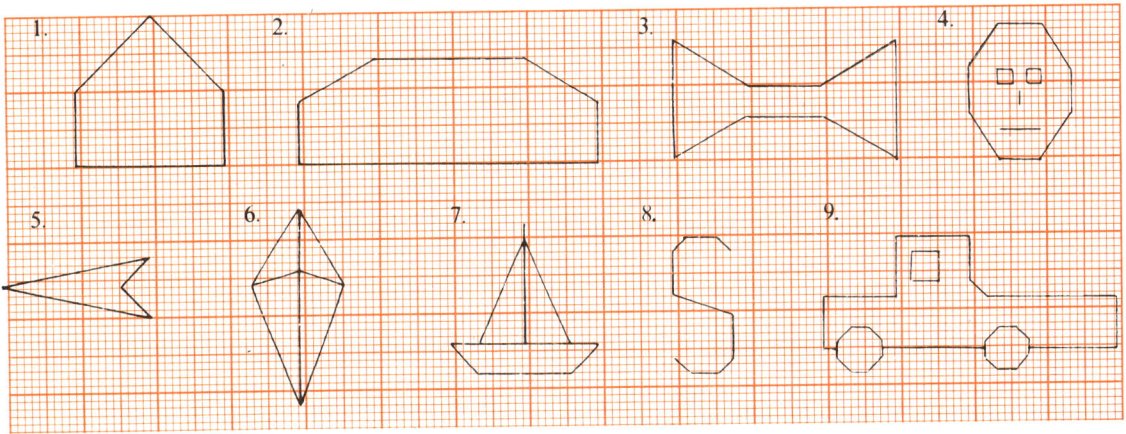

Reduce each of the figures above to half their size.

14.2.7   Cross-Sections

Exercise 14U

1. A cube of cheese is cut in the following manner. What is the shape of the cut face of the cheese?

(a)

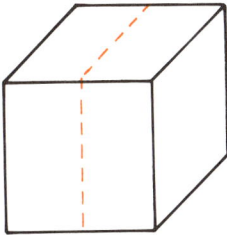

(b)

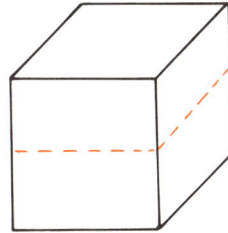

(c)

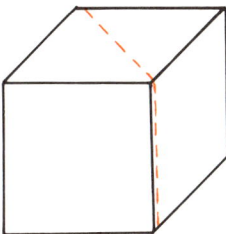

(d)

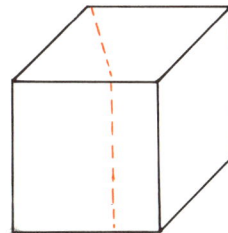

(e) 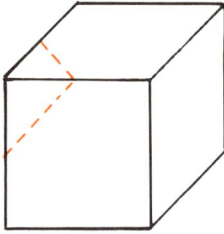   (f)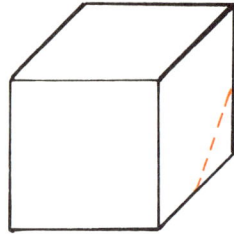

2. An orange is cut in the following manner. What is the shape of the cut face of the orange?

(a) 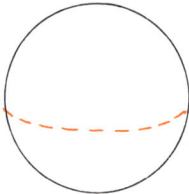   (b)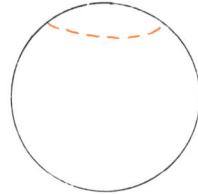

3. A cone is cut in the following manner. Draw the shape of the cut face of the cone.

(a) 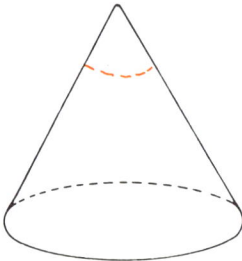   (b)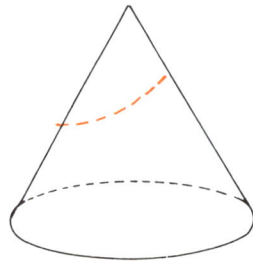

(c) 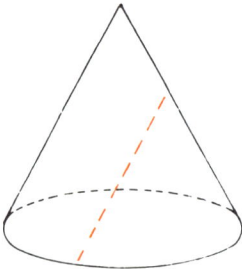   (d)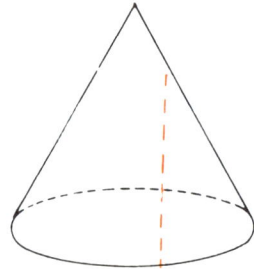

Index